Chemical Sciences in the Modern World

The Chemical Sciences in Society Series

Arnold Thackray, Editor

Fred Aftalion (translated by Otto Theodor Benfey). *A History of the International Chemical Industry,* 1991.
Peter J.T. Morris. *The American Synthetic Rubber Research Program,* 1989.
Seymour H. Mauskopf, ed. *Chemical Sciences in the Modern World.* 1993.

Sponsored by

Chemical Sciences in the Modern World

Edited by Seymour H. Mauskopf

University of Pennsylvania Press Philadelphia

Publication of this volume was supported in part by a grant from the Beckman Center for the History of Chemistry.

Earlier versions of several of the essays in this volume were presented at a conference on the Chemical Sciences in the Modern World organized by the Beckman Center for the History of Chemistry, held May 17–20, 1990.

Printed in the United States of America

Library of Congress Cataloging-in-Publication Data
Chemical sciences in the modern world / edited by Seymour H. Mauskopf.
p. cm. — (The Chemical sciences in society series)
Includes bibliographical references and index.
ISBN 0-8122-3156-2
1. Chemistry—History—Congresses. I. Mauskopf, Seymour H. II. Series.
QD11.C48 1993
540—dc20 93-30055
CIP

To Sidney M. Edelstein

scholar, bibliophile, and supporter
of the history of chemistry

Contents

Introduction: Chemical Sciences in the Modern World
Seymour H. Mauskopf xi

Part I: Practice 1

Theoretical

Philosophies of Chemistry Since the Eighteenth Century
Mary Jo Nye 3

Technical

Uses and Images of Instruments in Chemistry
Yakov M. Rabkin 25

Systems of Production: Drosophila, Neurospora, and Biochemical Genetics
Robert E. Kohler 43

Social

The Quiet Revolution of the 1850s: Social and Empirical Sources of Scientific Theory
Alan J. Rocke 87

Justus Liebig and the Construction of Organic Chemistry
Frederic L. Holmes 119

Part II: Production 135

The Evolution of the Chemical Industry: A Technological Perspective
John Kenly Smith 137

Chemistry and Biomedicine in an Industrial Setting: The Invention of the Sulfa Drugs
John E. Lesch 158

Defining Chemistry: Origins of the Heroic Chemist
Robert Friedel 216

Part III: Public Interface 235

Challenge to Preserve

Documenting Modern Chemistry: The Historical Task of the Archivist
Helen W. Samuels 237

Two Useful Tools for Documentation Strategies: Historical and Documentation Research
Joan Warnow-Blewett 254

Challenge to Public Outreach

History of Chemistry and the Chemical Community: Bridging the Gap?
William B. Jensen 262

The Museum, Meaning, and History: The Case of Chemistry
Robert Bud 277

Challenge to Public Policy

Between Knowledge and Action: Themes in the History of Environmental Chemistry
Christopher Hamlin 295

The Chemogastric Revolution and the Regulation of Food Chemicals
Suzanne White 322

The Chemical Industries and Their Publics: How Can History Help?
E. N. Brandt 356

Part IV: Prospects 365

The Prospect from Here
Erwin N. Hiebert 367

Bewitched, Bothered, and Bewildered
John W. Servos 370

Women in Research Schools: Approaching an Analytical Lacuna in the History of Chemistry and Allied Sciences
Pnina G. Abir-Am 375

Bibliography 393
List of Contributors 405
Index 409

Introduction: Chemical Sciences in the Modern World

Seymour H. Mauskopf

In 1950, at the century's midpoint, *Fortune* magazine proclaimed the twentieth century to be the "chemical century" because chemical industries had become America's premier industrial sector.[1] Like other American industries, they subsequently underwent the vicissitudes of "maturity" and intense foreign competition, yet the value of the products of chemical and allied industries in 1989 was $98.8 billion.[2] Even more visibly, chemical products have, for better or for worse, transformed our persons, our lives, and our surroundings. From pills to perfumes to paints to plastic to pollutants, we encounter artificially produced chemical products at every turn of our lives. Indeed, our very nostalgia for "natural" products is an index of the omnipresence of "synthetic" ones, often made of substances that did not exist prior to their creation by chemical researchers. The major global environmental concerns of our time, the greenhouse effect and the problem of ozone depletion, are essentially the result of the vast increase in the production and use of chemical products; their solutions will doubtless require the deployment of yet more chemical know-how.

The achievements of the chemical industries in creating the myriad new made-to-order chemical substances have been possible because of the tremendous advances in the chemical sciences: in the comprehension of the nature and interaction of chemical reagents under wide-ranging and vastly complex physical and biological conditions and in the refinement of manipulative experimental techniques to study and enhance these interactions. The very success of chemical sciences in the twentieth century has given rise to new hybrid fields between chemistry and other sciences: physical chemistry, chemical physics, biochemistry, geochemistry. Some fields that have dropped the name association with "chemistry" arguably also fall into this class of hybrids: of these, solid-state physics and molecular biology are the most prominent.

Yet the history of modern chemical sciences and industries has remained curiously invisible both to the general public and to the professional historian. World War I is sometimes termed the "chemists' war," but it is doubtful whether many within or without the historical profession could explicate this appellation beyond the horrible specter of poison gas it summons up. Nor, until the very recent appearance of a sophisticated history of industrial research at the Du Pont Corporation by David A. Hounshell and John K. Smith,[3] were we much better placed to elucidate the developments that put Du Pont in a position to claim "better things for better living through chemistry" for its industrial activities. Even in the teaching of the historians of science, the history of chemistry figures at best episodically and rarely in a context beyond the mid-nineteenth century.

Paradoxically, the invisibility of modern chemical sciences and industries in modern historical accounts is probably a result of the very achievements of chemistry. Its extension into industry and its hybridization with other scientific domains over the past century have created an enormous and polymorphous scientific entity that defies easy encapsulation and daunts the scholar by its size, complexity, and technical challenges. Some other factors help account for this historical invisibility of chemistry. Chemistry did not participate directly in the great and dramatic transformative scientific "revolutions" of the beginning of this century associated with such celebrated names as Einstein, Bohr, and Freud. And, in contrast to physics, biology, and the behavioral and social sciences, chemistry has not seemed to have much import for the great philosophical or existential issues of our time. Hence it has not been integrated in cultural history the way these other sciences have.

However, the invisibility of chemistry to the gaze of the modern historian has also been a result of the relative paucity of scholarship up to recent times, undoubtedly for some of the reasons just stated. But we are now witnessing a vivification of this research field among historians of science. At the same time, there is growing encouragement for producing historical scholarship in attractive and usable formats as well as for developing and preserving archival resources from chemists, teachers, industrialists and policy analysts. The establishment of the Chemical Heritage Foundation and its Beckman Center for the History of Chemistry at the University of Pennsylvania is testimony to interest and concern on the part of all of these groups for the enhanced prosecution of the history of modern chemical sciences and industries in all its aspects.

This volume of essays and commentaries is sponsored by the Beckman Center to promote the renaissance of scholarship in the history of modern chemical sciences and industries. It is the first such concentration of studies in this field to appear in book form since the history of science assumed its professional and academic status a half century ago. In its pages, the fruits of research by the leading scholars of the field are set forth, sometimes in the form of detailed, original research, sometimes as outlines for future research in major domains of scholarship, and sometimes as provocative reflections on the opportunities and challenges facing the historians of chemical sciences and industries and their audiences.

There are several objectives in bringing these essays together. The most straightforward is to make visible the best and most sophisticated scholarship in this field to the wider community of historians of science, technology, and medicine and to their students. This volume should also serve the needs of historically interested chemical scientists and chemistry teachers for readily accessible historical scholarship on modern chemical sciences and industries. As one of the essays in this volume points out, a major obstacle to effective communication between historians and chemists has been the lack of historical studies published outside the few specialized journals devoted to the history of chemistry, which most chemists and chemistry teachers do not encounter. The essays should prove especially useful to those concerned with promoting the public understanding of science, an educational objective receiving increasing national attention. Also, as the studies in Part III suggest, this volume should provide a valuable resource for those engaged in science policy studies.

Finally, it is hoped that this publication will be a stimulus to greater communication and interaction with generalist historians, social scientists and philosophers of science, to the greater enrichment of historical scholarship on this vast, complex, and very important domain of scientific and industrial activity.

The volume is divided into four sections. The first three are organized around complementary aspects of the historical analysis of chemical sciences and industries: *practice, production,* and *public interface.* In a concluding fourth section, consideration is given to future *prospects* for the field.

I

The first part examines *practice.* Subsumed under this rubric are, in turn, the components of chemistry that delineate it as a scientific activity: "theoretical," "technical," and "social." The theoretical deals with the nature of theories about chemical knowledge; the technical with the experimental generation of new chemical knowledge; the social with the reception, discussion, and adjudication of new chemical knowledge and theories by the chemical community.

Mary Jo Nye, addressing the first of these components, explores the question whether there is peculiarly "chemical" philosophy of theorizing. She argues that there is; modern chemical theorizing has been distinct from that of physics since at least the 1830s. She analyzes how the growing complexity of chemical phenomena brought about this divergence and traces the resultant tension between the persistence of the goal to subsume chemical phenomena under grand, unified theories like those of physics and the equally strong recalcitrance of the phenomena to such theoretical simplification. This study is unprecedented in its attempt to explore historically whether there has been a peculiarly *chemical* theoretical viewpoint in modern times and to delineate what it is.

Two essays examine the technical component of chemical sciences, specifically, the nature of chemical instrumentation. Yakov Rabkin does so from a broad historical perspective extending back to the eighteenth century. Robert Kohler focuses on a modern case study of the creation of living "instruments" like drosophila and neurospora by the geneticist-biochemist George W. Beadle and his collaborators. What is novel and provocative is their approach: instruments are considered not as passive or objective means of carrying out research agendas but rather as virtual collaborators in the setting of such agendas and in production of new knowledge. Rabkin, indeed, argues that instruments may even set the agenda for chemical research.

Kohler develops a broad contextual perspective, which he terms "mode" or "system of production," for situating the genesis and deployment of Beadle's living instruments. By "system of production" he means not only the conventional experimental context of laboratory procedures and goals but also social and professional contexts: the working style of research groups, the career goals of the protagonists.

Kohler's study as well as the others in this group are closely relatable to recent sociological and even anthropological approaches to the study of "laboratory life." In these approaches, experimental science is viewed as possessing no intrinsic cognitive superiority over other human activities that make knowledge claims. None of the papers in this group pursues such relativistic implications, but they do suggest the kinds of rich and complex contexts in which detailed studies of laboratory research will be examined in the future.

In contrast to the "microscientific" level of analysis of the preceding papers on laboratory chemistry, the final two studies in this section, by

Alan J. Rocke and Frederic L. Holmes respectively, are couched in "macroscientific" or social terms; that is, they deal with the reception of laboratory results (and concomitant theories) by communities of chemists. Each deals with an episode in nineteenth-century organic chemistry. Holmes treats the chemist Justus Liebig's controversies with French chemists over the various "radical" theories in the period 1825–1840. He deploys his well-known "fine structured" analysis to study Liebig, using that scientist's voluminous correspondence. In a revisionist account of the development of chemical theory in the 1850s, Rocke delineates what he terms the "quiet revolution," which, he argues, brought about the acceptance of many major reforms usually associated with the Karlsruhe Congress at the end of the decade some five to seven years earlier. Using the curmudgeon Hermann Kolbe as his marker, Rocke shows how German chemists in particular were quick to adopt the refined type theory of Alexander Williamson and Charles Gerhardt and the concomitant atomic weight reforms during the mid-1850s.

Both these papers are noteworthy for raising and taking a position on the hotly debated issue of whether the generation and reception of scientific knowledge has special claims to rationality and objectivity or is, like other domains of human activity, more or less conditioned by personal, social, cultural, and political contingencies. Both Holmes and Rocke opt for an intrinsic rationality to the scientific enterprise even while recognizing that the interplay of the extraneous contingencies is also continuous and sometimes crucial to the outcome of scientific discourse.

Although the issue of rationally generated versus socially constructed science has been playing over the study of the genesis of modern chemistry in the Chemical Revolution for some time, its appearance here in the analysis of nineteenth-century chemical activity is novel. It could well be extended to the discussion of the direction taken by chemical theorizing in the nineteenth and twentieth century as delineated by Nye. Nye implicitly favors a "rational" explanation for the peculiarities of modern chemical theory; however, the same kind of scrutiny given Rocke's and Holmes's arguments for the primacy of rational considerations might be applied to Nye's analysis.

Conversely, the question posed by Mary Jo Nye about the peculiar nature of chemical theory might also be addressed to the technical and social aspects of chemical sciences. Is there (or has there been in the past) anything "unique" about chemical laboratory life? If so, does it derive from intrinsic characteristics of "chemical" phenomena or from more contingent social factors? Rabkin initiates such discussion regarding instrumentation; further consideration of these questions awaits sophisticated research and analysis of the whole range of modern chemical practice.

II

The second part of the volume is devoted to *production*, by which I mean modern industrial chemistry. Even more than chemical *practice*, this subject touches virtually all areas of modern history: technological, institutional, business, social, political, and cultural as well as scientific. However, the central element of modern industrial chemistry relevant to the historian of science and the modern chemist is unquestionably the institutionalization of chemical practice as industrial research for the commercial creation and production of new materials. Starting up in the latter part of the nineteenth century, it resulted in the "chemical century" of our times, so rapid was its growth, so prolific and pervasive its products.

The examination of the development of industrial research is, comparatively speaking, an infant branch of historical scholarship, although a very lusty one. The three studies of *production* in this volume illustrate its wide range of scholarship by examining very different but complementary aspects of industrial chemistry: commercial, scientific, technological, institutional, and cultural. The lustiness (and formative state) of this field are illustrated by both the breadth of these studies and their concern with developing new conceptual apparatus—sometimes appropriated from other historical disciplines—to inform and control their material.

John K. Smith provides an overview of the development of twentieth-century American chemical industries. Smith patterns this development around three challenges that perennially face chemical industries: "scale" (the move from small-scale laboratory creation to industrial production of new materials), "scope" (the commercial deployment of the range of products that can be developed from an industrial reaction), and "need" (the satisfaction of market needs by new products).

Smith traces the development of American chemical industries through their post-World War II "maturity" down to their present-day problematic status vis-à-vis foreign (especially German) competition. As might already be suspected from the nature of the concepts he deploys in his analysis, Smith sees the science-technology-commercial relationship as a complex and interactive one rather than the conventional unidirectional sequence from scientific research through development to the marketplace. This view is supported by recent scholarship on the role of science in technological and industrial development, as exemplified by the next essay.

Focusing on the "invention" of the sulfa drugs, John E. Lesch examines the fine structure of industrial production in the biomedical-chemical field. His exemplification of a strongly interactive relationship between science and technology is vividly expressed in his conception of the invention of the sulfa drugs as "rational engineering" rather than conventional laboratory research. Moreover, his model for comprehending this invention is adopted from that of Thomas P. Hughes for technological change, with the deployment of such concepts as "technological momentum" and "reverse salients." Also participating in Lesch's analysis are analogues of Smith's ideas of "scope" and "need." The sulfa drugs are seen as being on a continuum of diversified products stemming originally from artificial dyes that industry is bent on developing in marketable form.

Lesch's study is not only complementary to Smith's overview of chemical industrial development; it is also closely related to Kohler's analysis of laboratory "systems of production" (Part I) in its delineation of the structure and function of research at I. G. Farbenindustrie, where the sulfonamide drugs were developed. Lesch delineates the various "modes of production" (in Kohler's sense) of the scientists of different disciplinary orientations (biomedical and chemical) who collaborated in the development of this drug within the framework of their industrial corporation.

The analytical apparatus proffered by Smith, Lesch, and Kohler in this volume are novel in conception and application. Although they are understandably not yet integrated here, they bode well to be used in an integrated manner by subsequent students of chemical industrial research and science-based industries generally.

Robert Friedel provides the "setting" for the other two essays on *production* by focusing on what might be termed the cultural context of chemistry in an industrializing age. Returning to the nineteenth century and using mainly chemical textbooks and encyclopedia articles as his sources, Friedel traces the change in "image" of the chemist as the century progressed (and under the rise of chemical industries) from analyzer of already existent materials and discoverer of natural laws of chemistry to creator of new materials. This essay touches on and raises a wealth of issues germane to the subjects and objectives of this volume. One is the concern with delineating the distinctiveness of "chemistry," the theme of the very first paper by Mary Jo Nye. Although Friedel and Nye approach this theme from very different perspectives, they agree that something important was happening in this respect in the latter part of the nineteenth century; in both analyses, the change seems to be related to a move away from the searching out of a deep, "philosophical" basis for chemistry toward a more empirical and practical science. What is particularly provocative in Friedel's paper is his emphasis of the active role of material creation as coming to mark the new "image" and objective of chemistry.

Clearly, we are just at the start of studying the cultural context of modern chemistry and chemical industries. Friedel's paper offers a promising vista about how this kind of history will be researched and written.

III

The third part of the volume is devoted to *public interface*. Subsumed under this heading are three challenges facing the history of chemistry in its interaction with the public domain: the challenge to develop and preserve archives and artefacts across a wide range of public and private institutions for future scholarly use; the challenge to make the history of chemistry meaningful to the interested but historically untrained layperson, be it through historical writing or though the more complex medium of the science museum; and the challenge to enhance the understanding of policy issues involving chemical sciences and industries through the insight provided by historical research, both for policymakers and for the general public.

Strategies for meeting the challenge of archival development and preservation are outlined by Helen Samuels and Joan Warner-Blewett. These challenges stem from the institutional complexity of modern chemical activities and from the consequent quantity, diversity, and wide dispersal of archival material.

Archivists have begun to formulate means to cope effectively with these challenges. Samuels and Warner-Blewett outline approaches for dealing with the formation of complex institutional archives (pharmaceutical companies, in Samuels's case) and disciplinary based archives, such as those being generated at the American Institute of Physics. In the documentation strategies outlined in these essays, historians and other social scientists team up with archivists to provide guiding analyses of the entities to be documented.

The challenge of outreach to other and wider audiences beyond the disciplinary confines of the history of chemical sciences and industries is addressed in the next group of essays. William Jensen deals with outreach to the chemists through historical books and articles. For Robert Bud, the audience is the general public and the medium, the science museum.

A professional chemist with a far-ranging and active historical interest in his discipline, Jensen is well positioned to address vigorously what he sees as a problematic relation between the productions of contemporary historians of science and the interest of professional scientists. Perhaps his most salient point is that recent scholarship in the history of chemistry, although more interpretively sophisticated in the eyes of the historical community than the traditional general histories of the past (organized around the development of scientific concepts and narratively structured), is, by the same token, less immediately accessible and meaningful to the outsider chemists, even those with serious historical interest. Although Jensen does not offer any easy remedy to the dilemma this poses, he has certainly brought it into the open in a most insightful and provocative way.

Robert Bud, of the London Science Museum, considers the challenge facing the historically oriented science museum curator to make his galleries of scientific and technological objects and equipment meaningful to museum visitors. Arguing that these objects are too esoteric to the general public to evoke powerful "folk memories" when displayed in the traditional museum context of narrative historical sequence, Bud sees the necessity to devise a context of more interrelatedness and significance. He suggests a way through the use of "myth," through which both enduring relational patterns and ambiguities can be simultaneously probed. Bud has delineated a challenge for the historically astute museum curator analogous to that raised by Jensen: how to convey meaningfully the achievements and approaches of modern historical scholarship to the lay public to promote—and not put off—its understanding and appreciation of the historical dimension of science.

The essays concerned with the final challenge of *public interface,* namely, that of illuminating science policy debates and decisions through historical analysis, focus on environmental issues. Christopher Hamlin attempts to provide historical perspective to current discussion about the nature of modern environmental agendas, debates, and policy: whether they are, or indeed can be, "rational" (i.e., based on scientific knowledge) or whether they are really about social and political issues and merely legitimated by claims to scientific verity. This issue will now be familiar from the first section on *practice.* Hamlin examines three environmental chemical debates ranging in date from the late eighteenth to the late nineteenth century.

Hamlin's general conclusion is that polarities in viewing environmental policy debates and decisions as being about *either* the deployment of scientific knowledge *or* that of political, social, and economic power are much too simplistic. His own historical cases of environmental debates exhibit complex intertwining of scientific and social beliefs and interests, not unexpected when such debates were (and are) conducted "in inevitable conditions of epistemic uncertainty and of ignorance about the future."[4]

Suzanne White's study of the "chemogastric revolution" and the resultant move toward enactment of regulatory laws for the use of chemical additives in foods extends Hamlin's context of "uncertainty" to include "change." The "chemogastric revolution" of 1930–1950 was marked by the ascendancy of "convenience foods" such as frozen foods. Socially, this transformation in food products was dependent upon the profound change of American lifestyle associated with the movement to the suburbs. Materially, the development of convenience foods depended on the deployment of a host of new chemical additives. White traces the regulatory concerns and strains to which these changes gave rise, culminating in the enactment of laws controlling the use of food additives following the congressional investigation of the Delaney Committee in 1950.

The final essay looks at the function of history of chemistry in environmental issues from a provocatively different perspective, that of the chemical industries. E. N. Brandt argues for the utilization of history of chemistry to promote a more positive public image of chemical corporations. As examples, he cites adverse publicity that accrued to earlier products and practices of his own company, Dow Chemical: the notorious cases of napalm and of mercury disposal in lakes and rivers. Brandt argues that the professional historian is needed by chemical corporations to provide a balanced historical perspective for the public. This, Brandt believes, will ineluctably soften or even eliminate the moral stigma often attached to the chemical industries in the public mind.

Brandt's essay has obvious relations to Hamlin's on the participation of historical research in illuminating current policy debates and to Friedel's on the "image" of chemistry and chemists. More provocatively, Brandt's blunt advocacy of a particular interest function for historians raises important issues about the epistemic and moral "objectivity" of historical research, especially as historians move into positions in companies and government. Here they will be engaged not so much in mere "outreach" to nonhistorians as in actual interaction with them and advocacy of their interests. It is appropriate to conclude the section on *public interface* with Brandt's paper, for, wittingly or unwittingly, it points up some of the opportunities, challenges, and even hazards that will face future historians of chemistry as they interact with the various public domains and institutions beyond their traditional academic home.

IV

In the concluding section, three distinguished historians of chemistry give thought to future *prospects* for their field. Erwin Hiebert presents the most general "map." He addresses the practical problem of research project choice and also takes up the question of methodology or "new moulds of thinking." Returning full circle to Mary Jo Nye's delineation of the philosophical complexity of modern chemistry, Hiebert argues that this peculiarity of the science (as compared with physics) emphatically demands pluralism and open-endedness of approach by the historians studying it.

Building on the theme of outreach, John Servos addresses a challenge for historians of chemistry concerning an important audience not considered earlier: the undergraduate student. Decrying the lack of historical studies accessible and attractive to students, Servos calls upon historians of chemical sciences and industries to put thought and effort into the writing of such works. It might be added that there is another related audience for such writing: the general historian and the history student (undergraduate and graduate). Communication and dialogue between historians of chemistry and their generalist colleagues might also suggest new research directions for relating the history of chemical sciences and industries to general social and cultural history.

Pnina Abir-Am in fact suggests several such directions. These include attention to collaborative and interdisciplinary research, gender (and, it might be added, ethnicity) in the history of chemistry, the public role of chemists, and the relationship of chemistry to the biological sciences. Abir-Am's research agenda in fact delineates many of the special features of modern chemical sciences and, indeed, of modern science in general.

It is fitting to conclude in this prospective mode. As the essays in this volume make clear, much has already been accomplished in the recent renaissance of historical scholarship on modern chemical sciences and industries, but, as in other human activity, achievement only serves as a spur to reach for yet richer opportunities and more demanding challenges.

Finally, I should like to acknowledge how the creation of this volume was made possible by the unswerving commitment of the staff of the Chemical Heritage Foundation, and by financial support from CHF, from the National Science Foundation (through its program on history and philosophy of science under Ronald J. Overmann), and from the Alfred P. Sloan Foundation (Samuel Goldberg, program officer).

Notes

1. John K. Smith, "The Evolution of the Chemical Industry: A Technological Perspective," this volume, p. 152.
2. Most recent estimate (July, 1992); current dollars. *Survey of Current Business*, 71, 4 (April 1991): 27.
3. *Science and Corporate Strategy: Du Pont R & D, 1902–1980* (Cambridge: Cambridge University Press, 1988).
4. Christopher Hamlin, "Between Knowledge and Action: Themes in the History of Environmental Chemistry," this volume, p. 316.

I
Practice

Theoretical
Technical
Social

THEORETICAL

Philosophies of Chemistry Since the Eighteenth Century

Mary Jo Nye

Epistemology of Chemistry

Is there a "chemical philosophy" or a "philosophy of chemistry"? What might this term mean, and how may it have changed over time? Some years ago the Leeds physical chemist E. F. Caldin suggested the need for a serious analysis of the methodology of chemistry. He said such analysis would contribute to discussions of questions of scientific method, for example, whether theories are explanatory as well as instrumental and what may be the status of theoretical entities like atoms.[1]

Erwin Paneth noted in a 1931 lecture, reprinted in 1962, that physicists, like philosophers, have endeavored to classify the epistemological foundations of physics, but that chemists have not done so.[2] Bernard Cohen commented in 1970 that, despite the writings of Pierre Duhem, Hélène Metzger, and Emile Meyerson, scholars concerned with conceptual analysis in the history of science have not usually turned to chemistry, except to study a few isolated figures like Paracelsus, Priestley, and Lavoisier.[3]

Aside from the specific question of the status of the atom, the challenge for history and philosophy of chemistry still lies largely unanswered.[4] In the last decades philosophers and historians interested in epistemology and theory choice mostly have focused on physics, biology (especially evolutionary biology), and the social sciences, with only occasional mention of atom-spheres, benzene hexagons, and the periodic table.[5]

Important exceptions to disinterest in chemistry among philosophers and historians can be found especially in the French tradition. Examples are in the work of Gaston Bachelard and François Dagognet,[6] the reflections of the Belgian chemist Ilya Prigogine, and recent studies by students of Prigogine and philosopher Michel Serres, including Isabelle Stengers and Bernadette Bensaude-Vincent.[7] The latter have treated chemical texts as examples of socially constructed literary technique, grammar, and rhetoric.[8] Still, when English chemists recently searched the professional literature for a critical review of the practice of modeling in chemistry, they found nothing helpful and ended by writing themselves the enlightening book entitled *Chemistry Through Models*.[9]

By and large, a pejorative view of the methodological sophistication of chemical science has prevailed, notably in comparison to physics. The structure of scientific explanation in chemistry is deemed child's play, or kitchen work. In his *Confessions d'un chimiste ordinaire* (1981), the chemist and historian Jean Jacques mused that chemists rarely write about their discipline for the public like their biologist and physicist colleagues. "Chemistry," he said, is "like the maid occupied with daily civilization: she is busy with fertilizers, medicines, glass, insecticides . . . for which she dispenses the recipes."[10] Some seventy years ago, the Toulouse physicist Henri Bouasse enraged his Nobel Prize winning colleague, the chemist Paul Sabatier, by claiming that chemists only "faire le cuisine."[11]

In the late nineteenth century, the French chemists Marcelin Berthelot and Henri Sainte-Claire Deville contributed to the general view of their métier as a descriptive, empirical science. Recalling a course taught in this tradition, Robert Lespieau, of the next generation, remarked in 1913: "Four and a half hours was all the time devoted to generalities; if one had doubled this time, it would not have been detrimental to the seventeenth property of chlorous anhydride."[12]

Skirting the Scylla of naive empiricism, chemistry has not escaped the Charybdis of naive realism. John Dalton's atomism is in fact unrelentingly and naively realist. His illustrative plates in the *New System of Chemistry* (1808) are claimed there to exhibit clearly "the mode of combination of some of the more simple cases" of ultimate particles forming bigger ones. Dalton's compatriot Alexander Williamson defended realism, albeit a more sophisticated variety, at a Chemical Society debate in 1869.[13] The philosopher John Bradley claimed that many nineteenth-century chemists were naive realists:

> To Avogadro and Cannizzaro, as to Couper and Kekulé, the molecules and atoms considered in this great theory were real objects: they were thought of the same way as one thinks of tables and chairs.[14]

This philosophical view of chemistry has been shared by many physicists, for example, Henry Margenau, Professor of Physics and Natural Philosophy at Yale University, who wrote in 1950:

> Twenty years ago many chemists would have defended the theory of bond arms as a satisfactory *explanation* because they had become accustomed to thinking of it as unique and as ultimate.[15]

However, this view of chemical philosophy simply does not stand up to historical scrutiny.

It appears to be true that chemists have spent less effort than physicists in scrutinizing the philosophical foundations of their subject. The philosopher Bas Van Frassen has claimed that the closer *physicists* are to experimental work, the less interested they are in fundamental questions.[16] Chemists have always been closer to experimental work, more thoroughly involved in the laboratory than natural philosophers, and less satisfied with idealizations of phenomena. Yet, as we shall see, some of the founding documents of modern chemistry explicitly address questions of epistemology. We can reconstruct epistemological views, where they are not explicit, by looking carefully at lectures, textbooks, and journal articles from the nineteenth and early twentieth century.

Nineteenth-century chemists were not dyed-in-the-wool instrumentalists or radical skeptics, any more than they were naive empiricists or naive realists. And, of course, chemists have not all shared the same views about the aims and methods of their discipline. Nonetheless, we shall attempt to gain some perspective on the meaning of "chemical philosophy" or "philosophy of chemistry," and how consensus, or debates, among chemists about chemical epistemology have evolved in the modern period from the Enlightenment to the mid-twentieth century.

We begin with the aims of "Chemical Philosophy" in the eighteenth century, analyzing the kinds of principles, laws, and theories that came to define chemistry by 1830. After 1830, the agenda for chemistry remained fairly constant for five or six decades, until the development of thermodynamics and the experimental validation of the ion and the electron, which I take to be turning points in the practice of chemistry.

I argue that chemical explanation from the late eighteenth century has been consistently characterized by three kinds of theoretical approaches which are well known to analysts of scientific method. We will call them (1) "philosophical" or "realist"; (2) "positive" or "exact"; and (3) "pragmatic" or "conventional." Among these, "positive" theories are largely nonhypothetical, descriptive, and inductive; "philosophical" theories are hypothetico-deductive, with varying degrees of truth probability for their "picture" or "system" of the world. Both "philosophical" and "conventional" theories employ models and metaphors, and they share with the "positive" method the use of analogies and signs (like the index and the symbol). The "conventional" theory is neither "exact" nor "probably true" and, especially as used in chemistry, it has overlapping, complementary, but incommensurate parts. Conventional theory is instrumental.

We will see that eighteenth-century "chemical philosophy" initially was identified with the aims of "natural philosophy" for establishing a "probably true" or real picture of mechanical causes and effects in the world. These aims were recognized by Lavoisier, though not by all his colleagues, to be premature. By 1830 philosophical chemistry became a "conventional" or "positive" chemistry.

That scientific language is itself a system of knowledge production, as well as a system of knowledge statements, was made explicit by Lavoisier's school. Nineteenth-century chemical explanation included ordinary language statements, charts and tableaux, simple statements of quantitative relations, and pseudo-algebraic symbols that were descriptive and conventional. Chemical explanation also included visual and pictorial imagery that appeared to be concrete, physical interpretations of invisible objects in the real world.

But for most chemists these images represented the multiple *functions* of a chemical molecule, not positions or activities of atoms within a molecule in three-dimensional space. It was only toward the end of the nineteenth century that chemists returned to the early project of chemical philosophy for understanding mechanisms underlying chemical functions. The extraordinary fruitfulness of the nineteenth-century agenda is borne out not only by its correlation in the twentieth century with a newly successful chemical and quantum mechanics, but also by its continued usefulness in modern chemistry.

I will conclude with an inquiry into whether the new mechanics in twentieth-century chemistry reconstituted the goals of eighteenth-century chemical philosophy.

Chemical Philosophy: Aims and Methods

In the late sixteenth century, Andreas Libavius warned a former student not to associate with chemists who are not philosophers. Libavius was the author of *Alchemia* (1597) and the founder of modern chemistry, according to historian Owen Hannaway.[17] In the philosophical pursuit of chemistry, Libavius followed the humanist tradition of Lutheran intellectualism as taught by Melanchthon. Knowledge is to be gained by sense experience and reason, but it would be rash to claim absolute conviction for naturally acquired knowledge. Progress in science (*scientia*) will come from collective endeavor in which individual contributions are subject to the scrutiny of one's peers and measured against the collective wisdom of the past. The nature of chemical reasoning is analogical, invoking the juxtaposition of natures or ratios, and therefore it is by nature probabilistic. Following Ramus, Libavius stressed the origin of knowledge in practice, but that chemistry must be demonstrative science, abstracted from its applications.[18]

This distinction between "theoretical" and "practical" chemistry was one observed in textbooks throughout the eighteenth and nineteenth centuries. A tradition of "philosophical chemistry" answered Libavius's challenge for chemists to abandon alchemical magic and Paracelsian iatro-chemistry in favor of newly philosophical principles in chemistry. Jacques Barner's *Chymie philosophique* in the seventeenth century is an early example; later, more famous texts in chemical philosophy are those of John Dalton (1808), Humphry Davy (1812), and Jean-Baptiste Dumas.[19] Texts called "chemical philosophy" were fewer than those in "natural philosophy," and very few texts in "chemical philosophy" were written after 1840.[20] Let us see why this was the case.

In a mid-eighteenth-century chemistry course given in Paris by P. J. Macquer and A. Baumé, the instructors self-consciously began their exposition with a set of general principles rather than with facts, in order to "provide links between facts and make it easier for students to learn. At the same time [we] will indicate where principles are only suppositions, probabilities, and matter for further research."[21]

The word "principle" had two meanings for mid-eighteenth-century chemists: the ideal, abstract, and hypothetical "principle" versus the material, empirical "principle." The aim of the chemistry of Macquer and Baumé was the explanation of chemical change through laws of affinity describing power relations among the fundamental chemical principles. Macquer vacillated between the hypothetical and material meanings of "principle," that is, the Newtonian notion of hypothetical massy particle and the Stahlian notion of material elementary substance.

If chemical "principles" were understood by philosophers to be invisible or visible matter, "cause" was understood to be invisible force. Like so many other chemists in the eighteenth century, Macquer and Baumé assumed that the forces of chemical affinity are simply instances of physical forces.[22] They took Newton to be a student of these affinity forces, and they found congenial Peter Shaw's view that "it was by means of chemistry that Sir Isaac Newton has made a great part of his discoveries in natural philosophy." Indeed, Newton's interest in chemical affinity may have led him by analogy to calculate the easier problem of celestial or gravitational "affinity."[23] The notion of the identity of chemical and physical force was based in the principle of economy, expressed in the standard form that similar effects "must arise from the same law, if we are not to multiply causes."[24]

What were the characteristics of chemical philosophy in the early nineteenth century? For Humphry Davy, the popular lecturer at the Royal Institution in London, the aim of chemical philosophy was to "ascertain the causes of all [chemical changes], and to discover the laws by which they are governed." In the British tradition of utility and natural theology, Davy taught that chemical philosophy has "ends" as well as "aims," namely its applications and uses "for increasing the comforts and enjoyments of man, and the demonstration of the order, harmony, and intelligent design of the system of the earth." Regarding methodology, "[in] chemical philosophy . . . observation, guided by analogy, leads to experiment, and analogy, confirmed by experiment, becomes scientific truth."[25]

Jean-Baptiste Dumas was more explicit about principles and causes in his lectures on chemical philosophy to students at the Sorbonne in 1837:

> Chemical philosophy . . . has for its aim to reveal the general principles [de remonter aux principes généraux] of the science . . . [to give its history], to give an explanation of the most general of chemical phenomena, to establish the link between facts and the cause of the facts. Chemical philosophy makes abstract the special properties of bodies; it is composed of the general study of the material particles chemists call atoms and the forces to which these particles are submitted.[26]

In short, Dumas's chemical philosophy aimed at general abstract principles, which were identified with chemical atoms and chemical forces; and it taught the history of chemistry as a guide to the progress of philosophical truth. Dumas does not here use the word "theory," but the "philosophical" part of chemical work generally was understood to mean theory.[27]

In marked contrast to Dumas's philosophical chemistry, let us consider the mode of presentation of Antoine Lavoisier's *Traité élémentaire de chimie* (1789). Lavoisier self-consciously begins with observations, not first principles, stating that "chemistry is an incomplete science, not like geometry." The new nomenclature he and his colleagues have developed is claimed to be based in a "natural order of ideas," not in metaphysics.[28]

What are the aims of chemistry? For Lavoisier, they are identical to chemical operations:

> Chemistry, in submitting different natural bodies to experiment, has for its goal decomposing them and putting them in a state to allow examining separately the different substances which enter into combination. . . . We cannot be assured that what we think simple today really is so; all that we can say is that some substance is the present end at which chemical analysis arrives.[29]

The laboratory method of chemistry is analysis/synthesis, and the conceptual method of chemistry is analogy expressed in systematic language, for, as Condillac had demonstrated, "Languages are true analytical methods. . . . The art of reasoning is nothing more than a language well arranged."[30]

On the face of it, nothing could seem more diametrically opposed than the views of Lavoisier and Dumas on the aims of chemistry, at least as expressed in these programmatic statements of 1789 and 1837. Yet, ironically, Dumas was at a crossroads in the mid-1830s, and his interpretation of the explanatory aims of chemistry was just then shifting from an atom-and-force program of explanation to a structure-and-function one. In so shifting, he appears to move in the direction of Lavoisier's definition of chemistry and away from the notion of "chemical philosophy" he and Humphry Davy had espoused. In making this shift, Dumas also was shifting to descriptive and conventional methods of chemistry from a philosophical method, which aspired to theories of probable, if not immediate, truth.

While the twentieth-century definition of "philosophy" is broad and flexible, eighteenth- and early nineteenth-century notions of "philosophy" more narrowly identified it with unified systems of causal demonstration. Natural philosophy was nothing if not causal, and it was oriented to strongly probable, or certain, knowledge.[31] But Lavoisier's *Elementary Treatise* is self-consciously nonphilosophical from the start; it begins with facts; it is an open-ended enterprise with no appearance of the certainty of geometric demonstration; its symbols may look like algebra, but they are only "simple annotations, of which the object is to ease the labors of the mind."[32] Its method is one of analysis, namely, a material or operational analysis accomplished by the instruments of the laboratory and a conceptual analysis achieved by a method of classification. The method is one of compositional grammar of "principles" defined by physical parameters like volume, weight, melting points, and boiling points; and by chemical parameters like acid, alkali, and neutral (or saturated, unsaturated) orders of substances. It is a generic system.

To the question, what is the *cause* of a chemical reaction, Lavoisier's answer was descriptive and concrete, not geometric and abstract. Lavoisier did not rule out the possibility of an abstract and geometric "chemical philosophy" in the future, but for Lavoisier the time was premature. An avid investigator of the chemistry of organisms, as well as of mineral substances, Lavoisier for the present shared the view of his contemporary the British natural philosopher John Robison that the phenomena of fermentation, nutrition, secretion, and crystallization are not susceptible to simple mechanical reasoning.

> Whenever we see an author attempting to explain these hidden operations by invisible fluids, by aethers, by collisions, and vibrations, and particularly if we see him introducing mathematical reasonings into such explanations—the best thing we can do is to shut the book, and take to some other subject.[33]

In the early 1800s there was a decided difference of opinion among chemists about the merits of pursuing philosophical chemistry. For John Dalton, like the young Dumas, the *cause* of chemical reactions and behavior lay in elementary atoms and forces of attraction and repulsion lying in atoms and in the clouds of caloric surrounding them.[34] Similarly, for Jöns Jakob Berzelius,

> the cause of chemical proportions [in reacting compounds] must be that bodies are composed of particles which should be mechanically indivisible; this explains all phenomena . . . especially multiple proportions. . . . Probabilities lead to imagining these elementary bodies in spherical form.[35]

Justus von Liebig, too, saw the aim of chemistry as the search not only to consolidate the truth of chemical proportions, but also to study the causes of the regularity and constancy of these proportions. Liebig took the cause of chemical action to lie in Newtonian-type atoms and forces of the Berzelian variety, that is, spherical atoms and electrical affinity forces.[36]

Dumas was to be the pivotal figure in shifting the study of causes in chemistry from atoms-and-forces to structure-and-function. This shift away from *force* as causal agent undermined the "philosophical" status of chemistry, which chemical philosophers already feared was becoming increasingly "unphilosophical" because chemistry was becoming a science of conventions. Thus Berzelius wrote to Liebig that, because of recognition of the variability of equivalent values (for example, the combining proportion of hydrogen with nitrogen varies depending on circumstances), the fundamental concept of chemical "equivalent" "was no longer positive, that is, empirical, but merely conventional."[37] Reflecting on the phenomenon of chemical catalysis, a new kind of chemical "affinity" recognized by Döbereiner and by Dulong and Thenard in 1823, Thomas Graham concluded that it was "unphilosophical" to simply refer the phenomenon to an unexplained "force": "The doctrine of catalysis must be viewed in no other light than as a convenient fiction, by which we are able to class together a number of decompositions."[38] The eighteenth-century aims of "chemical philosophy" became subsidiary to a new agenda for scientific explanation around 1830.

Chemical Explanation in the Nineteenth Century: Structure Versus Mechanics

Dumas told students in his 1836 lectures that the principle of an indivisible atom was an indifferent hypothesis, and therefore a matter of convention. "What difference does it make for the facts of chemistry if [elementary chemical masses] were capable of being cut up infinitely by forces independent from chemistry?"[39]

What led Dumas to this conclusion was a failed program of chemical philosophy. For some years, he had developed and employed the technique of vapor-density analysis to calculate relative atomic weights of elements that are easily vaporized, using as working hypotheses Dalton's atomic hypothesis and Avogadro's hypothesis (that equal volumes of vapors under identical conditions contain equal numbers of particles). Dumas further assumed that all elemental gas particles are composed of two halves (diatomic molecules in modern terminology), but this assumption got him into difficulty with the vapors of mercury (Hg), sulfur (S_6), and phosphorus (P_4). He began losing confidence in the general atomic theory, just as he also was developing a new method of explanation for chemical change.

Dumas's decade-long attempt to explain how the bleaching of wax candles resulted in the emission of noxious chlorine vapors at an 1827 court soirée led to his publishing a paper (1838) on the chlorination of organic substances like acetic acid and a theory, not of chemical *forces*, but of chemical *types*.

> The chlorinated vinegar is still an acid, like ordinary vinegar; its acid power has not changed. It saturates the same quantity of base as before. . . . So here is a new organic acid in which a very considerable quantity of chlorine has entered, and which exhibits none of the reactions of chlorine, in which the hydrogen has disappeared, replaced by chlorine, but which experiences by this remarkable substitution only a gradual change in its physical properties.[40]

Dumas (1800–1884), educated in Geneva and Paris, became perhaps the most significant figure advocating a new pathway for chemistry, developing the theory of molecular type and adapting to the needs of chemical explanation an array of analogies, metaphors, and classification schemes borrowed from the natural history of living organisms. He began to stress differences between physical and chemical properties and laws, rather than similarity or complementarity, gradually winning converts to his view. The distinguished physical chemist and historian of chemistry J. R. Partington gave us the following vignette:

Wöhler, who met Dumas in a visit to Paris ("Babylon") in 1833, although calling him a "windbag" and "Jesuit" (Dumas was a Catholic), said he was a "very industrious fellow" (ein sehr fleissiger Kerl) and had "a good heart." . . . Berzelius on repeating Dumas' experiments, could only confirm them. Liebig and Dumas were soon reconciled . . . and at a banquet in Paris in 1867, over which Dumas presided, Liebig said he had given up organic chemistry "since with the theory of substitution as a foundation, organic chemistry needs only labourers."[41]

In his banquet comments, Liebig was disingenuous about his reasons for giving up organic chemistry. But he did give credit to the transformation of chemical philosophy into what we might call a paradigmatic "normal" science. The substitution and then the type-and-structure theories *were* extraordinarily successful in producing and predicting new chemical results, including the synthesis of tens of thousands of compounds in the next decades. As Marcelin Berthelot noted, chemistry became the first science to create its own object.[42]

Substitution of elements or groups of elements ("radicals") into the molecular type became the explanation for the production of compounds, which then could be understood as, say, ammonia or methane "derivatives" ("degenerates"). The theory spawned further explanatory schemes. The carbon "chain" was a "backbone" or "vertebral column" for organic molecules. Why does carbon give birth to so many compounds? The explanation combines the early nineteenth-century equivalency theory and the mid-nineteenth-century carbon-backbone theory: "carbon is a quadrivalent element and it can combine with itself."[43] In the early 1870s, the "explanation" of the rotation of polarized light by organic acids was given by J. Wislicenus, Achille LeBel, and J. H. Van't Hoff by representing elementary "tetravalent" carbon as a regular tetrahedron in space.[44] None of these theories makes use of the indivisibility of the atom, rectilinear or curvilear motion, or centers of force.

In addition, molecular structures, like the structures of organisms, were found to have many overlapping functions. The empirical formula for a substance might be expressed $X_aY_bZ_c$, but further representations of the substance were needed to set out parallel or incommensurate functions of alcohol or aldehyde, acid or base, dependent on circumstances of time and place. During the nineteenth century the abandonment of force causality, on the one hand, and the adoption of a multiple-explanation conventional methodology, on the other hand, led many chemists to forsake the language of "chemical philosophy" and to stress the differences between physics and chemistry. The distinction became only more marked as mathematical physics became more thoroughly based in abstract analytic methods.

Thus, drawing on Baconian and natural history traditions, some nineteenth-century chemists tended to emphasize the empirical and descriptive characteristics of their work, distancing themselves from the practice of speculative theory and finding algebraic representation irrelevant. Perhaps the most extreme statement of this view was that of Marcelin Berthelot, who exerted considerable authority over chemical education and chemical careers in France for the last four decades of the nineteenth century.[45]

An extreme anti-atomist, wielding the truncheon of academic failure, Berthelot dissuaded students from writing their examinations in the notation of atomic weights and symbols used almost everywhere outside France after 1860. Berthelot argued that the notation of equivalent weights is based on chemical analysis and analogies, whereas atomic weights are based on physical hypotheses, and that chemistry is a science in the tradition of natural history, employing a system of classification founded on analogies, types, and functions. Chemistry, he argued, is a positive science, whereas physics relies too much on speculative and metaphysical hypotheses.[46] In this spirit, too, Hermann Kolbe attacked Van't Hoff and Wislicenus, as he had criticized Gerhardt and Laurent for their "pencil and paper chemistry."[47]

Yet chemists attached to "paper chemistry" were usually careful to make limited claims for its correspondence with reality. Berzelius himself wrote that theory

> is only a manner of imagining [se représenter] the interior of phenomena, although in a certain period of development of science, it serves it altogether like a true theory. . . . [With new facts accumulated through the centuries] one will probably change the modes of imagining phenomena in science, without perhaps ever finding the truth . . . it sometimes happens that two different explanations can equally take place: it becomes necessary to study them both . . . the true *savant* . . . studies the probabilities, and gives preference to no opinion unless it is founded on decisive proofs. . . . if we change theory it must fit the known facts better.[48]

On these grounds, Berzelius then introduced his theory of the identity of electricity and chemical affinity.

Edward Frankland was more circumspect:

> I neither believe in atoms themselves, nor do I believe in the existence of centres of forces, so that I do not think I can be fairly charged with this very crude notion.[49]

Auguste Kekulé, who also was a brunt of Kolbe's attacks on "pencil and paper chemistry," defended the use of theoretical hypotheses on the grounds of utility rather than probability:

> The question whether atoms exist or not . . . belongs rather to metaphysics. In chemistry we have only to decide whether the assumption of atoms is an hypothesis adapted to the explanation of chemical phenomena . . . whether a further development of the atomic hypothesis promises to advance our knowledge of the mechanisms of chemical phenomena. . . . I rather expect that we shall some day find, for what we now call atoms, a mathematico-mechanical explanation, which will render an account of atomic weight, of atomicity, and of numerous other properties of the so-called atoms. . . .[50]

Alan Rocke claims that the advocacy of the method of hypothesis, which had been familiar to physicists for a generation, began to replace inductivist rhetoric in chemical circles just about the time that Kekulé was formulating his benzene theory. But the method of hypothesis was a familiar one to chemical leaders like Berzelius and Dumas, and Dalton hardly avoids the fact of his hypothetical reasoning by his, indeed, inductivist rhetoric.[51]

Rocke suggests that, like the atomic theory, the benzene theory was *both* extremely popular, because it was successful, *and* disbelieved in realist terms.[52] In both cases, and in the case of the carbon tetrahedron, there were two compelling problems that deterred most chemists from believing that they had a real, probably true theory. First, neither the chemical atom nor the benzene ring had its own mechanics, that is, a set of laws of motion for what was going on inside the chemical black box. Second, the chemical theories of the atom and the benzene "ring" corresponded to no single explanatory device but to overlapping, partially incommensurate explanations of chemical behavior. The atom and the molecule acted more like organisms than like billiard balls or smoke rings, and rigorous deductions from a simple cause did not work. Chemical explanation was not "philosophical." But the recognition of its "positive" and "conventional" character does not belie an undercurrent of chemists' continued interest in reviving philosophical aims in the future.

Mechanics, the Elusive Dream

Chemistry is more than a descriptive science, said Thomas Graham, the "dean" of English chemistry following Dalton's death and England's most physics-minded chemistry professor at midcentury. Chemistry is not just classification because it embraces the action of bodies on each other.[53] "To suppose that rest, rather than motion, is the normal state of the particles of matter, is at variance with all that we know of the effects of heat, light, and electricity."[54] Thus chemical causality still was understood by Graham, and by his younger German colleague Lothar Meyer, to encompass mechanics, but perhaps not a mechanics of "force."

> If we look for an explanation of this astonishing property by which some atom can be united only to one other atom, while another kind of atom can be united to two, three, or even four, five, six atoms . . . we find ourselves before the door to which chemistry has been knocking for a hundred years without finding an answer.[55]

In order to be "scientific," continued the Tübingen chemist Lothar Meyer, arguably one of the first modern "theoretical" chemists, we must find out the cause of variable valence and how it is that an atom changes its nature according to its immediate environment.

The answer, Meyer thought, lies in the kinetic theory of heat and matter. This physical theory had been given explicit chemical meaning by A. W. Williamson's inference from studies of the synthesis of diethyl ether that atoms in chemical compounds must be continually changing places.[56] Molecules are not empty boxes in translation or rotation, but little Pandora's boxes filled with active entities. The goal of chemistry must be the understanding of chemical phenomena using theories of motion, not just theories of species or types.

Edward Frankland, ex-Manchester chemist and professor at the Royal College of Chemistry, similarly characterized atoms within the molecule as having their own proper motions, vibratory or otherwise, in addition to motion common to the whole molecule. But, he stressed, there also are bounds beyond which no atom can pass without causing rupture and decomposition of the molecule. We must understand the nature of these bounds that define the species or type, as well as the motion against them.[57]

Nor can affinity simply be explained away as heat, insisted Adolphe Wurtz, a leading advocate of chemical and physical atomism in France.[58] There was a thermochemical approach in chemistry, against which Wurtz cautioned. Thermochemistry attempted to solve the conundrum of the "affinity" problem through the phenomena associated with heat. The Danish chemist Julius Thomsen and the French chemist Marcelin Berthelot each tried to define and measure chemical affinity inferentially by the force required to overcome the affinity uniting the parts of a compound. This force, they argued, was measurable by the heat evolved in spontaneous exothermic reactions.[59] But their programs foundered on Thomsen's conflict with organic chemists over the energy content of single and double bonds and on the demonstration by physicist Pierre Duhem and others that the law of maximum work was wrong.[60]

More successfully, J. W. Gibbs, J. H. Van't Hoff and Pierre Duhem developed mathematical equations to describe states of chemical and physical equilibria. In their application of these techniques to chemical reactions, the term "affinity" disappeared.[61] "Arbeit," then "Energie" became the fundamental solvable parameters. Thus, in the new thermodynamics and the new physical chemistry associated with Gibbs, Helmholtz, and Duhem on the one hand, and Van't Hoff, Arrhenius, and Ostwald on the other, the energy term, later "free energy," was to replace chemical "affinity" as the driving force of chemical reactions.[62] Spontaneity and irreversibility were to enter the domain of physics, undermining the classical mechanics of matter and force in which processes are, in principle, reversible. On the one hand, chemists found energy and entropy more congenial to chemical explanation than matter and force; on the other hand, physicists began to recognize that chemical laws, unlike the laws of physics, could not be derived from any single cause or natural force, but, as Oxford chemist Vernon Harcourt insisted, from the whole *course* of chemical change.[63]

Thermodynamics had great appeal for "positive" chemistry, and the kinetic theory of heat had appeal for "philosophical" chemistry. Thermodynamics, whether atomistic or energetic, was a method of explanation grounded in direct, precise measurement and expressible with mathematical rigor. The theoretical concepts of energy and entropy were less problematic than the notion of force, because they were less anthropomorphic and more abstract. Ever since the work of Fourier on an analytic theory of heat, experimental and mathematical tools had been steadily developed by physicists to establish *laws,* not *causes,* of heat phenomena, and to reduce questions involving heat to problems of observation and pure analysis.[64] Thermodynamics created a physical chemistry conceived by contemporaries as a bridging science and even as a fundamental and autonomous theoretical chemistry. Walther Nernst used the terms physical chemistry and theoretical chemistry interchangeably, for example.[65]

But thermodynamics was a mechanics of bulk properties and systems for which the fine detail of the individual molecule was of little importance. This kind of mechanics fulfilled the dream of a rigorous chemical mechanics that was descriptive and inarguably "true." It was ideal scientific explanation, in the view of positivists like Pierre Duhem.[66] But it did not provide the kind of explanation that inspired most chemists in their work. It did not tell them what they wanted to know about the structure of chemical molecules and reaction mechanisms. Thus the return to mechanics, following a long post-

Bertholletian detour, proved a road that was smooth and clean, but inadequate for approaching many chemists' goal of chemical explanation.[67]

Theoretical Chemistry and Philosophy of Chemistry

In a reminiscence of his early work in theoretical chemistry, Joseph Hirschfelder recalled the "Golden Age of the 1920s" when quantum mechanics was first proffered as a key to understanding the inner mechanics of the chemical molecule and the details of the electron valence bond. But, he recalled, "the honeymoon came to an end at the close of the 1930s with the realization that although nature might be 'simple and elegant,' molecular problems were definitely complicated. . . . At this point, the theoretical physicists left the chemists to wallow around with their messy molecules while they resumed their search for new fundamental laws of nature."[68] Theoretical chemist E. Bright Wilson, in talking with S. S. Schweber about differences between physics and chemistry, commented: "I love my molecules."[69]

During the nineteenth century the chemists had learned the detailed behavior of their molecules. Through experience they learned to interpret the same symbol in different ways. When we write a double bond between two atoms, for example in the cases of diphenylethylene, ethylene, and fulvene, noted organic chemist Alfred Stewart, we do not really take the bonds to resemble one another chemically. Chemists know that there is an increase in unsaturation, or reactivity, to bromine or oxygen, as we go from one of these compounds to another. But these lengthening chains of implication about the bond need some principles of order. Chemists know that a "bond" must be looked at not as a fixed unit but as a sum of an infinite number of small forces. Now, he counseled, "we must go to the physicists and from them we can borrow as much of their theory as seems likely to help us with our own branch of science."[70]

In the early 1900s, the aid anticipated from physicists' theory was expected to come from the electron theory. Nineteenth-century chemists realized, but resisted, the notion that "affinity" or "valence" might constitute shared disciplinary terrain with physics. This terrain could be shared only if nineteenth-century chemists gave up the structure theory and returned to primitive notions of chemical affinity as a form of electricity or gravitation. They were not ready to do this during the nineteenth century because eighteenth- and early nineteenth-century conceptions of gravitation and electricity remained too simplistic to explain the complexity of chemical combinations and reactions.

The electron valence theory that provided the most help to chemists was the creation of G. N. Lewis and Irving Langmuir, working independently of each other following tentative applications of electron ideas to molecules by J. J. Thomson and a few other physicists in the early 1900s. For Lewis, who spelled out some of his views on the methods of chemistry in the book *The Anatomy of Science,* "force" was a concept that had fallen out of use because it purported to supply a single answer to the problem of causation.[71] The methods of the chemist and the physicist differ. The physicist takes a small number of measurements, works out an equation or a model following laws close to classical mechanics or electromagnetism, and arrives at a more complex model, which allows calculation of the unknown. The chemist follows a method that is "convergent rather than divergent," a method that is less exact and involves much more numerous data, much of it not metrical in character. Rough generalizations, like Mendeleev's periodic table, suffice for chemists who do not require rigorous logic.[72] Chemists use "little of the formal and logical modes of reasoning. Through a series of intuitions, surmises, fancies, they stumble upon the right explanation."[73] More recently, the Canadian materials scientist D. K. C. MacDonald similarly reiterated the point:

> A physicist tries to *reduce* . . . theory to as few parameters as possible and thinks a theory is respectable if it offers reasonably specific means for possible refutation. . . . The chemist regards theories as far less sacrosanct and is prepared in extreme cases to modify them continually as new pieces of experimental evidence come in. But a physicist is uneasy about this because he finds it difficult to see how a theory in this sense can ever be proved wrong.[74]

In twentieth-century chemistry, two basic approaches have been taken to explain the character and behavior of individual molecules. One is the theory of "reaction mechanisms" and the "curly arrow" pioneered by Arthur Lapworth, Robert Robinson, and Christopher Ingold. The other is the quantum theory of atomic and molecular oribtals, pioneered by Linus Pauling and Robert Mulliken, among others. The first is a strongly visual and pictorial theory employing the static structural and stereochemical images of late nineteenth-century chemistry, in combination with "the most successful fiction of the conceptual model, the curly arrow" to represent and suggest dynamic links between structural formulas.[75] The latter is an abstract and mathematical theory employing the probabilistic electron wave mechanics of twentieth-century physics.

Most chemists have argued that the daily work of organic chemistry is best served by the old pictorial theories; wave mechanics has corroborated nearly all the suppositions of the geometrical structure theory, for example, the tetrahedral building plan of the carbon atom, the bent structure of a water molecule, and fractional bond lengths in conjugated organic compounds.[76] The quantum mechanical model is the most "theoretical" and "fundamental" of the kinds of models chemists use, and it puts them closest to the old aims of chemical philosophy for logical rigor and "geometric" certainty in their knowledge.

But, while in principle all chemical properties can be predicted from a knowledge of the structure and behavior of particles configuring the chemical molecule, the old problem of the "empirical" versus the "rational" formula persists. So complex are the possible interactions among these electron particles that one must use either an exact mathematical model of a simplified system or a simplified model of the complete system. The adoption of "semiempirical" methods is "a threefold compromise between rigor, experiment and intuition." Some of the integrals in the Schrödinger equation are replaced by empirical parameters from reference compounds or other information chemists already have at hand.[77] And theoretical chemists' quantum mechanical calculations employ a fixed, or "clamped," nucleus approximation in which nuclei are treated as classical particles confined to "equilibrium" positions. Without such an approximation, quantum mechanics yields no recognizable molecular structure.[78]

If mechanics has always been an aim of scientific philosophy, then twentieth-century chemistry has revived its philosophical character. But chemists, perhaps more than physicists, remain self-conscious about the play back and forth between the phenomena revealed in operations of the laboratory and the symbols employed in the operations of mathematics. Precision, not rigor, is characteristic of chemical methodology; and parallel representation, not singular principle, is characteristic of chemical explanation. Whereas many twentieth-century physicists regarded conventionalism, complementarity, and indeterminacy as concessions of failure in their philosophical enterprise, chemists were not surprised that electrons and atoms, like molecules, were hard to predict.

It is a pleasure to acknowledge support for my research from the National Endowment for the Humanities, the National Science Foundation, and the University of Oklahoma. This paper was written while I enjoyed residence at the Rutgers Center for Historical Analysis at Rutgers University in New Brunswick, New Jersey.

Notes

1. E. F. Caldin, *The Structure of Chemistry in Relation to the Philosophy of Science* (London and New York: Sheed and Ward, 1961).

2. F. A. Paneth, "The Epistemological Status of the Chemical Concept of Element," *British Journal for the Philosophy of Science* 13 (1962): 1–14, 144–160, on p. 2 (translation of a lecture originally given in 1931).

3. I. Bernard Cohen, preface to Arnold Thackray, *Atoms and Powers: An Essay on Newtonian Matter-Theory and the Development of Chemistry* (Cambridge, Mass.: Harvard University Press, 1970), p. xxiii.

4. On the atom, see Alan J. Rocke, *Chemical Atomism in the Nineteenth Century: From Dalton to Cannizzaro* (Columbus: Ohio State University Press, 1984); and Mary Jo Nye, "The Nineteenth-Century Atomic Debates and the Dilemma of an 'Indifferent Hypothesis,'" *Studies in the History and Philosophy of Science* 7 (1976): 245–268.

5. John H. Brooke and Alan Rocke have addressed issues other than atomism in the epistemology of chemistry. Brooke, "Laurent, Gerhardt, and the Philosophy of Chemistry," *Historical Studies in the Physical Sciences* 6 (1975): 405–429; Rocke, "Hypothesis and Experiment in the Early Development of Kekulé's Benzene Theory," *Annals of Science* 42 (1985): 355–381; Rocke, "Kekulé's Benzene Theory and the Appraisal of Scientific Theories," in *Scrutinizing Science,* ed. Arthur Donovan et al. (Dordrecht: Kluwer Academic Publishers, 1988), pp. 45–161; and J. H. Brooke, "Methods and Methodology in the Development of Organic Chemistry," *Ambix* 34 (1987): 147–155.

Also, Hannah Gay, "Radicals and Types: A Critical Comparison of the Methodologies of Popper and Lakatos and their Use in the Reconstruction of Some Nineteenth-Century Chemistry," *Studies in the History and Philosophy of Science* 7 (1976): 1–52. On Kuhnian versus Popperian models, H. W. Schütt, "Guglielmo Körner (1839–1925) und sein Beitrag zur Chemie isomerer Benzolderivate," *Physis* 17 (1975): 113–125.

Philosophical criteria for theory choice are explored in Arthur Zucker, "Davy Refuted Lavoisier Not Lakatos," *British Journal for the Philosophy of Science* 39 (1988): 537–540; and H. Zandvoort, "Macromolecules, Dogmatism, and Scientific Change: The Prehistory of Polymer Chemistry as Testing Ground for Philosophy of Science," *Studies in the History and Philosophy of Science* 19 (1988): 489–515.

6. See Gaston Bachelard, *La pluralisme cohérent de la chimie moderne* (Paris: Vrin, 1932) and François Dagognet, *Tableaux et langage de la chimie* (Paris: Seuil, 1969); *Écriture et iconographie* (Paris: Vrin, 1973).

7. Ilya Prigogine, *From Being to Becoming—Time and Complexity in the Physical Sciences* (San Francisco: Freeman, 1980); Ilya Prigogine and Isabelle Stengers, *Order Out of Chaos: Man's New Dialogue with Nature* (New York: Bantam Books, 1984; first appeared as *La nouvelle alliance métamorphose de la science* (1979); Isabelle Stengers and Judith Schlanger, *Les concepts scientifiques: invention et pouvoir* (Paris: Éditions La Découverte, 1988); and Bernadette Bensaude-Vincent, *À propos de "méthode de nomenclature chimique." Esquisse historique suivie du texte de 1787* (Paris: Centre de Documentation des Sciences Humaines, 1983).

8. In addition to Stengers, Bensaude-Vincent, and Rocke, mentioned above, see Michelle Goupil, "Transferts mutuels de vocabulaire: psychophysiologie et physicochimie," in *Transfert de vocabulaire dans les sciences,* ed. Martine Groult, Pierre Louis, and Jacques Roger (Paris: Éditions du Centre National de la Recherche Scientifique, 1988), pp. 125–133.

9. Colin J. Suckling, Keith E. Suckling, and Charles W. Suckling, *Chemistry Through Models: Concepts and Applications of Modelling in Chemical Science, Technology, and Industry* (Cambridge: Cambridge University Press, 1978).

10. Jean Jacques, *Confessions d'un chimiste ordinaire* (Paris: Seuil, 1981), p. 5.

11. See Mary Jo Nye, *Science in the Provinces: Scientific Communities and Provincial Leadership in France, 1860–1930* (Berkeley: University of California Press, 1986), p. 286, n.100.

12. Robert Lespieau, "Sur les notations chimiques," *Revue du Mois* 16 (1913): 257–278, on p. 259. Also, on Deville and Berthelot see Mary Jo Nye, "Berthelot's Anti-Atomism: A 'Matter of Taste'?" *Annals of Science* 38 (1981): 585–590.

13. John Dalton, *A New System of Chemical Philosophy* (Manchester, 1808). For Williamson, see A. W. Williamson, "On the Atomic Theory," *Journal of the Chemical Society (London)* 22 (1869): 328–365; followed by "Discussion of Dr. Williamson's Lecture on the Atomic Theory," ibid., 433–441.

14. John Bradley, "On the Operational Interaction of Classical Chemistry," *British Journal of the Philosophy of Science* 6 (1955–1956): 32–42, on p. 32.

15. Henry Margenau, *The Nature of Physical Reality: A Philosophy of Modern Physics* (New York: McGraw-Hill, 1950), p. 99, n. 1.

16. Bas Van Frassen, "Interpretation in Science and the Arts," paper given at CCACC Conference on "Realism and Representation," Rutgers University, November 10–12, 1989.

17. Owen Hannaway, *The Chemist and the Word: The Didactic Origin of Chemistry* (Baltimore: Johns Hopkins University Press, 1975), pp. 75–79. H. W. Schütt claims for Robertus Vallensis the status of author of the first chemical text and history of chemistry, *De veritate et antiquitate artis chemicae* (1561); in Hans-Werner Schütt, "Chemiegeschichtsschreibung—'Zu welchem Ende'?" *Chemie in Unserer Zeit* 22 (1988): 139–145, on p. 140.

18. Hannaway, *The Chemist and the Word*, pp. 95–96, 109, 121–123, 141–142.

19. Jacques Barner, *Chymie philosophique* (1689), a system of acids and alkalis, written by the physician to the king of Poland; Antoine F. Fourcroy, *Philosophie chimique ou vérités fondamentales de la chimie moderne* (Paris, 1792; this is a reprint of the article "Axiomes" from vol. 2 of the *Encyclopédie méthodique*, article on "Chymie"); John Dalton, *A New System of Chemical Philosophy* (note 13); Humphry Davy, *Elements of Chemical Philosophy* (London, 1812); Jean-Baptiste Dumas, *Leçons sur la philosophie chimique* (Paris, 1837).

On the "chemical philosophy" of the Paracelsian tradition, see Allen G. Debus, *The Chemical Philosophy: Paracelsian Science and Medicine in the 16th and 17th Centuries*, 2 vols. (New York: Science History Publications, 1977); for a later period, see Arthur Donovan, *Philosophical Chemistry in the Scottish Enlightenment: The Discoveries of William Cullen and Joseph Black* (Edinburgh: Edinburgh University Press, 1975).

20. The most famous work of "chemical philosophy" in the second half of the nineteenth century is Stanislao Cannizzaro, "Sketch of a Course of Chemical Philosophy" (1858; reprint, Edinburgh: Alembic Club reprint no. 18, 1947), in which he argues the identity of the chemical and physical atom.

Part II of Benjamin Silliman, Jr.'s *First Principles of Chemistry, for the Use of Colleges and Schools* (Philadelphia, 1847) is entitled "Chemical Philosophy," encompassing laws of combination, nomenclature, affinity, crystallization, and chemical effects of electricity. Josiah P. Cooke published in 1860 the *Elements of Chemical Physics* (Boston: Little, Brown & Co., 1860), which was intended to be the first volume of an extended work on the "Philosophy of Chemistry" (see p. vi).

21. A. Baumé and P. J. Macquer, *Plan d'un cours de chymie expérimentale et raisonnée avec un discours historique sur la chymie* (Paris: Herissant, 1757), pp. 1–7.

22. Ibid., p. 8.

23. Peter Shaw, *A New Method of Chemistry,* 2 vols. (London, 1741), p. 173 n.; quoted in Shaffer, p. 65. On Betty Jo Dobbs's interpretation in *The Hunting of the Greene Lyon* (Cambridge: Cambridge University Press, 1975), see Prigogine and Stengers, *Order Out of Chaos* (note 7), p. 64. For the application of Newton's ideas in the Queries to the *Opticks,* see John Keill, *Introduction to Natural Philosophy* (1708; translation of 4th Latin ed., 1720) and John Freind, *Praelectiones Chymicae* (1709); or later, Georges Louis LeSage, *Essai de chymie mécanique* (1758).

24. Adam Walker, *A System of Familiar Philosophy in Twelve Lectures* (London, 1799), p. 144. In Newton's *Principia* the wording is similar.

25. Humphry Davy, *Elements of Chemical Philosophy* (note 19), p. 1. For natural theology and chemical philosophy, see William Prout's Bridgewater Treatise, *Chemistry, Meteorology and the Function of Digestion Considered with Reference to Natural Theology* (London: Pickering, 1834).

26. J. B. Dumas, *Leçons de philosophie chimique* (note 19), pp. 1–2.

27. See A. W. Hofmann, "The Life Work of Liebig in Experimental and Philosophic Chemistry," Faraday Lecture, March 18, 1875 (London: Macmillan, 1876), pp. 29, 35–36.

28. Antoine Lavoisier, *Elements of Chemistry* (1789), trans. Robert Kerr (Edinburgh: William Creech, 1790; reprint, New York: Dover, 1965), pp. xix–xx, xxvi; and Antoine Fourcroy, *Philosophie chimique* (note 19). J. A. Deluc notes that Fourcroy's "chemical philosophy" is better understood to mean the "new chemistry" of Lavoisier. See Deluc, *Introduction à la physique terrestre . . . précedé de deux memoires sur la nouvelle théorie chimique* (Paris: Nyon, 1803), p. 167.

29. Antoine Lavoisier, pp. 193–194 in vol. 1, *Traité élémentaire de chimie,* 2 vols. (Paris: Cuchet, 1789); in Dover edition (note 28), pp. 176–177.

30. Lavoisier, quoting Condillac, in the Dover edition (note 28), pp. xiii–xiv; on analysis and synthesis, 33; on analogy, 16–17, 61, 146, 156, 168.

31. See Stephen J. Toulmin, *Cosmopolis: The Hidden Agenda of Modernity* (New York: Free Press, 1990), for a very recent analysis of this seventeenth-century program.

32. See Charles C. Gillispie, *The Edge of Objectivity: An Essay in the History of Scientific Ideas* (Princeton, N.J.: Princeton University Press, 1970), pp. 242–245; and A. L. Lavoisier, "Considérations générales sur la dissolution des métaux dans les acides" (1782) in *Oeuvres de Lavoisier* (Paris, 1862), 2: 509–527.

33. John Robison, quoted in Keith Nier, *The Emergence of Physics in Nineteenth-Century Britain as a Socially Organized Category of Knowledge: Preliminary Studies* (Ph.D. thesis, Harvard University, 1975), p. 88.

34. Dalton, *A New System of Chemical Philosophy* (note 13), p. 141.

35. Jöns Jakob Berzelius, *Essai sur la théorie des proportions chimiques et sur l'influence chimique de l'électricité,* [1819] introd. Colin Russell (New York:Johnson reprint, 1972), pp. 21, 23.

36. See Justus von Liebig, *Introduction à l'étude de chimie* (Paris, 1837), p. 110.

37. Berzelius to Liebig, in J. Carriere, ed., *Berzelius und Liebig: Ihre Briefe von 1831–1845* (2d ed., 1898), p. 206.

38. Thomas Graham, *Elements of Chemistry: Including the Application of the Science in the Arts,* 2d ed., 2 vols. (London: Ballière, 1850, 1858), 1:234.

39. Dumas, *Leçons* (note 19), pp. 233–234. See later, Henri Poincaré in *Science and Hypothesis,* W. J. Greenstreet, trans. (New York: Dover Publications, 1952), p. 152.

40. J. B. Dumas, "Memoire sur la constitution de quelques corps organiques et sur la théorie des substitutions," *Comptes Rendus* (1839): 609–622.
41. J. R. Partington, *A History of Chemistry* (London: Macmillan, 1964), 4:339.
42. Marcellin Berthelot, *La synthèse chimique* (Paris: Baillière, 1876), p. 275.
43. Charles Friedel, *Cours de chimie organique, professé à la Faculté des Sciences, Paris* [1886–1887] (Paris, 1887), p. 2.
44. Ibid., p. 37.
45. For an excellent, if unsympathetic, biography of Berthelot, see Jean Jacques, *Berthelot: Autopsie d'un mythe* (Paris: Belin, 1987).
46. Marcellin Berthelot, *Leçons sur les méthodes générales de synthèse en chimie organique* (Paris, 1864), pp. 453–454 n. 5; and 521–523.
47. See H. Kolbe, *Journal für praktische Chemie* 14 (1877): 268; 15 (1877): 473, cited in Partington, *History of Chemistry* (note 41), p. 503; and Rocke, *Chemical Atomism* (note 4), p. 235. For Kolbe's general views on this subject, see "Meine Betheiligung an der Entwickelung der theoretischen Chemie," *J. prak. Chemie* 23 (1881): 305–323, 353–379, 497–517; 24 (1881): 374–425. Kolbe was editor of the journal.
48. J. J. Berzelius, *Essai sur la théorie des proportions* (note 35), pp. 18–19.
49. Edward Frankland, in "Discussion," pp. 302–305 following Benjamin Brodie's paper, "On the Mode of Representation afforded by the Chemical Calculus, as Contrasted with the Atomic Theory," *Chemical News* 15 (1867): 295–302; both reprinted in *Classical Scientific Papers,* ed. David Knight (New York: American Elsevier, 1968), p. 250 from original text, p. 302 in Knight.
50. Auguste Kekulé, "On Some Points of Chemical Philosophy," *Laboratory* 1 (1867): 303–306; reprinted in Knight, 255–258, p. 304 in original, p. 257 in Knight.
51. Rocke, "Kekulé's Benzene Theory" (note 5), pp. 156–157.
52. Ibid.
53. Graham, *Elements of Chemistry* (note 38), 1: 217.
54. Ibid., 2: 600–603.
55. Lothar Meyer, *Die Modernen Theorien der Chemie und ihre Bedeutung für die Chemische Statik,* 3d ed. (Breslau, Maruschke und Berendt, 1877), pp. 158–159.
56. Meyer, *Les théories modernes de la chimie et leur application à la mécanique chimique,* trans. from 5th German ed. by Albert Bloch (Paris: Carré, 1887), 1:viii.
57. Edward Frankland, "Contributions to the Notation of Organic and Inorganic Compounds," in *Experimental Researches* (London, 1877): 4–25, on pp. 4–5.
58. Adolphe Wurtz, *Introduction à l'étude de la chimie* (Paris: Masson, 1885), pp. 8–9.
59. See Helge Kragh, "Julius Thomsen and Classical Thermochemistry," *British Journal for the History of Science* 17 (1984): 255–272; Marcellin Berthelot, *Essai de mécanique chimique* (Paris, 1879), p. 259.
60. Kragh, "Julius Thomsen," pp. 264–266; Pierre Duhem, *Le potential thermodynamique et ses applications à la mécanique chimique et à la théorie des phénomènes électriques* (Paris, 1886).
61. It reappeared in the work by the Belgian physicist Théophile De Donder, *L'Affinité* (Brussels: Lamertin, 1927; rev. ed. Paris: Gauthier-Villais, 1936).
62. See comments by Pierre Duhem in *Le mixte et la combination chimique* (Paris: Naud, 1902), pp. 168, 171, 185ff., on chemical mechanics, founded on thermodynamics, as the key to chemical questions.

63. Vernon Harcourt, cited by Christine King, "Experiments with Time: Programs and Problems in the Development of Chemical Kinetics," *Ambix* 28 (1981): 70–82, on p. 70.

64. See Robert Kargon, on Fourier's physics and its influence, in "Model and Analogy in Victorian Science: Maxwell's Critique of the French Physicists," *Journal of the History of Ideas* 30 (1969): 423–436, on pp. 427–428.

65. On physical chemistry, see John Servos, *Physical Chemistry from Ostwald to Pauling: The Making of a Science in America* (Princeton, N.J.: Princeton University Press, 1990); and Walther Nernst, preface to the German ed., *Theoretical Chemistry: From the Standpoint of Avogadro's Rule and Thermodynamics,* trans. Charles S. Palmer (London: Macmillan, 1895), p. xii.

66. Pierre Duhem, *La théorie physique: son objet et sa structure* (Paris, 1906).

67. See Lothar Meyer's explicit remarks to this effect, in *Die Modernen Theorie der Chemie* (note 55), pp. 162–163.

68. Joseph O. Hirschfelder, "My Adventures in Theoretical Chemistry," *Annual Review of Physical Chemistry* 34 (1983): 1–29, on p. 1.

69. Quoted in S. S. Schweber, "The Young John Clarke Slater and the Development of Quantum Chemistry," *Historical Studies in the Physical and Biological Sciences* 20 (1990): 339–406, on p. 404.

70. A. W. Stewart, *Recent Advances in Organic Chemistry,* 1st ed. (London: Longman, Green, & Co., 1908), pp. 263, 266–267.

71. Gilbert N. Lewis, *The Anatomy of Science* (Washington, D.C.: American Chemical Society, 1926; reprint, Freeport, N.Y.: Books for Libraries Press, 1971), pp. 99–102.

72. Ibid., pp. 168–169. On convergence, also see Suckling et al., *Chemistry Through Models* (note 9), p. 7.

73. Ibid., p. 6.

74. D. K. C. MacDonald, "Physics and Chemistry: Comments on Caldin's View of Chemistry," *British Journal for the Philosophy of Science* 11 (1960): 222–223.

75. See Suckling et al., *Chemistry Through Models,* pp. 117–120.

76. See J. J. Mulckhuyse, "Molecules and Models," pp. 257–275 in *The Concept and the Role of the Model in Mathematics and Natural and Social Sciences,* vol. 12, *Synthese,* 1960, p. 258.

77. See Suckling et al., *Chemistry Through Models,* pp. 65, 135.

78. See R. G. Woolley, "Must a Molecule Have a Shape?" *Journal of the Chemical Society (London)* 100 (1978): 1073–1078; discussed in Stephen J. Weiniger, "The Molecular Structure Conundrum: Can Classical Chemistry be Reduced to Quantum Chemistry?" *Journal of Chemical Education* 61 (1984): 939–944.

TECHNICAL

Uses and Images of Instruments in Chemistry

Yakov M. Rabkin

Introduction

Scientific instruments[1] have a special relationship with chemistry. Some even argue that chemical instruments came to be used well before chemistry became a science.[2] Unlike physics, biology, or geology, all of which have had significant observational components, chemistry developed as an experimental discipline. Its emergence therefore depended heavily on the use of scientific instruments.

A historian of analytical chemistry claims, perhaps not unexpectedly, that "analytical chemistry is the mother of modern chemistry."[3] A broader history of modern chemistry by Aaron Ihde also ascribes to analytical chemistry

> a position of primary importance since only through chemical analysis can matter in its variety of forms be dealt with intelligently. The stimulus given to chemistry by new analytical approaches, either qualitative or quantitative, has been repeatedly observed. In general, however, analytical chemistry has never achieved recognition in keeping with its importance because the application of new techniques has resulted in new descriptive or theoretical knowledge that completely overshadows the technique which made the knowledge possible.[4]

Indeed, the traditional emphasis historians of science have put on theory as the motor of scientific development tends to obscure the roles of instrumentation that are at the root of progress in chemical analysis. Consequently, the instrument has acquired the appearance of a tool manufactured expressly for the chemical investigator intent on making an ultimate theoretical breakthrough. This imagery is related to the commonplace subordination of technology to science in much of the existent literature on the subject.[5]

Industrial history may provide a different perspective on the history of chemical instruments. It deserves to be carefully examined with this objective in mind.[6] Chemistry's old organic links with chemical industries are well recognized by the historians of chemistry. But this recognition and the subsequent urge to enhance the theoretical status of chemistry may in fact reinforce the dominant image of chemical apparatus as purely instrumental. Indeed, the history of modern chemical technology has been developed by scholars other than historians of chemistry. The latter probably share the myth that technology is derivative, "relatively speaking, unworthy of scholars' energies and interests, which are much better focused on the originative, creative subject, science."[7] One way of transcending this tradition in the history of chemistry is to show that instruments, often produced *for* the chemical industry, but certainly not *by* it, play a role whose confines and nature are more than instrumental. This may present a refreshing angle on the science-technology relationship.

Historians of other disciplines, notably physics and astronomy, have given greater prominence to the roles played by instruments in scientific research. In fact, physics seems to be almost the only discipline whose rapport with instruments has been undergoing investigation in the literature on the history and sociology of science.[8] However, it would be unwise to extrapolate findings about physics, whatever the diversity within the discipline itself, on the history of chemistry. No two scientific disciplines are alike,[9] and chemistry, because of its fundamental relationship with the instrument, must be understood in its own right. One should distinguish between the history of chemical instruments and the history of instruments associated with other scientific disciplines. One may wonder, for example, about the relevance of a recent study of experimentation on elementary particles, perhaps the most elusive physical phenomenon, to the historical understanding of chemistry.[10]

An instructive example concerns visual representations of Kekulé's formula for benzene. In 1865 he postulated, reportedly in the wake of a dream, that the molecule consisted of six carbon atoms in a ring. Many years later the field emission microscope showed shadows of the molecule that proved Kekulé right. The physical theory underlying the initial image had to be discarded eventually in favor of a totally different one, but for chemists the image was still true. "It made no whit of difference. People [more precisely, chemists] kept on regarding the micrographs of the molecules as genuinely correct representations."[11] This difference in the way chemists and physicists relate to the experiment and its interpretation should serve as a note of caution to historians of chemistry about cross-disciplinary extrapolations. This episode seems to support the view that "the instrument presents a phenomenon; this insures the reliability and validity of the information supplied by the instrument, even when there is no way to know just what property the instrument is presenting to us."[12]

There is another perhaps more mundane distinction between chemical and other scientific instruments that reinforces the validity of this "separatist" approach. Throughout the formative centuries of modern science, astronomical and physical instruments attracted the attention and interest of royalty, aristocracy, and bourgeoisie, whose houses they often adorned:

> De tous les objets de collection, les instruments scientifiques semblent avoir la faveur exclusive du sexe fort. Cela tient peut-être au fait que si la femme a la faculté de donner le jour à des mécaniques parfaites parce que vivantes et pensantes, l'homme a trouvé dans cette sorte de collectionomanie, un substitut quelque peu imparfait mais séduisant à ce besoin obsessionnel: Créer, créer sans cesse.[13]

But chemical instruments never acquired the same obsessive appeal. They were rarely collector's items because few amateurs could handle a chemical instrument the way they did an astrolabe. Consequently, the study of chemical instruments cannot rely on the instruments kept in private and public collections, as, for example, is often the case with astronomical instruments.[14] Because of the idiosyncrasies of individual collectors, collections of surviving physical or astronomical instruments offer a dubious basis for understanding the roles of instruments in the development of these scientific disciplines. Popular science also played a role in the dissemination of chemical apparatus. But magicians and other manipulators who would use chemical instruments for popular entertainment were of too humble estate to indulge in collecting such apparatus. Thus, luckily perhaps, our knowledge of chemical apparatus has to derive mainly from the history of chemistry rather than from antiquarian collections.

The point of this work is not to glorify the instrument-maker, and certainly not to rehabilitate an econocentric view of science, as if it were motivated purely by the exigencies of material production. Rather, it is to improve our understanding of the role of instruments in the formulation of research objectives and their orientation. This objective is distinct from attempts to reveal the influence of instruments on "the political economy of experimental practice and the culture of laboratory."[15]

Consequently, one purpose of this research is to explore how the diffusion of an instrument may affect the character and scope of chemical investigation. It will be argued that scientific instruments, even when these are produced at the express request of scientists and in accordance with their specifications, do not remain passive tools in the hands of the active researcher. Physics offers abundant examples of how a new measuring device can "boost" a field of research.[16] The acquisition of equipment, particularly when it constitutes a major investment, can shape the experiment and therefore the entire scientific process. While in some instances one may indeed wonder whether the scientist "uses" the instrument or the instrument conditions the thinking of the scientist, the purpose is not to argue for any strong, and usually misleading, program, of instrumental determinism or autonomous technology.[17]

A subsidiary purpose of this work is to analyze the impact of the chemical industry as the user of instruments on their diffusion and subsequent contribution to chemical research. The advent of serial, mass-produced scientific instrumentation increased the ease of exploitation. This led to a certain alienation of the scientist from the actual design of the instrument, particularly in the twentieth century. However, even in earlier centuries the production of instruments, mainly for astronomy and physics, was often affected by nonresearchers, popularizers of science, or instrument collectors: the phenomenon may not be quite so recent as we might assume. Another related point concerns the role of instrumentation in the emergence and development of the chemical profession. Because the use of instrumentation is indeed a major factor in the development of chemistry, we shall also examine possible connections between the instrument-building capacity of a country and its relative standing in the world of science. This query may help us better understand factors of inequality in the world distribution of scientific activity.

Roles of Instrumentation in Chemical Research

Chemical instrumentation originally developed in response to a simple question of everyday life: "What is in this object, and how much?" This question was asked by goldsmiths, tanners, potters, in fact by anyone who would receive coins in payment. Assaying of precious metals was, of course, among the more prestigious uses of chemical analysis from biblical times. The increase in precision of instruments such as balances was a crucial technological factor for the development of chemistry. The emergence of modern chemistry occurred at the very time when precision balances made their appearance in Britain and on the Continent.[18] Indeed, it has long been affirmed that chemistry "has been created chiefly by the careful use of the balance."[19] Antoine

Lavoisier and Dmitri Mendeleev, to name just two giants of chemistry, repeatedly emphasized the importance of quantification, only attainable through the use of measuring instruments, for the development of their science. This fundamental link between chemical research and its instruments has survived momentous changes in the history of chemistry. While theoretical studies in chemistry have been at times hard to distinguish from physics, the diffusion of physical instruments into chemistry has so far not produced a confusion with physics but rather reinforced the experimental, instrument-intensive character of chemical research.[20] The observation that "while theories may be conceived as a species of belief, it is difficult to conceive of experiment as a species of belief"[21] is particularly appropriate for the history of chemistry.

The birth of modern chemistry is usually associated with Antoine Laurent Lavoisier (1743–1794). His introduction of precision measurements offers a good perspective on the early interaction between the instrument and the conceptual development of chemistry.

"The prime instrument in Lavoisier's quantification was the balance," writes Trevor Levere. A significant leap in accuracy occurred around the middle of the eighteenth century. Was the new precision balance developed *for* Lavoisier or were those advances of measuring technologies made for other reasons and other customers? To quote Levere again,

> there was a dynamic process in which the needs of science sometimes required the development of new apparatus—Lavoisier's use of the balance provides a prime example—but in which the development of new standards and tools of craftsmanship also indicated and sometimes dictated the development of science.[22]

While the former observation is unexceptional for the mainstream history of chemistry, the latter part of the quotation confirms that even in this most fundamental episode in the history of chemistry the producers of instruments could actually *dictate* the development of science. Lavoisier was also shown to be "guided by his instruments in imposing logic on his results."[23] In other words, the instrument played a role that went well beyond the purely instrumental.

Lavoisier himself acknowledged this new role of chemical instruments when he found the main reason for the delay in the reception of his results in the lack of affordable instrumentation needed for proper verification of his results. Nobody would "make precise instruments in order to test theories."[24] The instrument was essential for both the conceptualization and the acceptance of scientific data developed by Lavoisier. Moreover, even Lavoisier, a man of considerable wealth, found the instruments, particularly the gasometer, developed earlier in the century by Stephen Hales, to be very expensive. The credit Lavoisier accorded to his costly instrumentation foreshadows the modern phenomenon of formulating the research program in accordance with the expensive apparatus acquired earlier. Indeed, Lavoisier can be considered the founder of modern chemistry in more than one sense.

Unlike the balance, the microscope is primarily an observational instrument. It was developed by Antonie van Leeuwenhoek (1632–1723) without chemical applications in mind. However, it too opened new opportunities to chemists. Microanalysis, that is, identification of minute amounts of substances, is based on the investigation under the microscope of crystals formed with different reagents. While the microscope had come to be used mainly for medical and biological research, its application also enabled chemists to investigate the nature and mechanisms of catalysis (vanadium was thus ascertained to be a catalyst for several organic reactions). The availability of the microscope oriented chemical research of catalysis in a direction that no available theory would have otherwise suggested.[25] The microscope became a standard tool for approaching new substances and phenomena in chemistry. Reasons for using the microscope, and later its electronic offspring, became basically akin to Galileo's reasons to use the telescope, namely to find out what a new instrument might reveal. Such use, which is traditional in astronomy, is quite common in chemistry as well, in spite of eventual rationalizations of experiments that may appear in print.

Electrochemistry, which appears to be a case by itself in the history of chemistry, owes its very existence to the invention of the electric pile by Alessandro Volta (1745–1827). The pile consisted of plates of different metals and offered a convenient source of electricity that soon attracted the attention of chemists. Many experimenters used the voltaic pile to investigate the phenomenon of electrolysis and ascertain properties of various compounds.

Much of this experimentation began without a theory or even a clear purpose. This is how within a few years the voltaic pile enabled Humphry Davy (1778–1829) to discover the electric character of chemical bonding, which had a profound effect on the development of chemistry. It also led to the discovery of a number of new elements. Davy's protégé Michael Faraday (1791–1867) is reputed to have been "content to uncover the secrets of nature, leaving it to others to develop the major theoretical ideas and to reap the practical benefits."[26]

The existence of this type of researcher agrees well with the role of scientific instruments as a motive force in the development of chemistry. To quote Baird and Faust, "one aspect of scientific progress [is] the accumulation of new scientific instruments."[27] Experiments with electrolysis enabled Faraday to formulate his idea of electrochemical equivalents, that is, a relationship between the amount of electricity and the quantity of substance decomposed. This discovery would be of great importance for the advances of atomic theories several decades after his death.[28] While in the history of organic analysis advances in instrumentation were largely made at the request of research chemists (e.g., Lavoisier, Gay-Lussac, Berzelius, or Liebig) for purposes of promoting chemical knowledge, the progress of instrumentation in electrochemistry had little to do with the needs of chemical science. The case of electrochemistry shows how a new instrument, the voltaic pile (which may not be an instrument in other contexts), was developed for purposes quite distinct from chemical research and made a broad and durable impact on the evolution of chemical knowledge. Similarly, most advances in spectroscopy, which led in the nineteenth century to the discovery of dozens of new elements, came from outside the expanding bounds of chemistry. Physical phenomena that constitute a legitimate object of research in physics have often become instruments in the context of chemistry. This situation emphasizes the distinction made earlier in this essay between the two disciplines brought out on the example of Kekulé's formula for benzene.

An important exception to this pattern can be found in the work of Robert Wilhelm Bunsen (1811–1899) and Gustav Robert Kirchhoff (1824–1887), who adapted the spectroscope for the express purpose of investigating chemical elements. In turn their instrument, which "embodied no new principles but brought together the necessary collimating and viewing tubes on a single stand for convenient operation,"[29] contributed not only to the development of chemical analysis but also to the discovery by astronomers of the chemical composition of different planets. Reasons for the development and functions of the use of the instrument turned out to be different and should be distinguished by the historian. The development of the classification of elements, a major contribution to the theoretical foundations of chemical research, ensued from instrumental advances in electrochemistry and spectroscopy.

Instruments at the Chemistry-Industry Interface

The very emergence of modern chemistry in the eighteenth century is intimately linked with a new generation of instrument-makers who were satisfying diverse customers: navies in need of precision chronometers to determine longitude at sea, amateurs doing experimental demonstrations, and, ultimately, research scientists. But by concentrating on science and ignoring industry we miss important markets for chemical instruments, and so also miss important stimuli for the development of instrumentation that, as we have just seen, affect the nature and direction of chemical research.

The pivotal role of industrial uses in the progress of chemical instrumentation is rarely reflected in the history of chemistry. There are, however, two important exceptions: the microscope and the polarimeter, often dubbed the saccharimeter. Needs of medical and biological research are credited for frequent improvements in the microscope, while the development of manufacturing and international commerce in sugar is rightly considered the main motive force behind the refinement and the routinization of the use of the polarimeter.[30] Whatever the significance of these two cases, the historiography of chemistry usually highlights requests of scientists as the stimulus for the design and amelioration of chemical instrumentation.

My earlier work suggests that not only instruments, but in fact "extraneous" forces that regulate their development and production, affect the conduct of scientific research. This view was largely based on an analysis of infrared spectroscopy and later expanded to other instruments used in twentieth-century chemistry.[31] The spectroscopy example clearly shows how the war effort, and more specifically the impact of military procurement on petroleum refining, affected the design and, later, the diffusion of the new instrument into chemical research. It challenges the traditional view of the instrument as a tool built in response to the needs of theory-driven scientists.

The history of refractometry offers a similar instance of diffusion of a new scientific instrument from industry to research laboratories. The phenomenon of optical refraction has been identified since the seventeenth century. This physical phenomenon finds itself at the basis of a new analytical instrument, in that refractometry is a simple and reliable method of chemical process control. It was produced by Zeiss Jena and then by its competitors elsewhere in Germany in the second half of the nineteenth century. Consequently, the instrument became routinely used in chemical research laboratories doing work on correlations between properties and structures of the growing number of organic compounds.[32]

Another example of an industry-driven instrumental technique widely used in chemistry is titrimetry. The efficiency of a bleaching plant depends on the precision in determining the hypochlorite concentration. At the end of the eighteenth century, François Antoine Henri Descrozilles (1751–1825) put together a titration assembly using the relatively simple and previously known pipette and burette. (Descrozilles later developed the volumetric flask, which further perfected the techniques of volumetric analysis.) After a few years of largely industrial use the method of titrimetry spread to chemical research laboratories. This was done by Joseph Louis Gay-Lussac (1778–1850), better known for his theoretical contributions to the science of chemistry. He modified earlier apparatus, introduced more finely calibrated burettes and other technical improvements, and was the first to use the term titrimetry in connection with this method of chemical analysis.[33] After him titrimetry became a standard analytical method used in many branches of chemical research. Although it may be hard to assess the total impact of titrimetry on advances in chemistry, it did routinize determinations of the composition of substances and thereby contributed to a better understanding of reactions between them.

Instruments and the Chemical Profession

Didactic and social roles of instruments deserve particular attention. Laboratory training and access to chemical instruments were essential for the professionalization of chemistry in the middle of the last century.[34] Practical work with chemical instruments came to occupy an ever-growing place in the curricula of chemical training. In the popular image and in the more sophisticated perception of university peers, the road to professional life in chemistry was seen to pass through extensive exposure to laboratory life.

Chemical instrumentation has become, perhaps, the main pillar of that identity and of the general culture of chemistry that distinguishes chemists from fellow natural scientists. Moreover, it leads to a broader distribution of chemistry graduates among government, industrial, and university laboratories than is the case for graduates in other natural sciences. For example, chemists have become active in the public health movement, arguing for tighter controls on domestic fuels, foods, and drugs. Unlike most other scientists, chemists use instruments in various institutional contexts, and this versatility is largely rooted in their experimental and instrumental skills.

Most prominent schools of chemistry are grounded in experimental work and excel in devising and refining new apparatus for chemical research. Teaching laboratories, by definition the natural abode of scientific instruments, are often built through private initiative and financing by enthusiasts of chemistry. In the mid-nineteenth century, "central" nations such as Germany and France fared only slightly better than the then peripheral United States and Russia in building teaching laboratories. Attached to a university, a hospital, or an industrial company, laboratories offered positions to technical staff, some of whom used experimental work to become research chemists in their own right. The case of Jean-Baptiste André Dumas (1800–1884)[35] may well illustrate this point.

Hired as a *répétiteur* at the École-Polytechnique, Dumas had only the most rudimentary equipment at his disposal. It might have been sufficient for teaching future engineers, but Dumas "built up a supply of apparatus and chemicals suitable for his own research. Then, following in the footsteps of those scientists who, both in Geneva and Paris, had been so generous in providing him with facilities to carry out his experiments, he too invited young men to work in his modest laboratory, to assist him in his research.[36] Subsequent improvement in status enabled Dumas to acquire more sophisticated instruments and attract more students. In Germany Justus Liebig's laboratory was a breeding ground for both research and industrial chemists, whereas growing numbers of French chemists with advanced training joined laboratories in chemical and related industries.[37]

Reliance on experiment and equipment combined in Dumas's laboratory with a strong motivation to advance chemical theory. In this sense Dumas can be compared to Lavoisier, whose experimentalist bent was certainly not a goal in itself. By devising new laboratory procedures and techniques for research and analysis, and increasing considerably the data base of organic and analytical chemistry, foundations were laid for a rational science of organic chemistry.[38]

Professional chemists transferred to industry and government laboratories the techniques of chemical experimentation they acquired at the university. They frequently chose the same equipment they had used in their formative years. The diffusion of professional chemists into industry thus created not only a demand for experimental training in chemistry but, no less important, a demand for chemical instruments. Industry, alongside medical and governmental institutions, began to constitute an important market for chemical instrumentation. Thus began the routinization of chemical apparatus, which in turn facilitated its acquisition and application for chemical research.[39] The links between industry and university—mainly of circulation of chemists and of contract research performed for industry at university research laboratories—increased the similarity of chemical equipment used at industrial and university laboratories. Thus chemical instruments played multiple roles in the professionalization of chemistry in the last century.

Instrumentation as a Factor of Inequality

Chemical instrumentation has also been a major factor in the international distribution of chemical research. Two brief examples illustrate this point and suggest avenues for further exploration. One has to do with the transfer of the scientific center from France to Germany in the nineteenth century. The other is the decline of Russian chemistry after the October revolution of 1917.

Germany emerged as the undisputed leader of world chemistry by the mid-nineteenth century. The loss by France of the central position in chemical research profoundly injured national pride and led such a notable of French chemistry as Charles Adolphe Wurtz to begin his major compendium of chemical knowledge with the affirmation: "La chimie est une science française."[40] Several decades later, on the occasion of the Universal Exhibition in Paris in 1900, the decline of French chemistry was underlined by the obvious domination of the instrument-making scene by German companies.

In the nineteenth century, French instrument-makers, following the lead of French science, were largely concentrated in the capital. Their fame was established in the early part of the century. The disruption of commercial links with Britain, then the leader in scientific instrumentation, gave impetus to serious qualitative advances in the instrument-making in France.[41] French instrument-makers appeared innovative: for example, they were the first to make use of aluminum anemometers, electrometers, and sextants.[42] But their ingenuity alone could not ensure that France would retain its position. Since the 1870s Germany has held the lead, and while the French may maintain the quality of craftsmanship, commercially France's instrument-making industry has badly lagged behind:

> Les nouvelles conditions du marché ne sont pas comprises, et l'artisan français prend de plus en plus la mentalité de 'l'artiste', entendant par là le praticien de l'un des beaux-arts. Certes—et ceci contribue à expliquer cela—la construction française donne encore à cette époque des réalisations spectaculaires, mais ce qui lui manque c'est la solidité en profondeur sur le plan commercial, acquise à la même époque par les constructeurs allemands.[43]

It is the industrialization of instrument-making that decisively accounted for the disparity. While some apologists of the time tried to demonstrate that French instrument-makers were on their way to triumph, contemporary historians compared this attitude "to the optimism feigned before terminally ill patients."[44] Praise for "the national temperament" of the French who could allegedly distinguish between "real" and "nominal" precision further illustrates the despair of French instrument-makers, faced, at the turn of the century, with an avalanche of mass-produced German instruments.

But the growth of German science alone cannot explain the spectacular advance of that country's instrument-makers in the late nineteenth and early twentieth century. The emergence of science-intensive industries, chemical industries first among them, created a new and substantially more significant market for chemical apparatus. Many of these instruments were first produced for and used by industry, and only later reached the confines of scientific research. Unlike their French colleagues, whom industrialists among their compatriots usually disdained as theoreticians incapable of offering practical expertise, German scientists were well integrated into the mass production of scientific instruments. Their advice was sought by such industrial giants as the optical concern Zeiss Jena, a leader in the production of both sophisticated scientific instruments and substantially more numerous optical devices for the industrial markets. This phenomenon not only illustrates the relative decline of French science but further confirms the point made earlier in this work with respect to infrared spectroscopy and refractometry: industrial uses of instrumentation lead to its eventual diffusion into scientific research.

Albeit never claiming chemistry as a Russian science, Russia was a major center of chemical science in the nineteenth and early twentieth century.[45] The names of Mendeleev, Butlerov, and Chugaev illustrate the importance of Russian chemistry. The October revolution of 1917 caused relatively few Russian chemists to flee abroad. Moreover, the Soviet government put particular emphasis on science as a key to "building a new society." However, Russian chemical research declined steadily in terms of its impact on world science. While restrictions of freedom and severe mismanagement impeded normal development of Soviet science, inadequate supply in instrumentation was another major obstacle with profound consequences for the country's research potential.

Unable to mass-produce sophisticated instruments, advanced experimental chemistry in the Soviet Union used occasional and badly maintained imports and local custom-made instruments. The low priority accorded to chemical research in the Soviet defense program translated into poor instrumentation available to chemists (as distinguished from physicists) in the Soviet Union. Indeed, a comparison of Soviet and American participants in international scientific exchanges in midcentury reveals a largely unilateral pattern of influence, from American to Soviet chemists. Moreover, the most frequent factor identified by Soviet chemists as crucial for the success of their stay in an American laboratory was the level and availability of scientific instrumentation.[46] More theoretical, less technology-dependent disciplines in the USSR by and large fared better than experiment-bound disciplines like chemistry.[47] Even within chemistry theoretical research proved more promising and yielded more significant results. For example, N. N. Semenov received a Nobel prize for his work on chemical kinetics.[48] However, the decline of Russian chemistry continued throughout the Soviet period. The resulting marginalization of Soviet chemists, as distinguished from mathematicians or physicists, was largely caused by inadequate instrumentation.

The examples of France and the Soviet Union suggest that deficiencies in instrumentation constitute a decisive factor of marginalization in the world of science. They also suggest that mass production of scientific apparatus is an essential distinction between central and peripheral science.[49]

Concluding Remarks

A blue-ribbon report in the 1960s on chemistry in the United States[50] stated that there exist four stages in the development of a new instrument:

(1) discovery of suitable means of observing some phenomenon;
(2) exploration of this phenomenon with special, homemade instruments or commercial prototypes;
(3) widespread use of commercial instruments; and
(4) routine applications of the instrument to control industrial production as well as research.

This scheme seems to be at variance with much of the evidence. It has been shown that the integration of physical instruments into chemical research has rarely been due to a demand on the part of the chemist. Rather it occurs through vigorous production of advanced instruments on the part of the industry. This scheme is indeed surprising. The company that proposed these four stages in the report has itself had experience when stages 3 and 4 occur in the reverse order; moreover, stage 4 is by far the most decisive factor in the development of new instrumentation.[51]

Beyond regular publicity, the instrumentation industry mounted adult education campaigns, beginning with infrared spectroscopy in the 1940s, with the express purpose of introducing the researcher into the new world of instrumentation. An analogy with the general consumer market would not be far-fetched: advances in, say, stereo equipment or personal computers rarely constitute a response to consumer demand but rather derive from strategies of the manufacturers. When the marketing approach is right, the habits of the consumer will change, sometimes drastically, not because of an inner felt desire for a new product but because of its availability and prestige. Chemistry, traditionally linked with advances in instrumentation, naturally follows this path of development and undergoes changes in orientation and conceptualization that are often not of the chemist's own making.

A recent edition of the authoritative *Chemical Instrumentation* illustrates this point:

> . . . the modern methods of analytical chemistry that are used in analysis and research can best be mastered if chemical instrumentation is studied in its own right. Accordingly, . . . this book turns to a systematic treatment of instrument *design* and instrumental *methods*. The adequate consideration of the first theme has dictated the strong undercurrent of physics, engineering, and physical chemistry that is evident on inspection of this book. These disciplines provide the fundamentals that are needed to understand design and function. The pursuit of the second theme, modern methods of analytical chemistry, has required a running discussion of physical properties and behavior throughout. Throughout the book the emphasis is on fundamentals.[52]

It appears that in this book the chemist is introduced to a novel domain whose fundamentals ought to be mastered. It is also telling that this university book on instrumentation is coauthored by experts from instrument-making companies. Indeed, it has been suggested that instrument makers constitute a third full-fledged category of scientists, alongside with experimenters and theoreticians.[53]

Another analogy between the consumer and the research chemist consists in the growing acceptance of the new product or instrument as a black box whose inner working may be irrelevant, beyond one's technical proficiency, or both. The chemist is urged at least to "understand the instrument on the macroscopic level."[54] Consequently, while stages 1 and 2 may indeed occur in the development of a new instrument, it often happens beyond the confines of the chemical laboratory. It is only if the instrument is meant mainly for the scientific market that stage 2 may involve tests in a chemical laboratory. Otherwise, stage 4 may precede stage 2 in the above scheme.

The images of chemical instruments show a greater continuity than their actual uses. Major advances in chemistry of the twentieth century have resulted from experiments carried out with instruments designed for studies of physical phenomena and later adapted to the needs of chemistry. Although this circumstance does not blur the borders between the two disciplines, the prominence of physical approaches to chemical realities represents a continuing trend that is at least a hundred years old. Physical phenomena have become instruments in chemical research. They profoundly affect chemical practice, introducing new ideas and—perhaps more important—new, physical discourse that supersedes the conceptual apparatus inherited from the last century. Thus instruments, developed mostly outside the chemical laboratory, produce a massive impact on the way chemical research is conceptualized, conducted, and reported.

Chemical science relies more and more on instrumentation that not only measures a variety of physical properties but directly reveals molecular and atomic structures. For example, the passage from the optical to the electron microscope reinforced the observational element in the practice of chemistry to such an extent that it made the tradition of analytical chemistry based on intrusion and destruction largely irrelevant. Advances in computers, optics, and electronics regulate the progress of chemical analysis. The transition to "the new analytical chemistry" represents a discontinuity no less striking than the transition from the horse to the locomotive. And, similarly, those responsible for the advent of the new instruments had no greater knowledge of (or commitment to) the traditional methods of chemical analysis than railway engineers had with respect to horse breeding.

However, chemical research maintains its distinctiveness with respect to physics. While chemists are just beginning to see atoms and molecules on the screens of their apparatus, physicists face quite a different problem of distinguishing between real and artifactual phenomena. The concerns and agendas of the two groups are not the same, a fact that casts doubt on the ability of the innumerable historical accounts of physical experimentation to help us understand the rarely documented story of the role of instruments in chemical research.

"Despite the slogan that science advances through experiments, virtually the entire literature of the history of science concerns theory."[55] Albeit borrowed from the history of physics, this paradox holds true for chemistry. The habitual image of instrumentation in the mind of the chemist and layperson alike remains that of a tool at the service of the chemical researcher. This happens even though instrumentation has played a crucial role in the foundation of modern chemistry, its professionalization, and its international development. The standard literature on chemical instrumentation suffers from an even greater degree of amnesia than textbooks of chemistry that make occasional references to the history of chemistry. My use of historical evidence here suggests that chemical research should be viewed in the context of its instrumentation, which may often introduce approaches linked with its origins, usually quite distant from the chemical laboratory. Instruments and their production profoundly affect the very identity of chemical research and remain crucial in the ongoing stratification of scientific nations. These considerations should inform the history of chemistry in a more consistent manner than so far attempted.

The author acknowledges helpful comments from Vladimir Ganin, Sy Mauskopf, Steven Shapin, and Spas Spasov.

Notes

1. The range of what may be considered an instrument is very broad, from the inanimate magnifying glass and the balance to live Drosophila and the inmates of National-Socialist concentration camps. It is the use rather than the nature of an object or an organism that defines its instrumental character in a given context.

2. The use of chemical technologies also predates the emergence of chemistry as a science. Moreover, even as chemistry developed its explanations often lagged behind the progress of essentially chemical technologies (see, e.g., Yakov M. Rabkin, "La chimie et le pétrole: les débuts d'une liaison," *Revue d'Histoire des Sciences* 30(1977): 303–336).

3. F. Szabadvary and A. Robinson, "The History of Analytical Chemistry," in *Wilson and Wilson's Comprehensive Analytical Chemistry,* ed. G. Svehla (Amsterdam: Elsevier, 1980), 10:63.

4. Aaron J. Ihde, *The Development of Modern Chemistry* (New York: Harper & Row, 1964; reprint, New York: Dover Publications, 1984), p. 277.

5. See Derek De Solla Price, "Is Technology Historically Independent of Science?" *Technology and Culture* 6(1965): 550–568; and Yakov M. Rabkin, "Science and Technology," *Fundamenta Scientia* 2, 3–4(1981): 413–423.

6. Yakov M. Rabkin, "Technological Innovation in Science," *Isis* 78(1987): 31–54.

7. Donald de B. Beaver, "The Secrets of Science Unlocked," *Science, Technology and Human Values* 13, 3–4(1988): 377.

8. This impression is reinforced by a recent review of current sources on the subject of scientific instruments, even though it was written by a historian of chemistry: Jeffrey L. Sturchio, "Artifact and Experiment," *Isis* 79(1988): 369–372.

9. Gérard Lemaine et al., *Perspectives on the Emergence of Scientific Disciplines* (The Hague: Mouton, 1976).

10. Peter Galison, *How Experiments End* (Chicago: University of Chicago Press, 1987).

11. Ian Hacking, *Representing and Intervening: Introductory Topics in the Philosophy of Natural Science* (Cambridge: Cambridge University Press, 1983), p. 200.

12. Davis Baird, "Instruments on the Cusp of Science and Technology," preprint.

13. Maurice Rheims, préface, in Henri Michel, *Instruments des sciences dans l'art et l'histoire* (Bruxelles: Albert de Visscher, 1973), p. 8.

14. J. V. Field, "What is Scientific About a Scientific Instrument," *Nuncius* 3,2(1988): 3–26.

15. Sturchio, "Artifact and Experiment" (note 8), p. 371.

16. See, e.g., a recent study of the thermopile, an instrument of consequence for the development of infrared spectroscopy: Edvige Schettino, "A New In-

strument for Infrared Radiation Measurements," *Annals of Science* 46(1989): 511–517.

17. It is well to remember that the late Derek De Solla Price, who time and again emphasized the importance of instruments for the development of scientific thinking, stayed away from strong programs. Even though it may appear ironically irreverent not to cite the father of citations, private conversations with him have had a greater influence on me than have his publications. After all, a sure sign of a real scientific classic is that the classic is no longer cited.

18. Simon Schaeffer, "Scientific Instruments and Their Public," in Roy Porter, Jim Bennett, Simon Schaeffer, and Olivia Brown, *Science and Profit in 18th-Century London: published to accompany a special exhibition at the Whipple Museum of the History of Science* (Cambridge: Whipple Museum of the History of Science, 1985).

19. W. Stanley Jevons, *The Principles of Science* (London: Macmillan & Co., 1879), p. vii, quoted in Trevor H. Levere, "Lavoisier: Language, Instruments and the Chemical Revolution," in *Nature, Experiment and the Sciences: Essays on Galileo and the History of Science,* ed. Trevor H. Levere and William R. Shea (Amsterdam: Kluwer, 1990), pp. 207–223.

20. More on the early relations between chemistry and physics can be found in Arnold Thackray, *Atoms and Powers: An Essay on Newtonian Matter-Theory and the Development of Chemistry* (Cambridge, Mass.: Harvard University Press, 1970).

21. Davis Baird and Thomas Faust, "Scientific Instruments, Scientific Progress and the Cyclotron," *British Journal for the Philosophy of Science* 41(1990): 147–175.

22. Levere, "Lavoisier" (note 19), p. 209.

23. Ibid., p. 217.

24. Hacking, *Representing and Intervening* (note 11), p. 242.

25. Szabadvary and Robinson, "History of Analytical Chemistry" (note 3), p. 163.

26. Ibid., p. 136.

27. Baird and Faust, "Scientific Instruments" (note 21), p. 3.

28. A similar example of Einstein as a "parasite, feeding haphazard on long dead experiments" is brought up with respect to Armand-H. Fizeau and his experiment on the velocity of light (see Hacking, *Representing and Interviewing* (note 11), p. 237.

29. Ihde, *Development of Modern Chemistry* (note 4), p. 235.

30. Ibid., p. 295.

31. See Rabkin, "Technological Innovation" (note 6) and a presentation at the annual History of Science Society meeting in Cincinnati in December 1988.

32. Boris V. Ioffé, *Refraktometrichiskie metody v khimii* (Leningrad: GNTIKhL, 1960), pp. 5–6.

33. Szabadvary and Robinson, "History of Analytical Chemistry" (note 3), pp. 177–180.

34. J. B. Morrell, "The Chemist Breeders," *Ambix* 19(1972): 1–46.

35. Leo J. Klosterman, "A Research School in Chemistry in the Nineteenth Century," *Annals of Science* 42(1985): 1–80.

36. Ibid., p. 8.

37. Dumas married a daughter of the director of the Sèvres porcelain factory. This was another kind of connection between academic and industrial worlds related to chemistry that remains to be explored.

38. Klosterman, "A Research School," p. 80.

39. This idea was developed more fully for a later period in the United States in Rabkin, "Technological Innovation in Science" (note 6).

40. C. A. Wurtz, *Dictionnaire de la chimie pure et appliquée* (Paris: Hachette, 1868). This claim provoked a major controversy among German chemists, as analyzed by Alan Rocke in a paper presented at the international symposium "Interaction Between Jewish and Scientific Cultures" in Montreal in June 1990.

41. Jacques Payen, "Les constructeurs d'instruments scientifiques en France au XIXe siècle," *Archives Internationales d'Histoire des Sciences* 36(1986): 84–161.

42. Anita McConnell, "Aluminium and Its Alloys for Scientific Instruments, 1855–1900," *Annals of Science* 46(1989): 611–620.

43. Payen, "Les constructeurs," p. 86.

44. Ibid., p. 152.

45. Alexander Vucinich, *Science in Russian Culture* (Stanford, Calif.: Stanford University Press, 1970).

46. Yakov M. Rabkin, *Science Between the Superpowers* (New York: Priority Press, 1988), Chap. 2.

47. See, e.g., Thane Gustafson, "Why Doesn't Soviet Science Do Better Than It Does," in *The Social Context of Soviet Science*, ed. Linda L. Lubrano and Susan Gross Solomon (Boulder, Colo.: Westview Press, 1980), pp. 31–37.

48. N. N. Semenov, *Tsepnye reaktsii* (Leningrad, 1934). See also an historical essay on the subject: V. A. Kritsman, *Razvitie kinetiki organicheskikh reaktsii* (Moscow: Nauka, 1980).

49. An outline of the center and periphery in science can be found in E. Shils, "Center and Periphery," in *Center and Periphery: Essays in Macrosociology* (Chicago: University of Chicago Press, 1975), p. 5. See also T. Schott, "International Influence in Science: Beyond Center and Periphery," *Social Science Research* 17(1988): 219–238.

50. National Academy of Science/National Research Council, *Chemistry: Opportunity and Needs* (Washington, D.C.: NAS/NRC, 1965), p. 92.

51. This point is documented in my *Isis* article of 1987 (note 6).

52. Howard A. Strobel and William R. Heineman, *Chemical Instrumentation: A Systematic Approach* (New York: John Wiley & Sons, 1989), p. vi.

53. Baird and Faust, "Scientific Instruments" (note 21), p. 36. A connected idea that architects also impose their values through the laboratory design may be a bit far-fetched.

54. Ibid., p. x.

55. Galison, *How Experiments End* (note 10), p. ix.

Systems of Production: Drosophila, Neurospora, and Biochemical Genetics

Robert E. Kohler

Historians of science have only recently realized how much they have to learn about how scientists actually conduct research in laboratories or in the field.[1] There are many reasons for this dearth of knowledge—the paucity of sources, academic deference to theory and theorizers, and a scientific moral economy that systematically veils or mythologizes the process of production. The situation is likely to change very rapidly in the coming years. The language of scientific experiment and "practice" has become fashionable. Sociologists and historians have come forward to define what "practice" means and to prescribe how it should be treated. We may look forward to lively and interesting contention among a diversity of genres and ideologies.[2]

In this article, I deploy terms like "modes" or "systems" of production, which signify accepted configurations of instrumentation, routinized procedures, and social infrastructure (working groups, user networks, and so forth). Modes or systems of production give communities of practitioners their distinctive identities and ecologies. I do not present a micro-analysis of experiment; the art and craft of experimental or field "practice" is an essential part of systems of production, but only a part. Rather, my aim is to reassemble the whole material and social system in which a group of investigators produced, published, competed, and made careers.

The language of production and system, though suggestive of certain theories and ideologies, is not used here as code for any systematic theory. I use market terms metaphorically, to suggest fundamental similarities between science and other social and cultural activities, especially those concerned with the production and distribution of material goods. Science is different from manufacturing and commerce in ways too obvious to point out here. But there are similarities, too, which are urgently in need of exploration. Both are fundamentally collective, marketplace activities, driven and constrained by the production process and by the activities of consumer or user constituencies, and both have acquired an astonishing, even alarming, productive and moral power.

So too with the metaphor of system, which I use to invite empirical studies of what scientists mean when they speak of "experimental systems." That phrase is part of the ordinary language of laboratory work. When asked about their work, experimental biologists or chemists usually talk about their experimental "materials" or "systems," by which they mean something more than single instruments but less than the productive machinery of a laboratory or research field. They are fundamental units of scientific production, of which practitioners have tacit knowledge, and which historians need to discover.

Instruments offer convenient points of entry into these complex systems of production. Here "instruments" signify not just air pumps, accelerators, combustion tubes, or spectrometers, but also colonies of Drosophila or rats, ecologists' quadrats and plats, geological expeditions, and social surveys.[3] It would have seemed odd, even a few years ago, to regard living organisms as instruments; we recognize today that the standard rat and the mutant Drosophila are constructed artifacts. "Standard" animals, produced by generations of inbreeding, are unambiguous examples of engineered instruments.[4] With found objects like sea urchin eggs or primates, the artifice resides less in physical redesign than in the accumulated knowledge of what they will do in the artificial conditions of a laboratory experiment. Organisms and instruments come to embody libraries of craft knowledge and routine practices. That, by definition, is how they become systems of production. Indeed, whole fields can be conceived in terms of competing and evolving instruments and systems of production.

Consider the case of biochemical genetics. In the 1950s, the K-12 strain of the bacterium *Escherichia coli* was the system of choice because it was the most productive. In the 1940s, the pink bread mold, *Neurospora crassa,* and in the 1930s, the fruitfly *Drosophila melanogaster*; each seemed the most promising in a field of competing organisms. Why did workers discard proven experimental systems for new ones? The two systems I shall consider were invented by one individual, the geneticist George W. Beadle. In 1935, Beadle and Boris Ephrussi, hoping to create an experimental genetics of development, adapted the method of transplantation to Drosophila. Its new embryological use did not, however, fundamentally change Drosophila as an instrument of genetics. The accumulation of genetic craft and knowledge in Drosophila caused it to be used in familiar ways. But a more fundamental change occurred in 1941, when Beadle and Edward L. Tatum brought into the laboratory the hitherto little-used Neurospora. Although designed to be an improved genetic instrument, Neurospora unexpectedly proved more productive of biochemical knowledge. Drosophila was abandoned, and the practice of developmental or physiological genetics was transformed. Thus, searching for more convenient and productive forms of known experimental systems, Beadle and his coworkers came to invent new systems of production, the astonishing productivity of which transformed the practice of both genetics and biochemistry.

That is not the whole story. Many others were designing organisms and practices to unite genetics with embryology and physiology: intersexes of the gypsy moth *Lymantria dispar* (Richard Goldschmidt); eye color in the meal moth *Ephestia kühniella* (Alfred Kühn, Ernest Plagge); eye color in the freshwater shrimp *Gemmarus* (Julian Huxley); skeletal deformations in lethal mutants of mice (J.B.S. Haldane, Hans Grüneberg); plant pigments (Muriel Wheldale Onslow); cysteinuria in Dalmation hounds (Erwin Brand); and coat color pigments in guinea pigs (William E. Castle, Sewall Wright), to mention only the main competitors.[5] Comparison of these competing modes of production would be revealing. Here, however, I want only to show how one unusually creative experimentalist designed systems, chose (or fell into) problems, utilized resources, and made a career that took him unexpectedly into different kinds of practice.

Beadle and Ephrussi, Apprentice Drosophilists

Both Beadle and Ephrussi were recent converts to Drosophila genetics when they invented the practice of transplantation in 1935. Beadle was a maize geneticist, trained at Cornell University (Ph.D., 1931) by Rollins A. Emerson in the classical methods and problems of chromosomal mechanics and transmission. His ten papers on the mechanics of synapsis and cytokinesis in "sticky chromosome" mutants reveal a virtuoso performer with an elaborate and well-worked experimental instrument. Boris Ephrussi did his apprentice work in other classical veins of experimental biology, the embryology of the sea urchin egg and mammalian tissue culture, which he hoped to apply to the study of development. Both Beadle and Ephrussi, in becoming credentialed practitioners, made a substantial investment in standard practices. What then brought them to Thomas Hunt Morgan's school at the California Institute of Technology and to the unfamiliar world of Drosophila genetics?

For Beadle, the change was shaped by the social machinery of career-building. In 1930, the best and brightest were expected to do postdoctoral work, and so Beadle applied for and received a National Research fellowship. He would have stayed at Cornell and continued his work on "sticky genes" in maize had not the chairman of the fellowship board insisted that he go instead to Caltech (his second choice) and learn something about a different system. Beadle later recalled his initial bewilderment at the arcane and unfamiliar jargon of Drosophila mutants. At first he continued his research on maize with the group's plant geneticist, Sterling Emerson, the son of Rollins Emerson. Sterling, however, was teaching himself Drosophila genetics, and he and Beadle learned together. Their coauthored papers, the first for both on Drosophila, dealt with the mechanics of "sticky" chromosomes. Beadle moved into a new mode of practice by the most familiar and easy route, apparently not with the intention of giving up maize permanently for Drosophila.[6]

Ephrussi changed in a more deliberate way. Having decided to apply tissue culture to the development of mice with lethal developmental mutations, he realized that he must first master genetics. He came to Caltech in January 1934, with a fellowship from the Rockefeller Foundation, to learn genetics from those who had defined the field. Like Beadle he took a course of least resistance, working at first with "lethal" mice, which he brought with him from Paris, and taking, as his first work with Drosophila, the closely related problem of the "scute-8 lethal" mutant. This problem was suggested to him by Alfred Sturtevant, one of the original inhabitants of the Columbia "fly-room," and the man who served as the unofficial intellectual shepherd of the Caltech group. His influence on both Beadle and Ephrussi was profound. Within a month of his arrival, Ephrussi began experiments with tissue culture and familiarized himself with Drosophila.[7]

Sturtevant was very interested just then in genetic gynandromorphs, or "mosaics." These were female flies in which some cell lines lost a sex ("X") chromosome in the process of development, resulting in organs or patches of tissue that were genetically different from surrounding tissues. Mosaics had been observed since the earliest examinations of Drosophila, but they appeared unpredictably and were not amenable to systematic experiment. Around 1930, however, Sturtevant constructed a multi-allelic stock of *D. simulans,* which threw off X chromosomes frequently and early in the developmental process, resulting in whole organs of mosaic type. He transformed a genetic oddity into a laboratory instrument, with which he could investigate experimentally what genes did in the process of development.

In the most interesting cases, the effects of mutant genes were suppressed by adjacent wild-type tissue, presumably by supplying the mutant tissues with diffusible chemical substances. Sturtevant found several such cases of "nonautonomous" development. The mutant gene for "vermillion" eye color, for example, was expressed in mosaics as if a wild-type gene; eyes genetically vermillion had wild-type color. Likewise, tissues carrying the "scute-8 lethal" gene survived in mosaics.[8] Mosaics offered a purely genetic alternative to what embryologists did by physical transplantation or by tissue culture: namely, to see how genes acted in development by placing them in genetically varied environments. Despite the differences in language and bench technique, the method for studying Ephrussi's "lethal" mice and the "scute-8 lethals" in mosaic flies were not so different. Ephrussi quickly became an enthusiastic Drosophilist.[9]

One reason that Beadle and Ephrussi took so easily to the practices of Drosophila genetics was that Morgan's group, transferred almost completely intact from Columbia University to Caltech in 1928, was designed to assimilate visitors.[10] It had to be in order to accommodate the stream of scientists who came to learn Drosophila practice at the source. Under Morgan's aegis, the Drosophila group—Sturtevant, Calvin Bridges, Jack Schultz (the last Ph.D. of the Columbia period), and Theodosius Dobzhansky (who came as a fellow of the International Education Board in 1928 and stayed)—continued the collective work style of the fly-room at Columbia.[11] Although they had distinctive personal interests and interpersonal conflicts, the members of the core group at Columbia did not stake out and defend personal territories but worked together to make sure that they were at the center of every new development in Drosophila genetics. Cooperation was not preached as an official doctrine, but everyone made his first priority "getting on with the work." When Schultz decided to study the "minute" character, Bridges and Sturtevant began to save mutants for him. Students were not molded by one professor but by the whole school. Morgan deputized the "care and feeding of students" to Bridges.[12] The participants of the Drosophila group were all supported by a single source of funding, an annual grant from the Carnegie Institution of Washington, and Morgan customarily published the results of the group's work collectively in the Institution's *Yearbook*.

This collective style of work continued at Caltech and was embodied in the physical arrangements of the workplace. There were no private offices, telephones, or closed doors, only common work tables and a common space for an easy exchange of news and ideas. No one had a personal assistant and everyone worked side by side at the bench. Thus everything discouraged individuals from withdrawing into specialized lines of research. Beadle recalled that the Caltech fly-room was a most stimulating place to learn and to work.[13] Personal competition and conflict did arise, of course: Dobzhansky's possessiveness over specific problems and students caused conflict in the late 1930s at Caltech, in the same way that Herman Muller's personality had presented problems at Columbia.[14] But Morgan's authority sufficed to prevent personal ambitions from making the group into a factionalized academic department.

The effects of this organization can be seen in the high proportion of papers coauthored by Sturtevant, Bridges, Dobzhansky, and Schultz—from one-third to one-half—and by the ability of the Caltech group to stay in the vanguard of virtually every important trend in genetics. That population genetics and biochemical genetics both had their origins in the collective work of the Caltech group in the mid-1930s is itself a remarkable achievement. The Caltech work-collective lost this special character after Bridges's death in 1938 and Morgan's retirement a few years later, but while Beadle and Ephrussi were there together in 1933 and 1935, the Morgan group was *the* place to nurture new scientific practices.[15] Sturtevant and Schultz were instrumental in Beadle and Ephrussi's invention of transplantation in Drosophila.

Putting the Genes in the Whole Organism

Beadle and Ephrussi did not intend to make genetics more biochemical, but rather to reconnect it with embryology by refashioning Drosophila into an instrument for both. Transplantation was a standard practice in the study of the differentiation of germ layers, the regeneration of limbs in amphibians, and the development of embryonic buds or "imaginal disks" in insects. A researcher would dissect the disks from a donor larva, transplant it into a host larva, and observe the character of the developed organ. The new wrinkle with Drosophila was that genetic constitution, not morphological position, was taken to be the key variable. Drosophila had been designed and constructed to be an instrument of genetics, just as sea urchin and Ascaris embryos had gradually become specialized instruments for embryology. Flies were ill-suited for embryological practice, as invertebrate eggs were for genetic research. These laboratory creatures had not been constructed deliberately to keep the two disciplines separate, but separation happened as distinct bodies of knowledge and usages developed around them. Because of their specialized instruments and modes of practice, the two disciplines, like diverging species, could no longer interbreed.

Unlike natural species, however, they could regain that ability. In the 1930s, geneticists and embryologists began to reconsider the mutual advantages to both available through collaboration.

Beadle and Ephrussi's attempt to recombine genetics and development was not unique. Curt Stern tried transplantation with Drosophila pupae in 1934 and thought for a time that he had succeeded.[16] Geneticists, especially, predicted that the next big breakthrough would be in the genetics of development: in the chains of developmental chemical reactions, presumed to link the primary action of genes to morphological characters. This belief took diverse forms: every research specialty or biological interest group seemed to have its particular claim. Mainstream Drosophila or maize geneticists, who had invested their careers in the mysteries of genetic transmission and gene mapping, placed their hopes on the chemistry of the genetic material and the physiology of chromosomal mechanics. T. H. Morgan took this position in 1926, when he gingerly urged geneticists to go beyond the safety of counting ratios and move to the physiology of mutation and segregation, although he could not say how that could be done. On the other side, developmental biologists cautiously welcomed a rapprochement with geneticists while calling for more embryological data.[17] Geneticist Richard Goldschmidt used a language of physiology and development to assert his alternative vision of a mainstream genetics without a material gene.[18] Physiologist Frank R. Lillie called for a new synthesis of genetics and development around his work in reproductive physiology.[19]

Many kinds of biological practitioners hoped to ride the fashion to "put the genes back in the whole organism" to fame by giving a new twist, or at least a new name, to their experimental practices. The Caltech Drosophilists too felt the fashion. But at the time there were no patently productive genetic methods for examining experimentally how genes acted in developmental processes. Why, then, did Beadle and Ephrussi go beyond mainstream practices to a new one, designed specifically to "put the genes back in the whole organism"? The incentives probably did not come from the work they were doing at the time, but from the general conversation and consciousness-raising within the Drosophila group.

Nothing in Beadle or Ephrussi's apprentice work in Drosophila suggests an internal reason for their shift in productive strategies. As a graduate student at Cornell, Beadle had taken courses in biochemistry and had witnessed sympathetically Emerson's fruitless efforts to persuade biochemists and plant physiologists to use genetic methods. But it does not follow, as Lily Kay asserts, that Beadle "had already [by 1932] been captured by the biochemical puzzle of gene action, especially in relation to enzymology."[20] His nine publications on Drosophila from 1932 to 1935 give no hint of any interest in the physiology of gene action, much less biochemistry or enzymes. However, Beadle was not a geneticist who disdained all connection with neighboring disciplines. He was open to the ideas that Sturtevant and Schultz discussed in the Caltech fly-room.

The Drosophila group took a lively interest in the problems of genetics and development. The continual brainstorming there may well have inspired Beadle and Ephrussi to invent an experimental way to study the role of genes in development.[21]

Everyone seemed to share Morgan's belief that the future of genetics lay in its integration with other disciplines. Unlike Morgan, however, they were actively trying to put that belief into practice at the bench. Since the mid-1920s at least, Bridges and Sturtevant had worked on problems having to do with quantitative effects, like "bar" eye, gene dosage results, multiploids, and intersexes, which invited physiological interpretation. So too did Schultz and Dobzhansky, especially after 1928. Judging from the group's publications, the late 1920s and early 1930s were the apogee of the group's interest in connecting genetics with physiology or developmental biology.[22] By the late 1930s, each member of the group had returned to his own particular problem: Dobzhansky on the genetic structure of wild populations, Sturtevant on the comparative genetics of Drosophila species, Schultz on the chemistry of deoxyribonucleic acid, and Beadle on the genetics of eye-color pigments. Thus research agendas in the fly group were most fluid and exploratory in precisely the years that Beadle and Ephrussi were there, participating in the daily shop talk.

Sturtevant and Schultz inspired their efforts. The method of transplantation did by physical manipulation what Sturtevant's mosaic method did with fancy genetics, as Beadle and Ephrussi were always ready to acknowledge. Schultz, the closest of the group to Beadle and Ephrussi, was its only member to have done research in physiology and biochemistry. While at Columbia, Schultz had spent time in the biophysics laboratory of Selig Hecht learning to isolate and to characterize spectroscopically the eye pigments of Drosophila. By reclassifying the numerous eye-color mutants in terms of chemical and developmental function rather than in terms of their position on the chromosomes, Schultz hoped to reveal how genes functioned physiologically. He did not publish much of this work; it did not quite live up to expectations, and Schultz always preferred starting new projects to finishing old ones.[23] The problem of connecting genetics and development experimentally nonetheless remained very much on his mind between 1933 and 1935. He pointed out in a lecture in June 1934 that advanced researchers required an experimental system in which the problem could be attacked both genetically and embryologically at the same time, and he urged Drosophilists to exploit their strategic advantage by expanding into embryology. Embryologists at New York University and elsewhere had recently revealed that experimental embryology could be done with Drosophila eggs. "It is for us to make use of these opportunities," Schultz exhorted his listeners, "we have a complete story to unravel, because we can work things from both ends at once."[24] Schultz's lecture is a window on the brainstorming within the Caltech group. It offers a glimpse of what inspired Beadle and Ephrussi to forego the predictable results of their ongoing research and to spend, or "to waste, if necessary," six months trying to find a way to apply the practices of developmental biology to Drosophila.[25]

Sturtevant and Schultz's bright vision of a developmental genetics seems to have cast a pall over Beadle and Ephrussi's labor over "sticky genes" and "scute-8 lethal." Beadle recalled much later that his work with Sterling Emerson had not succeeded as he had hoped in revealing the nature of the gene.[26] Beadle thought that the prospects for a more direct attack on gene physiology made standard cytogenetic practice seem slow and unpromising. Ephrussi's discontent was more acute. Three months after returning to Paris, Ephrussi wrote to Schultz that he could not get back to business as usual. He was desperately "home seek" for the lively scientific life of the Morgan group. He missed its openness and free-wheeling intellectual style. His efforts to recreate Morgan's seminars at the Rothschild Institute had been obstructed by the "particularism" of Parisian scientists; they refused to take an interest in anything outside their own disciplines. He felt "asphyxiated." Embryology seemed tame after being at Caltech. He was quitting work on lethal mutants in mice, his enthusiasm chilled by the sharp criticism he had received from his friends at Caltech, and he was at a loss to know what to do with tissue culture. What he really wanted to do, he knew, was "la vrai génétique," but his laboratory was not set up for such research, and he did not know where to start. He recalled with chagrin how he failed to take advantage of friendly conversations with Morgan at Woods Hole to get some ideas for specific genetic projects. When would Beadle write again? Why did his other friends not write? What was he missing? Had he been forgotten so quickly?[27]

Discontent with mainstream practices, plus the promise of developmental genetics, made it easier for Beadle and Ephrussi to take a gamble on new modes of genetic production. Thus, with financial help from Morgan, Beadle arranged to spend the first half of 1935 in Ephrussi's laboratory at the Rothschild Institute. There he would try to invent experimental methods for doing developmental genetics in Drosophila.[28]

Inventing Transplantation

In order to realize their new kind of scientific production, Beadle and Ephrussi had to engineer the adaptation of one of the familiar techniques of experimental embryology to Drosophila larva. Tissue culture was an obvious choice, given Ephrussi's expertise and the available technical resources of the Tissue Culture Laboratory. Another obvious possibility was transplantation, but Ephrussi was not an expert in this technique. Beadle later recalled that he and Ephrussi first tried and failed to culture larval tissue, and only then turned to transplantation. However, in a letter written one month after he arrived in Paris, Beadle wrote that they had not tried tissue culture and were trying transplantation.[29] Most likely, Ephrussi had tried to culture Drosophila tissue sometime before Beadle arrived and concluded from the experience that transplantation might be more promising.[30]

The leading French insect embryologist, Charles Perez, warned them that transplantation would almost certainly fail with Drosophila. Embryologists avoided flies because their imaginal disks all looked alike, making it almost impossible to pick out disks of specific organs. Drosophila would be even worse because of its small size. (Hoping that the mutant "giant" might be a little easier, Beadle asked Milislav Demerec to send him a culture from the stock of Cold Spring Harbor.)[31] Beadle and Ephrussi decided to try transplantation anyway. Since they were interested in combining embryology with genetics, the potential reward of success with Drosophila, with its rich genetic history already available, outweighed the risk on the embryological side.

A month into the work, Beadle and Ephrussi were just beginning to get the knack of picking out disks and injecting larvae without killing them, though still with no positive results.[32] But one morning—it must not have been long after Beadle wrote to Demerec—he and Ephrussi came into the laboratory and found a newly hatched fly with a third eye clearly visible just beneath the skin of its abdomen. They knew then that they had unwittingly dissected out an eye disk. Could they do it again on purpose? They retired to the Capoulade, a nearby coffeehouse, to plan a series of experiments. The first thing to do was to test the vermillion eye-color mutant. Fortunately, Beadle had already asked Demerec to send a culture of vermillion flies. It was the obvious first step because Sturtevant's experiments with vermillion mosaics showed Beadle and Ephrussi in advance what a positive result would look like. They knew that the vermillion gene developed nonautonomously: therefore, an eye disk from a vermillion larva, transplanted into the larva of a wild-type fly, should develop into a wild-type eye. And it did.[33]

Beadle and Ephrussi realized at once that they had hit on a general method for producing knowledge of host genes that acted in the process of development. Transplantation did everything that Sturtevant's mosaic method could do, but with none of the limitations of a genetic system that produced mosaics at random and that was useful only for sex-linked genes. In the new method, any combination of genes could be tested by manipulating the genetic constitution of transplant and host. Sturtevant had touted the practical advantages of mosaics over the grafting method of traditional embryology. Beadle and Ephrussi tipped the balance the other way. They had not set out consciously to make Sturtevant's method fully experimental, but that is what they did.[34] Morgan and others quickly pronounced it to be as important as Hermann Muller's invention of mutation by X-ray or Theophilus Painter's discovery of the giant salivary chromosomes of Drosophila—quite a compliment. Sturtevant eagerly spread the news and displayed a three-eyed fly at a meeting of geneticists at Woods Hole.[35] Drosophila had been transformed into an instrument for experimental developmental genetics.

The system that Beadle and Ephrussi invented was an exacting bit of laboratory handicraft. Transplantation required close cooperation between two workers, seated opposite each other at a narrow table behind two binocular microscopes. One microscope was fitted with a dissecting stage and a microdissection apparatus to pick out imaginal disks; the other had a micropipette to suck up the minute disk and inject it into the host larva. The pipette itself was a work of art: a capillary tube drawn down to 0.01 millimeter, cut off at just the right angle and constricted just enough to keep the disk from being swallowed by the microsyringe. One person washed donor larvae and dissected out the disks while his partner washed the host, etherized it—at just the right rate to get the larva to extend—and placed it in position for injection. The dissector then sucked the disk slowly into the micropipette and injected it into the side or ventral surface of the host larva, which his partner held in position with a blunt needle. This last and crucial operation required a good deal of craft knowledge to insure success:

> The precise way of holding a larva, the manner of inserting the pipette, the speed at which the tissue and saline should be injected and the rate and direction of withdrawal of the pipette and its relation to the rate of flow of liquid are all matters that can best be learned by experience. The most frequent difficulty and the hardest to learn to overcome is "blowing out" of the larva, i.e., flowing out of part of the internal organs.[36]

With experience, however, it was possible to do 30 injections per hour and 100 to 200 per day. On a good day, 60 percent of the larvae would develop into adult flies, though the average was closer to 30.

Determining the phenotype of the developed transplants also required craft skill. It was relatively unusual for eyes, say, to develop on the surface of the host; more usually they developed deep within the abdomen attached to the genital organs. Also, the eyes developed inside out, with the facets on the concave surface and with antennae attached along with parts of the head. To get the eye out for inspection required microdissection and cleaning. Furthermore, eye colors did not develop quite normally inside the body, so elaborate controls were necessary. To show that a disk became a vermillion eye in a claret host, for example, three control experiments were done in addition to the test to provide reference colors of wild type, claret, and vermillion.[37] There was in addition the routine business of maintaining several hundred different genetic stocks, making food and washing bottles, collecting eggs, culturing larvae, and so on—all of which had to be done with precision. For instance, as the experiments progressed, the experimenters realized that they needed to monitor the exact age of test larvae. Drosophila was as complex a piece of technology as any instrument to be found in laboratories of chemistry or physics.

Transplantation, a Genetic Strategy

The addition of transplantation to the repertoire of Drosophila practices did not alter the fact that the creature was a genetic instrument, embodying a vast accumulation of genetic know-how. Most of the thirty-odd papers on transplantation that Beadle and Ephrussi published between 1935 and 1938 concerned genetics, not embryology or biochemistry. Beadle and Ephrussi had geneticists' priorities and wrote for an audience of geneticists. For both, the genetic knowledge and craft skills built into Drosophila defined the problems to which it was applied. Geneticists knew from recombination analysis that genes were generally "pleiotropic": that is, individual genes had multiple effects, and the expression of each individual gene was modified by many others. Geneticists painstakingly constructed different combinations of genes and tried to infer patterns of interaction among them. The same kind of analysis could also be done with transplantation—indeed, much more easily. Eye-color mutants constituted an exceptionally rich accumulation of genetic knowledge of this sort.[38]

Beadle and Ephrussi systematically worked through the twenty-six known eye-color mutants to see what turned up. Because Paris lacked a Drosophila laboratory, it did not belong to the network through which stocks were swapped and shared. Beadle had to ask Demerec to rush some mutant cultures by fast boat or even air mail, a novelty at the time, so that he and Ephrussi could finish their preliminary survey before Beadle had to leave Paris.[39] He and Ephrussi did not test every one of the 676 (26 × 26) possible combinations, but tested each mutant in a reciprocal transplant with the wild eye type to discover if any mutants other than vermillion developed non-autonomously. Such cases, obviously, were the most amenable to experimental manipulation and the most likely to reveal details of where, when, and how genes acted in the development of eye pigments. Only two clear-cut cases of nonautonomous development occurred, vermillion and cinnabar, plus one ambiguous case, claret, and a dozen or so cases of genes slightly modifying the expression of others (Figure 1).[40]

The most striking and unexpected result came with vermillion (v) and cinnabar (cn). To see whether these two genes were involved in the same developmental process—the synthesis of the red and brown eye pigments—Beadle and Ephrussi arranged a reciprocal transplant of a cn disk in a v host and a v disk in a cn host. If the products of the v and cn genes—call them the v^+ and cn^+ substances or "hormones"—were the same, then the expression of neither v nor cn genes should be altered by the reciprocal host. If v^+ and cn^+ were chemically different, however, the transplanted eye should be wild type in both cases, since the host could make up for the genetic and chemical deficiencies of the eye disk. Because the experimental process of a reciprocal transplant was precisely analogous to the geneticists' standard recombination for allelism, Beadle and Ephrussi anticipated only these two possibilities. Genes were allelic or they were not; hence they inferred v^+ and cn^+ were either identical or independent. Beadle and Ephrussi used transplantation like a genetic instrument; hence their surprise when a cn disk in a v host gave a cn eye but a v disk in a cn host gave a wild-type eye.

The simplest explanation of this asymmetrical result was that v^+ and cn^+ were chemical intermediates in a single chain of chemical reactions, $ca^+ \rightarrow v^+ \rightarrow cn^+$, the third intermediate ca^+ being inferred from similar evidence with claret flies.[41] Thus, the transplantation technique allowed Beadle and Ephrussi to deduce a biochemical relationship that could not be revealed through a purely genetic analysis. Drosophila had become something more than a genetic instrument. In hindsight, the $ca^+ \rightarrow v^+ \rightarrow cn^+$ chain can be recognized as the first step in the transformation of a genetic mode of production into a biochemical one. But only in hindsight, for it took some time for Beadle and Ephrussi to adapt fully their experimental practice to their new instrument.

Beadle and Ephrussi devoted an extraordinary amount of labor and ingenuity to examining the effects of moderating genes on the $ca^+ \rightarrow v^+ \rightarrow cn^+$ system. They constructed a more sensitive double recessive

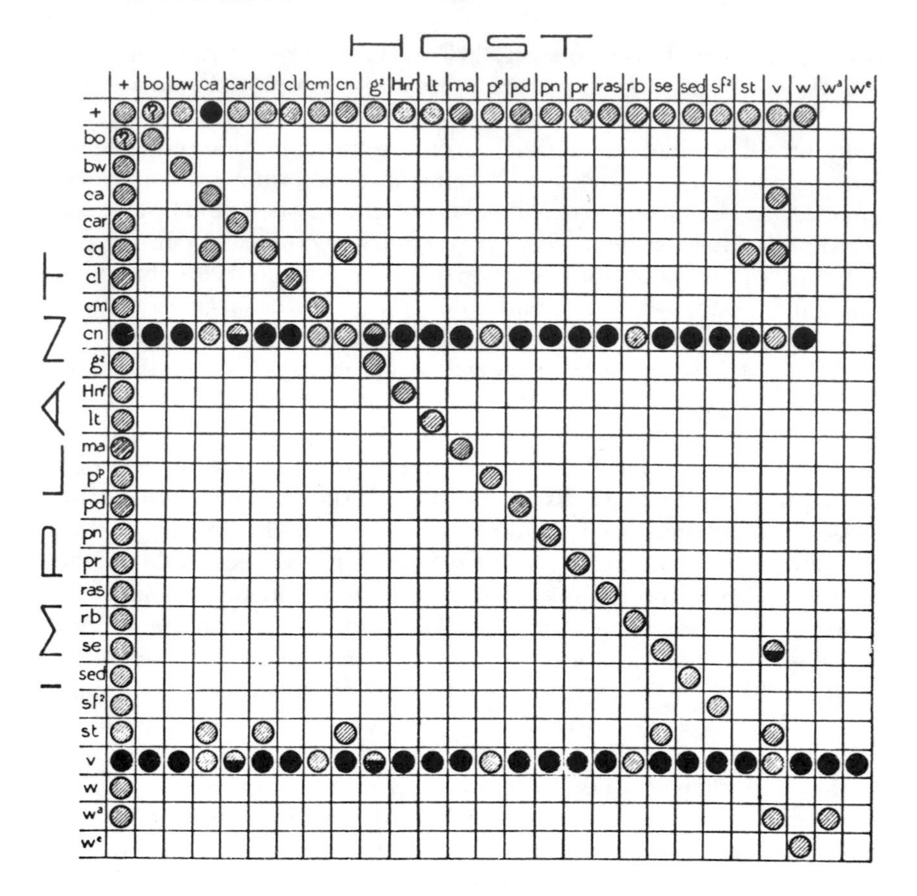

Figure 1. Diagrammatic representation of the results of eye transplants. Shaded circles indicate autonomous development of the pigmentation of the implant. Black circles indicate nonautonomous development of pigmentation. Circles half black and half shaded indicate nonautonomous development of such a nature that the resulting implant is intermediate in color between two controls. Source: Beadle and Ephrussi, "The Differentiation of Eye Pigments in Drosophila as Studied by Transplantation," *Genetics* 21 (1936): 225–247 (on 230).

test fly that could register the slightest moderating effects of various genes on eye color. They did thousands of transplants to explore systematically the permutations and combinations of a genetic system that grew more and more complicated. By the time they gave up this line of work, after three years, Beadle and Ephrussi had constructed a system in which no fewer than fifteen genes participated in eye color, in

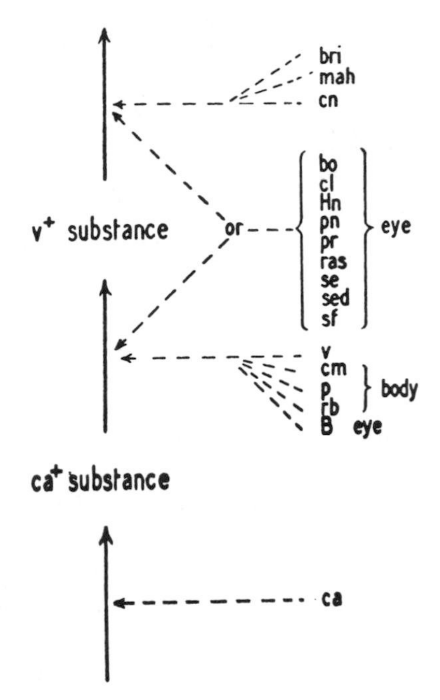

Figure 2. Diagram indicating assumed relations of various genes to the three diffusible substances. Dotted lines indicate the step to be interfered with in the various mutant types, in body, eye, or both. The particular alleles indicated in the diagram are not necessarily the ones used in the experiments. Source: Ephrussi and Beadle, "Development of Eye Colors in *Drosophila:* Diffusible Substances and Their Interrelations," *Genetics* 22 (1937): 76–86 (on 85).

addition to vermillion and cinnabar (Figure 2).[42] Claret, it turned out, was simply a modifier of v and cn.

The complex interrelations between the twenty-six eye-color genes were Beadle and Ephrussi's chief experimental resource. They naturally took for granted that a knowledge of how genes functioned would be won by exploiting genetic complexity. This assumption was built into the very foundations of the Morgan school's beliefs and practices. In the early 1910s, it provided a welcome alternative to the failing "presence absence" theory; in the 1920s and 1930s, it stood as a crucial bulwark against critics of the gene theory like Richard Goldschmidt. Without such a concept, genetics could not easily be integrated with development and evolution. By constructing more complex genetic systems, Drosophilists sought to reveal how genes worked. The greater the variety of interactions they spied in an experimental system, the more hopeful geneticists were of catching a glimpse of some new pattern, some essential feature of the genetic machinery. For Drosophilists, a challenge to pleiotropy constituted a challenge to the very enterprise, a devaluation of their systems of production and of their accumulated expertise.

Beadle and Ephrussi adopted these beliefs and practices when they chose Drosophila for their experiment in developmental genetics. It was precisely the knowledge of complex genic interactions that encouraged Drosophilists like Sturtevant, Schultz, and Beadle to suppose themselves better placed than embryologists to find out when, where, and how genes acted in development. As Schultz put it: "The analysis of the behavior of genes in combination with each other forms a bridge between the description of development in the different mutant races—what may be called Mendelian embryology—and the study of gene action proper."[43] For Schultz, pleiotropic effects were vital clues in sorting out the effect of individual genes on developmental processes. Beadle and Ephrussi's work on transplantation build on that conviction.

The genetic development of the transplantation method enabled Beadle and Ephrussi to rework rapidly a vast body of existing genetic knowledge. Their remarkable productivity (thirty papers in just three years) had dramatic effects on their careers. In 1935, Ephrussi was promoted from research assistant to associate director of the Tissue Culture Laboratory, the resources of which he increasingly diverted to his work with Beadle. He obtained the services of a senior tissue culture technician and of a new staff to maintain genetic stocks and produce a steady supply of larvae for transplantation. In 1937, Ephrussi was named *Maître de recherches* of the Centre National de la Recherche Scientifique and director of the new Laboratory of Genetics at the École des Hautes Études. This was a meteoric ascent for a French scientist. In 1934, Ephrussi had dreamed only of escaping the Old World and returning to Pasadena. In 1936, he was hatching plans to bring Pasadena to Paris—Schultz and Dobzhansky for a year, Beadle for another summer of collective work, and others. Those plans never materialized, though Ephrussi returned to Caltech as a visiting professor in 1935.[44]

The transplantation work also attracted new patrons. The Rockefeller Foundation's grant to the Rothschild Institute was crucial for the rapid expansion of the Drosophila work, which strained the internal resources of a laboratory dedicated chiefly to work on tissue culture. Warren Weaver and his chief man in Europe, Wilbur Tisdale, were especially eager to further Beadle and Ephrussi's careers. Rockefeller money sustained transatlantic collaborations and enabled Ephrussi to prise matching support from the Caisse Nationale des Recherches Scientifiques. In 1936 and 1937, Ephrussi created a Drosophila laboratory with a core of four full-time workers, including Simon Chevais—a graduate from the Sorbonne—and four technicians. The transplantation work attracted bright young biologists to the new field of developmental genetics, most notably Jacques Monod.[45]

As for Beadle, the transplantation work earned him offers from Cold Spring Harbor and Harvard in 1936 and a counteroffer from Morgan for a promotion to assistant professor. Beadle turned down Harvard at first, because their "enthusiasm for Drosophila work was a bit weak," but accepted a renewed bid.[46] Just a year later, he accepted an offer of a full professorship at Stanford and with it the promise of strong support for Drosophila work. A grant of $3,000 from the Rockefeller Foundation was arranged, and with that Beadle hired Edward L. Tatum, a bacterial biochemist, to work full-time on the chemistry of the eye-color hormones.[47] For Beadle, Stanford had the great attraction of a group of outstanding young experimental biologists assembled by Charles V. Taylor.[48]

During the preceding decade, Taylor, a cell physiologist and parasitologist, had replaced an older generation of morphologists and systematists with the best young researchers he could find in the various branches of "chemicophysical biology." Taylor envisioned a unified program of experimental biology. By 1935, he had secured Cornelius B. Van Neil in chemical microbiology, Douglas M. Whitaker, Morgan's son-in-law, in physiology and development, Victor C. Twitty in transplantation and regeneration, Lawrence R. Blinks in plant physiology, and Arthur C. Giese in photobiology. Stanford had excellent laboratory facilities, a research endowment at the Hopkins Marine Station, and plans to get the Rockefeller Foundation to assist in building a new research institute for Taylor's group. For Taylor, Beadle was the key person to tie the group together, working as he did at the point where genetics, developmental mechanics, and physiology converged.[49]

The community into which Beadle entered in 1937 had the cooperative spirit of Morgan's school, but not the shared interest in one organism and criss-crossing collaborations that had made the Caltech Drosophila group unique. Like most academic departments, the Stanford group was deliberately diverse. Beadle was the only geneticist. He had no colleagues like Sturtevant or Schultz, free-floating intellectuals with a hand in everyone's projects. The course of Beadle's work was influenced far less by his biological colleagues than by his biochemical assistant, Edward Tatum. So too with Ephrussi, whose only close collaborator at the Rothschild Institute was Yvonne Khouvine, a bacterial biochemist in the Institute's chemistry section. Ephrussi hoped collaborations would develop with the Rothschild biologists, but none did.[50]

Set in motion by a vision of a new mode of production in genetics and development, the transplantation work acquired a dynamic of its own as Beadle and Ephrussi gained knowledge of what it was possible to do. As they worked through the genetic resources of Drosophila, the limitations of transplantation for embryology became more apparent, and the work turned from its initial focus toward biochemistry. The arrival of biochemists Tatum and Khouvine indicated a change in practice around 1937.

The limitations of Drosophila transplantation as a general method for experimental biology first became apparent in work on morphological characters, the sort that embryologists liked to examine. With his training in *Entwicklungsmechanik,* Ephrussi was eager to use transplantation for studying the morphological development of the eye and the hybridization of species that could no longer interbreed. Beadle and Ephrussi transplanted disks of wings, legs, and other organs in the hope of revealing how genes acted in the development of morphology. Ephrussi planned to revive work on lethals to study the genetics of early embryonic development.[51] The results of these forays were meager, however. Most organs developed autonomously, unaffected by the host tissues. A group of Drosophila embryologists at New York University under Ruth B. Howland had the same experience, and their early enthusiasm for transplantation and genetics also waned.[52]

Transplantation experiments on eye-color genes also revealed to Beadle and Ephrussi that genetic complexity did not necessarily lead to fundamental insights. They began to appreciate just how complex a physiological and developmental system they dealt with, and how fraught it was with opportunities for error. They never ceased to hold that the simple $v^+ \rightarrow cn^+$ chain was a key to understanding when, where, and how genes acted in development. But knowledge of the complexity of the developing eye system forced them to become more cautious in their public interpretations. For a time, they worried that even the v^+ and cn^+ substances, like claret, might turn out to be only different amounts of one thing.[53] Further experiments shed little new light on other steps in the $v^+ \rightarrow cn^+$ reaction chain or on the primary actions of genes.

The period of intense, cooperative work on transplantation ended in 1937. Beadle and Ephrussi, now working independently, moved in the direction of biochemistry. Between 1937 and 1940, Beadle and Tatum gradually transformed Drosophila from an instrument of developmental genetics into an improvised and awkward tool for biochemical work.

New Practices: Transfusion and Feeding

After 1937, Beadle's research on the interactions of eye-color genes gave way to investigations of the complex physiology of development. New experimental practices appeared. Transplantation of imaginal disks gave way to transfusion—transferring body fluid from one larva to another—and to feeding chemical substances directly to larvae as additions to their normal fare of sugar, protein digest, and dried yeast. The feeding method, though it evolved from transfusion and had similar uses, turned out to be a distinctly new departure. It was an experimental practice not of experimental embryologists but of nutritional biochemists.

Beadle and Ephrussi introduced transfusion into their repertoire of practices to supplement transplantation, to confirm inferences about the transport of diffusible hormones, and to guard against physiological artifacts. They began to use the feeding method because it promised to be faster and simpler than transfusion for a quantitative assay of eye-color hormones, since the task of isolating the v^+ substance might require thousands of measurements. At first, biochemical methods were adjuncts to an essentially genetic practice. Gradually, however, they transformed the problem of genes and development in unforeseen ways. Using Drosophila as an instrument of nutritional biochemistry made chemical problems more accessible and appealing, increasing the importance of biochemists in the production process. It also revealed the practical shortcomings of Drosophila as a biochemical instrument. In time, experience with feeding inspired Beadle to fashion Neurospora into a new instrument that combined biochemistry and genetics in a way that he and Ephrussi had never anticipated.

Beadle and Ephrussi conducted their first experiments on transfusion of "lymph" in 1935 and 1936 simply to confirm their belief that host larvae altered the development of pigment in implanted eye disks by supplying diffusible "hormones." They found what they expected: a v or cn fly developed a normal wild-type pigment if, at the pupal stage, it was injected with fluid extracted from wild-type pupae. Transfusion also played a minor role in Beadle and Ephrussi's exploration of the physiology of pigment development, confirming that v^+ and cn^+ were indeed released from specific organs at specific times, transported, and transformed into pigments in the developing eye. Transfusion was a modest improvement on transplantation as a mode of genetic analysis.[54]

Transfusion acquired a new and more important use as a method of biochemical assay as an unanticipated consequence of the reversal in transfusion of the relation between implant and host. In transplantation, the host modified the implanted disk; with transfusion, the host was itself the test organism, revealing the effects of transfused chemical substances.[55] Beadle and Ephrussi used Drosophila in the same way that nutritional biochemists used rats or mice to assay for traces of vitamins or essential food factors. As Ephrussi later phrased it, transfused flies functioned as chemical "reagents" for detecting tiny amounts of biologically active substances. It was, he recalled, "the first real indication that it should be possible to isolate the active principles."[56] Thus an apparently minor variation in practice inaugurated a major change in the perception and pursuit of the problem. Having an instrument for biochemical diagnosis became, in time, a powerful incentive to study biochemistry with it.

Ephrussi had first approached chemistry in mid-1936 with Yvonne Khouvine, transfusing randomly selected amino acids into v larvae in the hope that one would turn out to be the v^+ substance. This shortcut to identifying active substances, although a standard practice of nutritional biochemists, was shooting in the dark. When none of the compounds that Khouvine and Ephrussi tested altered the color of test flies, they turned to other things.[57] Beadle began chemical work a few months later at Harvard, teaming up with an old friend from his Caltech years, Kenneth Thimann, a biochemist with a broad range of interests in the physiology of microorganisms and plants. Their collaboration was serendipitous since Thimann had also just come to Harvard in the fall of 1936.[58] Beadle continued these experiments with Lloyd W. Law, who, after finishing his Ph.D. in genetics at Harvard in 1937, followed Beadle to Stanford as a postdoctoral fellow.[59]

Law's results convinced Beadle that the feeding method could be made into a quantitative "reagent." Everything seemed in place for a classical biochemical isolation of the v^+ "hormone." Tatum's experience in designing bioassays for essential amino acids would serve to make Law's crude assay into a precision instrument for measuring v^+ activity in extracts—an essential prerequisite for isolating the pure substance. Told of the feeding method by Beadle during a visit to the U.S. in August 1937, Ephrussi returned to chemical work, using Tatum's new method. His first experiments using feeding started with Khouvine in October.[60] In the process, Drosophila would be transformed from a genetic to a chemical instrument.

Between Development and Biochemistry

In late 1937, Beadle and Ephrussi talked of resuming their collaboration to work on the chemistry of eye-color hormones. They were no longer mobile post-docs, however; each bore institutional responsibilities that could not be delegated, and their chemical co-workers did not share their enthusiasm for intergroup collaboration. Khouvine doubted that cooperation at such a long distance was practicable. Not wanting to push "the chemists," Beadle and Ephrussi agreed to work independently without a common plan, but to swap research news and to send each other copies of manuscripts before publication. During the next few years, Beadle supplied Ephrussi with Tatum's latest bioassay technology. Ephrussi provided Beadle with *Calliphora* pupae, the best source of hormones. They discussed differences in experimental results and interpretations through an almost weekly exchange of letters.[61] Ephrussi noted that this was "actually a form of cooperation, and perhaps the most convenient under the present conditions."[62] The two groups nonetheless followed different paths. Beadle was slower and more deliberate in exploring experimental systems. Tatum, a perfectionist in experimental practice, set out to design a precise, quantitative, and reliable bioassay. Ephrussi, in contrast, liked to cast about for dramatic chemical breakthroughs. Both styles of work had their price: Tatum ended up being scooped on the chemical structure of the v^+ substance; Ephrussi bogged down in some complicated and not very interesting nutritional physiology.

In hindsight, the most important result of this work was the transformation of Drosophila itself into a new kind of laboratory instrument. Tatum took an instrument of developmental genetics and refashioned it into a reagent for quantitative biochemical analysis. He constructed double mutants, combining v or cn with "apricot," which, because it had virtually no eye pigment, revealed the slightest deepening of color. By combining various mutant genes, he constructed a set of nine standard reference flies in order that the results of tests with extracts could be compared with a known spectrum and given a numerical value. Here indeed was a reagent, a living biochemical instrument designed to measure semiquantitatively the amount of v^+ or cn^+ substances.[63]

Tatum's bioassay required much practical craft skill, as Ephrussi discovered when he tried to make it work. "I should be *very, very* grateful to you," he wrote Beadle, "for giving me a short description of your adopted procedure—(temperature? how to raise them [larvae]? At what age are the different controls used for comparison? Can you keep one set of control flies for some time? etc.)—as fast as possible."[64] There were many little ways to go wrong, and it required detailed instructions to make Drosophila work as a chemical assay as reliably in Paris as in Palo Alto.

The convenience of Beadle's feeding method encouraged Ephrussi to repeat his earlier experiments with amino acids on a much larger scale. Some remarkable things turned up almost immediately. Most of the test larvae died of starvation on a diet of sugar and a few amino acids, including all the controls. Two of those that did hatch, however, showed a distinctly deeper eye color: sure evidence, in standard biochemical practice, that those amino acids were chemically related to the v^+ hormone. Ecstatic, Ephrussi repeated the experiments with still larger numbers of larvae, hoping to retrieve live controls. More positive results appeared, but with a different set of amino acids! Ephrussi could not repeat his earlier results, nor could he find any specific connection between amino acids and pigment formation. Ecstasy turned to puzzlement, and puzzlement to despair. Only a few weeks after thinking he had cracked the chemistry of v^+ at the first try, Ephrussi worried that the entire project might be "a big mistake." It was "a hell of a mess," he wrote Beadle, "difficulties came from the most unexpected sides and mixed up everything." He had not worked so hard since the first months of their transplantation project, and never to so little effect.[65]

The more work Ephrussi did, the more complicated the situation became. Although he suspected something in the growth medium, he eventually agreed with Beadle and Tatum that starvation had caused the phantom effects. They had concluded that the amino acids ordinarily used to make protein were diverted in starvation to the synthesis of eye pigments.[66] It was a meager reward for labors that had consumed most of Ephrussi's time and energy for several years. Having experienced the amazing productivity of the transplantation work, he felt frustrated and miserable about the feeding method.

He blamed a lack of genuine cooperation with Beadle.[67] His dependence on Tatum's experimental practices did make it awkward for him to publish independently: he could not say that he had done a thorough study of starvation because he knew that Tatum would do the same.[68] But much of the trouble originated with Ephrussi's eagerness for a dramatic result and his incautious use of an unfamiliar laboratory practice. Drosophila was not an appropriate instrument for nutritional biochemistry. Too little was known about its physiology, leaving too much room for unsuspected traps like the starvation effect. Even Beadle hesitated to go public with a description of the feeding method.[69] Drosophila had no rich accumulation of biochemical phenomena that could be made accessible by a new technique, as its genetics had been by transplantation. Thus, while Tatum labored to refashion Drosophila into a reliable biochemical instrument, Ephrussi bogged down in a physiological morass.

The chemistry of the eye color "hormones" was slow work that stalled for months at a time. At one point, Calliphora refused to pupate, thus cutting off the supply of crude extract. Sometime later, apparently pure v^+ hormone would not crystallize. Ephrussi began to doubt that Khouvine was as capable a chemist as he had thought. By July 1939, he had decided to terminate their collaboration and arranged with an organic chemist to do the structural work of the v^+ substance. Tatum, meanwhile, had got crystalline derivative of the v^+ substance, but was having trouble identifying the active principle.[70] Outside competition also became a problem for the first time. Through a double slip-up in sterile technique, the two groups accidentally revealed that the v^+ hormone was a relative of the amino acid tryptophane.[71] But before they could capitalize on their lucky break the prize was snatched from their grasp by the eminent organic chemist Adolf Butenandt, and by Alfred Kühn, Beadle and Ephrussi's German competitor in the eye color research.[72] Knowing where to look, Butenandt had only to keep testing relatives of tryptophane until he found the active one.

The Germans' triumph made a bitter pill for Tatum especially, because he had invested so much in perfecting Drosophila as a biochemical tool. But it was a common experience in nutritional biochemistry, a notoriously volatile and risky game in which years of work could be devalued by a newcomer's lucky guess. The German scoop, an especially distasteful defeat in mid-1940, underlined that in adopting the feeding method Beadle and Ephrussi had moved into an area of scientific production where different, and not entirely congenial, rules prevailed. In the nascent field of developmental genetics, Beadle and Ephrussi had few competitors. Possession of the transplantation method enabled them to control the pace of production and limit competition long enough to skim the cream off a problem. Nutritional biochemistry, in contrast, was a densely populated field where victory went to the quick-footed.

Between the end of transplantation work in 1937 and the beginning of Neurospora in 1941, Beadle and Ephrussi did not define a single dominating line for research in Paris or Stanford. Both men looked again at embryology and genetics. In 1937, Ephrussi began a line of experiments on morphogenetic mutants, returning to the scute-8 lethal, but resuming his examinations of the eye-color hormones when the scute-8 studies stalled. He grew very excited when he found some months later that Calliphora extracts caused an increase in the number of eye facets in the "bar" mutant of Drosophila. He felt that he had found evidence of a "facet-forming" morphogenetic substance. The experiment itself was inspired by Sturtevant's earlier observation of nonautonomous development of "bar"-eye in mosaics.[73] In late 1938, Ephrussi plunged into work on the physiological basis of "gene dosage," combining the transplantation technique with Tatum's quantitative bioassay for eye pigment. It involved great labor, over 7,000 test larvae so far, he wrote Beadle in July 1939, but Ephrussi enjoyed working once more in the familiar field of experimental embryology.[74]

Beadle too picked up a variety of earlier interests, the most important of which concerned the problem of genetic dominance. By combining classical genetic analysis with Tatum's new reagent for measuring eye pigment, Beadle hoped to understand dominance in quantitative, physiological terms, in a manner that Schultz had attempted earlier through spectroscopic analysis of the pigments.[75]

Although they published little of their work on dosage and dominance, Beadle and Ephrussi had initiated an important change in their experimental practices. Leaving the biochemical side of the eye color problem to Khouvine and Tatum, Beadle and Ephrussi returned with new tools to the problem that had inspired the invention of transplantation: the problem of characterizing genes by their physiological and developmental effects. The two collaborators seemed to share the mood of anticipation, palpable in the late 1930s, that chemical knowledge of the gene was about to break wide open, and that the chain of reactions between gene and character was not as long and inscrutable as some had thought.[76] Neither man intended to abandon the experimental system of eye-color genes. But they explored new avenues with the experimental tools they had struggled to develop. They sought new opportunities among the major problems of mainstream genetics, applying biochemical methods to situations that had hitherto been posed in purely genetic terms, perhaps in the hopes of recapitulating their success with transplantation. Thus it was that Beadle unexpectedly conceived a distinctly new experimental instrument, Neurospora.

The Invention of Neurospora

Beadle's invention of the Neurospora method is a classic tale of an "aha" experience. Inspiration struck early in 1941 while he sat listening, or not listening, to Tatum lecture on comparative nutrition and biochemistry.[77] A simple idea stole into his thoughts: rather than starting with visible mutations and laboriously working out their complex chemistry, why not start with biochemical mutations and work out the genetics? It was the mirror image of the stratagem that he and Ephrussi had used in applying transplantation to Drosophila, but it entailed giving up Drosophila for a microorganism whose biochemical mutations could be easily identified as nutritional requirements.[78] With Ephrussi, Beadle had gambled on adapting a method of experimental embryology to an organism with a rich genetics; with Tatum, he gambled that a microorganism could be found for which a genetics could be constructed.

Neurospora was a bold move into a new mode of production, although Beadle may not have thought so at first. He did not expect to give up the eye-color work, and certainly not his new work on the physiology of genetic dominance. In fact, Beadle may have designed Neurospora more for Tatum than for himself, as an alternative to a general line of work that Tatum was about to develop in the nutritional biochemistry of insects. This interpretation differs from the usual view. Beadle and Tatum later recalled that the eye-color work had reached an impasse, thus provoking a radical change in practice. "Isolating the eye-pigment precursors of Drosophila was a slow and discouraging job," Beadle recalled. Tatum told Joshua Lederberg that the experience with Butenandt caused Beadle and himself to rethink their research strategy. Beadle later said that Tatum's wasted efforts to identify the cn^+ hormone prompted them to drop Drosophila and to take a chance with Neurospora—a "blessing in disguise."[79] On the basis of such testimony, historians might well conclude that the Drosophila work had reached a point of erratic and diminishing returns.[80]

Yet to most observers Butenandt's stunning success with the v^+ hormone seemed to quicken the pace of work, not to dampen it. Tatum lost no time in resuming work on the cn^+ hormone, and Beadle assigned a student to the red pigment. Ephrussi, though hampered in turn by French mobilization and German occupation, picked up where he left off as soon as he was safely relocated at Johns Hopkins in 1941. He looked forward to "the old cooperation" with Beadle and expected a complete chemical elucidation of the chain of reactions between tryptophane and the red and brown eye pigments.[81] Sturtevant predicted that chemical knowledge of the v^+ hormone would open up the physiology of hormone production and perhaps even make it possible to work backward through the chain of chemical intermediates to the gene itself.[82] Sturtevant worried that chemists could not be certain when the gene had been reached, but Ephrussi gently reminded him that "unpredictable experimental situations often . . . yield a little more than permitted by the theory."[83] Reviewing the prospects of the eye-color problem in January 1941, Beadle and Tatum stated boldly that the primary products of the v and cn genes were specific enzymes.[84]

It does not seem that Beadle's decision to invest in a new genetic instrument was born of frustration with the eye-color problem. Rather, the advent of Neurospora may have occurred because Tatum had been struggling to make Drosophila into a general instrument of biochemical research. Tatum worked much more on the nutrition of Drosophila than seems necessary for bioassaying eye-color pigments. He explicitly pointed to insect nutrition as an important and underworked field. Likening it to animal nutrition on the eve of the discovery of vitamins, he signalled his intention to lead in its development.[85] Tatum was strategically placed to exploit such an opportunity. He was familiar with the nutritional physiology of Drosophila from his work on the starvation effect and, unlike virtually every other biochemist, he had already mastered genetics. To make Drosophila into a tool for systematic nutritional work, he had only to find, by a process of elimination, what larvae needed to grow on a chemically defined, minimal medium. He had done the same many times over with propionic bacteria and other microorganisms.[86]

Beadle, an experienced strategist, might well have felt uneasy, knowing that Tatum intended to invest a lot of time in making Drosophila into an instrument of biochemical analysis. What were the possible returns on such an investment? What other characters were as well suited to biochemical analysis as eye pigments? Nutritional mutants of Drosophila would have been impossible to make, since larvae do not survive starvation and cannot be divided like a colony of bacteria. To an ex-bacterial biochemist looking for ways to strike out on his own, Drosophila might well have seemed the system through which genetics and biochemistry could best be united. Beadle, however, would have had good reasons to doubt that such a unification would ever repay a large additional investment on the biochemical side. Given the available options, Beadle understandably thought it would be easier to invent a genetics of fungi than biochemistry of insects.

Beadle had other compelling reasons to intercede in Tatum's program of research. Tatum's career at Stanford approached a critical point in late 1940. He still held the position of a research associate, dependent upon soft money after three years of devoted and very productive work on the eye-color hormones. Beadle pushed to give Tatum a faculty appointment, but some of the biologists hesitated to give a chemist a prized position among them. Beadle wanted to keep Tatum in his group, but he advised him to take a post elsewhere if one could not be arranged soon at Stanford.[87] Beadle felt a particular responsibility for Tatum, having enticed him away from the certainties of biochemistry into the no-man's-land between genetics and chemistry. Beadle was dramatically reminded of the risk when Tatum's father, a professor of pharmacology, took him aside and expressed concern about Edward's future "in a position in which he is neither a pure biochemist nor a bona fide geneticist."[88] Tatum needed to establish himself before taking large risks.

Tatum became an instructor in biology a few months later. If the main motive for switching from Drosophila to a microbiological instrument had been to give Tatum standing as an independent member of Taylor's group, the strategy worked. As *the* expert in chemical microbiology in the group, Tatum would visibly direct Beadle's research on the physiological functions of genes. At the same time, he would avoid the short term risks of hormone hunting and the long term risks of insect nutrition, and he could stake out a strategic position on the new frontier between genetics and biochemistry. Tatum benefited in more than scientific ways from the switch to Neurospora.

Constructing Neurospora

What was this creature, Neurospora, that displaced Drosophila in Beadle's workshop? In the wild, Neurospora is an unruly creature. It grows irrepressibly in tropical countries like Indonesia, where botanists first studied its natural history. There, because its spores are heat resistant, Neurospora was the first colonizer of terrain devastated by volcanic eruptions. The organism covered landscapes of blackened stumps and debris in a blanket of bright orange fuzz. Semidomesticated, it was used by the Javanese to brew *ontjom,* a favorite concoction made by leaving peanut mash, the residue of oil pressing, to collect a thick orange covering of the fungus. Yet Neurospora resisted the more rigorous domestication of the scientific laboratory. The Dutch botanist Friedrich Went, who first investigated the fungus in Java in the late 1890s, was disconcerted by its ability to grow right through the cotton plugs of his culture tubes. He could do nothing with it until he returned to the cool, well-regulated North. European strains of Neurospora were also troublesome: bakers dreaded infestations of the red bread-mold, *N. sitophila,* whose tenacious spores passed unharmed through baking ovens inside the loaves.

By 1930, however, Neurospora had been domesticated in botanical laboratories. Bernard O. Dodge, plant pathologist at the Brooklyn Botanical Garden, revealed that the creature had a sex life. The normally asexual, "haploid" growth filaments, "conidia," of two mating types fused and their chromosomes recombined. This discovery made Dodge, a Columbia Ph.D. and a regular visitor to Morgan's group, into an avid promoter of Neurospora as an organism for genetic research. Dodge tried to persuade Morgan that his fungus was far more convenient to study than Drosophila. Morgan, though not persuaded, agreed nevertheless to take a collection of Dodge's cultures with him to Caltech. There he persuaded a graduate student, Carl C. Lindegren, to work out the cytogenetics for his dissertation (1931). Neurospora became a junior partner to Drosophila at Caltech, where it came to Beadle's attention. When Beadle began to cast about for a suitable microorganism for doing biochemistry and genetics, Neurospora naturally came to mind.[89]

Neurospora required some engineering to make it a useful instrument for biochemical genetics. Thanks to Dodge and Lindegren, its genetics were unusually well documented for a fungus; they developed standard procedures to determine whether mutations occurred in the same or different alleles, and to measure linkage. Neurospora's physiology was not so well documented; it does not figure in the 1938 survey of fungal nutrition by Nils Fries, a source upon which Tatum relied.[90] Since Tatum did not know its dietary requirements, he could not be sure that it would grow on a minimal medium—an essential condition for making biochemical mutants. But being an opportunistic survivor, Neurospora was not a fastidious eater, requiring only a source of carbon and nitrogen, salts, and biotin (a B vitamin). Its propensity to grow also had to be controlled, since measurement of small amounts of metabolic intermediates required precise and reproducible growth rates. Initially, colonies were simply scraped off, squeezed dry, and weighed; subsequently, cultures were grown in special long "race tubes" so that the position of the leading edge of the mycelium could be measured over time to get a rate. Francis J. Ryan, a young National Research Council fellow who came to work on embryology with Whitaker, invented this trick. Ryan was so taken with the Neurospora work that he loitered around the lab until Beadle finally agreed to take him on despite lack of space.[91] Within only a few months, Neurospora had been fashioned into a serviceable instrument for biochemical genetics.

Producing and identifying mutants in Neurospora was much simpler than in Drosophila. Filaments irradiated with X-rays or ultraviolet light were allowed to fuse with unirradiated filaments of the opposite sex. From each of the resulting perithecia, or fruiting bodies, a single ascospore or spore pod was dissected out, split open, and its eight spores removed, each placed in order on an agar plate—four potential mutants and four wild types. Each spore was then removed with a bit of agar and transferred to a growth tube containing enriched growth medium. From the resulting colonies, bits were transferred to "minimal" medium and allowed to grow into new colonies. Mutants blocked in one or another biosynthetic pathway could not grow, however, and could thus be recognized instantly. To determine the interrupted metabolic pathway, the investigator added known chemical substances, one by one, to see which restored growth on the minimal medium. A group of mutants, say for the amino acid arginine, could then be analyzed by recombination genetics to see if they were altered in the same or different genes. It required only time, persistence, and some chemical intuition to work out the individual steps of biosynthetic pathways.[92]

Because this entire process used standard genetic or biochemical practices, Beadle and Tatum knew beforehand that it would work. But would biochemical mutants appear so infrequently that the hunters would be discouraged and quit before finding one? To prevent despair, they agreed to isolate 1,000 single-spore cultures before testing for mutations. It was an unnecessary precaution: culture number 299 turned out to be a mutant for pyridoxine (vitamin B_6), and 1,090 for thiamine, or B_1. Beadle wrote Lindegren elatedly that he "always knew they [Neurospora] were fine bugs to work with." Six months after the first experiments, Beadle wrote Ephrussi that "the mutants are turning up so fast we haven't got time to digest them yet." By November 1941, he and Tatum had a dozen "good mutants," that is, mutants that could be characterized biochemically. Their first paper was in press. By July 1942, the number of "good mutants" had quadrupled.[93]

The productivity of the Neurospora method astonished everyone. Beadle's first public presentation of the work in December 1941 was the hit of the scientific season. Sewall Wright considered it the most important discovery in genetics since the salivary chromosomes. Even H. J. Muller, who disliked acknowledging anyone's brilliance, was impressed. One colleague professed not to believe Beadle's claim to have produced so much so fast.[94] By September 1942, the Stanford group had tested 33,000 single-spore cultures and identified 83 biochemical mutants. By 1945, over 60,000 tests had turned up mutations in 100 distinct genes. Beadle and Tatum added more knowledge of biosynthetic pathways than two or three generations of biochemists had using traditional methods.[95] A mutant that especially delighted Beadle was one in the tryptophane series that revealed to him the identity of the elusive cn^+ hormone.[96]

To exploit the productive capacity of Neurospora required more workers and more money than Beadle could rightfully claim as his share of the Rockefeller Foundation's grant to Taylor's group. In December 1941, Beadle asked Warren Weaver for a grant specifically for Neurospora work. He estimated that it would take five or six full-time workers to staff the project, besides himself and Tatum: an experienced researcher to develop biochemical assays, plus a technician; two more technicians to maintain mutant stocks and do the routine genetic analysis; and two Ph.D. research assistants to characterize biochemical mutants and start work on the cell physiology of Neurospora. The whole operation would cost $11,000 per year.[97] Ordinarily, the Rockefeller Foundation did not make personal grants to individuals already supported by a group grant. But Weaver urged that an exception be made. The work was "one of the most important and exciting leads which had developed for a long time in the ultramodern field lying between genetics and biochemistry," Weaver wrote, and it had resulted "in a major part, from circumstances which we helped to create."[98] In the next few years, Beadle also tapped the Research Corporation; Nutrition Foundation; American Philosophical Society; Merck, Sharp and Dohme; and the U.S. Department of Agriculture.[99]

Within a year, Beadle's group had grown to almost a dozen members. The most important new additions were two recent Ph.D.'s whom Beadle recruited from Caltech: David Bonner, a structural organic chemist, and Norman Horowitz, an experimental embryologist and expert on respiration enzymes—a subject that Beadle wanted especially to develop. Both engaged in the search for biochemical mutants, along with Ryan; geneticist Adrian Srb (who came to work on Drosophila but found them all in cold storage), six graduate students, and two part-time research assistants. In 1943, they were joined by Herschel K. Mitchell and Mary Houlahan, who later married one another. All worked together in what Beadle was soon calling the "mutant hunting room"—a name and a mode of organized group work that owed much to his experience in the Caltech fly-room. The process of screening mutants and identifying them biochemically lent itself to assembly-line production. Results poured out of the mutant hunting room faster than anyone had anticipated. Horowitz recalled that every day brought exciting new results in that "scientific paradise." Back home, Francis Ryan felt like an exile from paradise, much as Ephrussi had in Paris. "On hearing of the 12,000–16,000 culture tubes weekly," Ryan wrote, "we in New York, the proverbial hub of the universe, feel provincial."[100]

Biochemistry Versus Genetics

The extraordinary productivity of Neurospora completely upset the ecology of Beadle's group—it colonized new terrain as vigorously in the laboratory as in the wild. Within months, work on the Drosophila eye-color system was sidelined, then abandoned. Beadle never meant that to happen. In August 1941, he might still have shared Ephrussi's hope that his amazing success with Neurospora would not cause him to give up Drosophila.[101] By November, however, Beadle reluctantly had to admit that even his new research on the physiology of gene dominance would have to be set aside. That was especially hard: he had constructed various combinations of eye-color mutants to test with Tatum's bioassay and was eager to get on with the work:

> There's a very interesting set of relations among the alleles (I have 6 or 7), but unless I can find a student to interest in it, I'm not sure I'll ever get back to it. The Neurospora work is constantly accelerating, and I find I'm more and more tied to it. This state of affairs seems to apply to all thc "fly lab" workers, so Drosophila seems to be at least temporarily out of luck at Stanford.[102]

Another casualty of Neurospora was the nascent collaboration with Ephrussi on eye-color research. Beadle gently suggested that Ephrussi might strike up a collaboration with Donald Costello, a zoologist at the University of North Carolina and a recent visitor at Stanford.[103] Ephrussi was politely unenthusiastic.

The productivity of the Neurospora method resulted in a shift from genetics to a predominantly biochemical mode of practice. In their first papers on Neurospora, Beadle and Tatum dwelt on genetic problems like the relation between genes and specific enzymes, the nature of dominance, and the mapping of mutations. They looked forward to finding mutants for nondiffusible substances, that is, for morphological and developmental characters of interest to biologists. In his first report to the Rockefeller Foundation in 1942, Beadle stated that the primary purpose of the Neurospora work was to learn about the relations of genes to biochemical characters, to test the idea of one-gene-one-enzyme. "A secondary purpose, not entirely anticipated at the outset," was to use the method for purely biochemical purposes—isolation, bioassay, and analysis of biosynthetic pathways. Beadle's policy until then had been to work out a bioassay only where a good chemical assay did not exist.[104]

Soon, the biochemical tail was wagging the genetic dog. In 1944, Beadle's report to the Rockefeller Foundation projected ten lines of work, nine of which had to do with biochemical pathways and only one with genetics. In 1945, Beadle laid out a plan to make Stanford a center of research on intermediary metabolism. Almost as an afterthought he added two lines of work in genetics: chemical mutation, and what seemed to be a case of cytoplasmic inheritance in strains of Neurospora that adapted to new diets without genetic change.[105] Of approximately twenty-eight papers written by the Stanford group between 1941 and 1945, fifteen were distinctly biochemical, four biological, and nine mixed (methods, reviews); twelve appeared in biochemistry journals.

Beadle tried to preserve the biological side of the Neurospora work. In 1944, for example, he persuaded Barbara McClintock, an old friend from his Cornell days, to spend some months trying to resolve the technical problems associated with cytogenetics in Neurospora. With Vera Coonradt, he tried to adapt Neurospora to research on the mechanism of crossing over.[106] He enjoined Tatum's Neurospora team to save mutants of biological interest: for example, strains that showed unusual colonial morphology or habits of growth—characters of the whole organism.[107] Beadle drew his biologist colleagues Geise and Whitaker into cooperative work on the cell physiology of Neurospora.[108] Beadle's preferences as an experimentalist were still those of a biologist.

Yet everything seemed to conspire to make Neurospora an instrument for producing knowledge of biochemistry, not biology. There was the instrument itself—the rapidly growing family of its mutant strains. Like the eye-color mutants of Drosophila, these biochemical mutants constituted a resource that powerfully shaped research practice. They invited Beadle and Tatum to construct metabolic pathways, just as the eye-color mutants of Drosophila had invited Beadle and Ephrussi to construct networks of genetic interactions and reaction chains. As much as the accumulated genetic knowledge in Drosophila made transplantation a genetic mode of production, so the biochemical richness of Neurospora made it a biochemical instrument and the basis of a biochemical mode of practice.

Neurospora biology, in contrast to its biochemistry, proceeded at a slower pace. Even McClintock, the master cytogeneticist, could not entirely succeed with Neurospora chromosomes. To his chagrin, most of the morphological mutants that Beadle had carefully saved turned out to be fungal contaminants, the consequence of imperfect sterile techniques and his ignorance of systematic mycology.[109] Such setbacks, and the effortless productivity of the biochemical work, were irresistible incentives for Beadle's group to move away from biology into biochemistry. Constructing linkage maps simply could not compete with analysis of biochemical pathways, either in novelty or in productivity.

Competition between experimental practices does not take place in a vacuum, of course, but in a marketplace of producers and consumers.[110] As Drosophila geneticists had constituted a large body of eager consumers for Beadle and Ephrussi's transplantation work, biochemists were the most numerous and ready users of what the Neurospora group produced. Neither group of users needed persuasion. Geneticists could incorporate directly into their own daily practice the new knowledge produced by transplantation. No biochemist could ignore the new knowledge of metabolic pathways that flooded out of the mutant hunting room at Stanford. Active interest by important communities of practitioners could not help but influence how the transplantation and Neurospora practices evolved.

Experimental embryologists, on the other hand, had never been convinced of the utility of Drosophila for them. Similarly, geneticists were never consumers of Neurospora work in the same way as biochemists. Though impressed by Beadle and Tatum's creation of a new specialized branch of genetics, geneticists remained skeptical of Beadle's radical one-gene-one-enzyme concept. Practitioners of Drosophila, mouse, or maize genetics could not use the new genetic knowledge for their sorts of production. Neurospora genetics was not indispensable to the practice of other geneticists, whereas the biochemistry of Neurospora was important to all nutritional biochemists, whatever organisms they happened to be working on themselves.

Circumstances connected with the war also impelled Beadle into biochemistry. Shortages of materials and the military draft made it increasingly difficult for scientists not working on war-related research to continue business as usual. As Lily Kay has shown, Beadle undertook work on practical problems that qualified under mobilization laws as relevant to war production. He helped develop bioassays for vitamins and essential food factors; but the genuine interest that he took in inventing new and useful biochemical tools soon flagged under the mounting volume of demands for routine assays. The routine had the virtue, however, that it kept his mutant-hunters out of the draft and the Neurospora machine running at full speed.[111] The same pressures also helped shift the balance of Beadle's work away from biology. That, indeed, may have been the war's most important and lasting effect on Neurospora practice. War work and war funding amplified an impetus toward biochemistry inherent in the Neurospora practice itself: the bio-instrument, the organization of the mutant hunting room, the career strategies of the mutant-hunters.

The rapid expansion of Neurospora work also changed Beadle's and Tatum's roles in the production process. Though he took real pleasure in scientific handiwork, Beadle increasingly had to leave the bench work to Tatum, Bonner, and Horowitz. It took all his time to manage a team of ten workers, cultivate a half dozen granting agencies, and promote his vision of biochemical genetics and the work of the Stanford school. As Tatum and his group took charge of the experiments, Tatum's overmastering interest in metabolism and biochemistry began to push to the fore. For him, genetics was a tool for doing biochemistry, not an end in itself. Even the new genetics of bacteria, in the invention of which he played a key role, did not tempt him to abandon Neurospora.[112] Beadle set the intellectual tone, much as Morgan had in the Drosophila group. The one-gene-one-enzyme theory bears Beadle's distinctive hallmark. But the practices and products of the Stanford school reflect the interests of Tatum and his team of biochemists.

Genetics remained an integral part of the daily practice of the Neurospora group as a routine stage of the production and characterization of biochemical mutants. Each new mutant was crossed with other mutants of the same phenotype, to see if they were altered in the same or in different genes, and with different mutants, to detect linkages and map chromosomes. Outcrosses with wild-type stock indicated when mutation occurred in a single gene. For Beadle this test was more than routine, because it confirmed the one-gene-one-enzyme theory. But he did surprisingly little to solicit independent evidence for the theory until the late 1940s, when Horowitz undertook to defend it against a rising tide of critics. Neurospora functioned primarily as a biochemical instrument, producing knowledge for a constituency of biochemists.

Conclusion

Focusing on experimental production offers many advantages to historians of science. It underlines how significant careers and trends in experimental science are constructed around the invention of new instruments and new modes of production. George Beadle, though best known for the one-gene-one-enzyme theory, was above all a highly original inventor of new experimental systems. Transplantation in Drosophila, Drosophila nutrition, and Neurospora, all three the creations of Beadle's group, carried the succession from developmental to physiological and biochemical genetics between 1934 and 1945. Like the remarkable achievements of the Morgan group in Drosophila genetics, those of Beadle's group need to be understood as changes in know-how and production.

Laboratory organisms should be treated as constructed artifacts, akin to physical instruments, and as tools for investigation rather than as objects to be investigated. Genetic instruments are extended systems, consisting of large and fecund families of mutants designed for particular purposes and freighted with ever-growing bodies of craft knowledge and skills. These systems include the routine technology of stockrooms and the user networks for exchange of mutants. The large families of structurally related and interconvertible chemical compounds (aromatics, terpenes, or steroids, for example) have a similar place in the practice of organic chemists, and analogous systems could no doubt be identified in every discipline. The skills and knowledge embedded in such instruments are a crucial, and much too little studied, element of scientific production.

A metaphor of production also makes clear how the imperative to produce may override intellectual commitments to particular problems or to theoretical positions. At crucial points in his career, Beadle chose the path most likely to produce results, even when his decision entailed an abandonment of problems dear to him. He also chose to produce in fields that had large and influential constituencies of users. This case study supports the conception, which Latour has to date laid out most systematically, of experimental scientists as small producers in markets made up not of passive consumers but of other small producers.[113]

Focusing on production and practice also illuminates the complex relations between instruments and problems, between programmatic and opportunistic production. Many have observed that this relationship runs both ways.[114] Sometimes people invent tools to resolve particular problems; new problems arise when an instrument displays unexpected capabilities. Also, specific problems may not enter in at all. New practices can be strategic gambits to defend or to enlarge the boundaries of schools, or to create new constituencies in neighboring disciplines.

Beadle and Ephrussi invented Drosophila transplantation not because they were fascinated by the eye-color problem but because they had been persuaded by Sturtevant and Schultz that geneticists should occupy a position in developmental biology. However, the research that Beadle and Ephrussi conducted with their new instrument was guided not by any program but by what the instrument did most productively. Their choice of problems was determined largely by the vast potential for genetic analysis that had been built into the creature by scores of geneticists. The nutrition method was borrowed from biochemistry to make Drosophila a more convenient "reagent" for biochemical analysis. With Drosophila nutrition Beadle discovered the limitations of Drosophila and the kind of physiological genetics it made possible. Beadle designed the Neurospora method to overcome those limitations in Drosophila, inadvertently driving Drosophila from the lab and making genetics an element in a new, and essentially biochemical, mode of production. We see a constantly changing dialectic between tools and uses, uses and audiences.

Finally, the metaphors of production and markets direct attention to how producers of knowledge compete and seek to control competition. Beadle's inventions were strategic gambits in the marketplace for genetic knowledge. With each one, Beadle and his co-workers managed to work for a few years in a highly productive field with minimal competition. So complete was their temporary monopoly, in fact, that the competition might not be apparent. With whom was Beadle competing? At first, he had to compete with other bright young Drosophilists who were striving to impress seasoned fly-men like Sturtevant and Morgan. But whereas most beginners just modified familiar experimental systems, Beadle created new ones, which enabled him to avoid close competitors.

Beadle and Ephrussi followed a similar strategy in their relations with the Göttingen group—Ernest Plagge, Alfred Kühn and others—who worked on eye pigments in other organisms. The groups found ways to avoid one another. As the eye-color work moved from genetics to biochemistry, the nature of the competition also changed. Alfred Butenandt proved a dangerous competitor for Beadle and Tatum. In the search for hormones, large chemical laboratories had a competitive edge. Even with Neurospora, the work on bioassays brought Beadle into a realm of cutthroat competitors. He was disappointed several times when collaborations with growth-factor and hormone chemists proved to be one-way exchanges of information. Accustomed to Drosophilists' cooperative folkways, Beadle was repelled by the chemists' "bitch the other guy if you can" attitude.[115]

Yet another kind of competition affected the groups of physiological geneticists who worked with different systems of production—primroses, shrimps, guinea pigs, and mice. Beadle did not try to exclude them. He had every reason to want confederates in a new and ill-defined field, as long as they were not direct competitors. Biologists seem deliberately to have avoided direct competition by working on different organisms. On the other hand, the lively controversy over Beadle's one-gene-one-enzyme theory, which contrasted so strikingly with the universal acceptance of his experimental work, shows that biologists who use different systems of production may nonetheless compete over theory. Theoretical disputes are one way in which the authority of different, but potentially competitive, modes of practice are negotiated.

A metaphor of production suggests concrete ways of investigating large questions of scientists' collective mores and behaviors. It reveals the need for systematic studies of the marketplaces for skills and knowledge and for comparative studies of groups that use similar and different means to produce knowledge. It remains to be seen how new modes of production spread, how the founders of new practices accommodate to a more crowded field of producers, and how those who have invested in other systems of production respond when novel systems like Drosophila, Neurospora, or *E. coli* force them to reevaluate their own methods.

An earlier version of this essay was published in *Historical Studies in the Physical and Biological Sciences* 22, 1 (1991): 87–130, and is reprinted by permission.

Notes

The following abbreviations are used: *AJB, American Journal of Botany*; *AN, American Naturalist*; *ARG, Annual Review of Genetics*; *BB, Biological Bulletin*; *BBFB, Bulletin Biologique de France et Belge*; *BHM, Bulletin of the History of Medicine*; *BMNAS,* National Academy of Sciences, *Biographical Memoirs*; *CRAS,* Académie des Sciences, *Comptes Rendus*; CS, Curt Stern Papers, American Philosophical Society, Philadelphia; *DSB, Dictionary of Scientific Biography*; GWB, George W. Beadle Papers, California Institute of Technology Archives, Pasadena, California (documents are identified by box and folder); *HSPS, Historical Studies in the Physical Sciences*; *JBC, Journal of Biological Chemistry*; *JEZ, Journal of Experimental Zoology*; *JGP, Journal of General Physiology*; *JHB, Journal of the History of Biology*; JS, Jack Schultz Papers, American Philosophical Society Library; LCD, Leslie C. Dunn Papers, American Philosophical Society Library; MD, Milislav Demerec Papers, American Philosophical Society Library; *PNAS,* National Academy of Sciences, *Proceedings*; *PR, Physiological Reviews*; *PRSL,* Royal Society of London, *Proceedings*; *PSEBM,* Society for Experimental Biology and Medicine, *Proceedings*; *QRB, Quarterly Review of Biology*; RF, Rockefeller Foundation Papers, Rockefeller Archive Center, North Tarrytown, New York (documents are identified by record group, series, box, and folder); *SSS, Social Studies in Science.*

1. Bruno Latour's program for a sociology of scientists "in action," for example, makes use mainly of his own observations of laboratory life and two participant accounts. Bruno Latour, *Science in Action: How to Follow Scientists and Engineers Through Society* (Philadelphia: Open University Press, 1987).

2. Some that I have found most stimulating are Steven Shapin, "Pump and Circumstance: Robert Boyle's Literary Technology," *SSS* 14 (1984): 481–520, and "The House of Experiment," *Isis* 79 (1988): 373–404; Shapin and Simon Schaffer, *Leviathan and the Air Pump: Hobbes, Boyle, and the Experimental Life* (Princeton, N.J.: Princeton University Press, 1985); Latour and Steve Woolgar, *Laboratory Life: The Construction of Scientific Facts* (Beverly Hills, Ca.: Sage, 1979); Peter Galison, "Bubble Chambers and the Experimental Workplace," in *Observation, Experiment, and Hypothesis in Modern Physical Science,* ed. Peter Achinstein and Owen Hannaway (Cambridge: Cambridge University Press, 1985), 309–373; and *How Experiments End* (Chicago: University of Chicago Press, 1987); James A. Secord, "The Geological Survey of Great Britain as a Research School," *History of Science* 24 (1986): 223–275; Roy Porter, "Gentlemen and Geology: The Emergence of a Scientific Career, 1660–1920," *Historical Journal* 20 (1978): 809–836; David Gooding, Trevor Pinch, and Simon Schaffer, eds., *The Uses of Experiment: Studies in the Natural Sciences* (Cambridge: Cambridge University Press, 1989).

3. Boelie Elzen, "Two Ultracentrifuges: A Comparative Study of the Social Construction of Artifacts," *SSS* 16 (1986): 621–662; Yakov M. Rabkin, "Technological Innovation in Science: The Adoption of Infrared Spectroscopy by Chemists," *Isis* 78 (1987): 31–54; Timothy Lenoir, "Models and Instruments in the Development of Electrophysiology, 1845–1912," *HSPS* 17 (1987): 1–54.

4. Bonnie Clause, "The Wistar Rat" (unpublished paper, University of Pennsylvania, 1988).

5. George W. Beadle, "Biochemical Genetics," *Chemical Reviews* 37 (1945): 15–96; Sewall Wright, "The Physiology of the Gene," *PR* 21 (1941): 487–527.

6. Beadle and Sterling Emerson, "Further Crossing Over in Attached-X Chromosomes of *Drosophila melanogaster,* " *Genetics* 20 (1935): 192–208; Beadle and Alfred H. Sturtevant, "X Chromosome Inversions and the Meiosis in *Drosophila melanogaster,*" *PNAS* 21 (1935): 384–390. Beadle talked in 1932 of returning to maize genetics: Beadle to Demerec, 9 Jan 1932 (MD).

7. Ephrussi to L. C. Dunn, 12 Dec 1932, 6 and 20 Feb, 17 Sep 1933, 21 Jan, 26 Feb, late Apr 1934, and Dunn to Ephrussi, 12 Mar, 2 May 1934 (all in LCD); Sturtevant to Demerec, 21 Jan [1934], and Demerec to Sturtevant, 3 Feb 1934 (both in MD).

8. Sturtevant, "The Use of Mosaics in the Study of the Developmental Effect of Genes," VIth International Congress of Genetics, *Proceedings* (1932), I: 304–307, and "The Vermillion Gene and Gynandromorphism," *PSEBM* 17 (1920): 70–71; E. B. Lewis, *DSB,* s.v. "Sturtevant."

9. Richard M. Burian, Jean Gayon, and Doris Zallen, "The Singular Fate of Genetics in the History of French Biology, 1900–1940," *JHB* 21 (1988): 357–402, on 391–393; Boris Ephrussi, "The Absence of Autonomy in the Development of the Effects of Certain Deficiencies in *Drosophila melanogaster,*" *PNAS* 20 (1934): 420–422, and "The Behavior in Vitro of Tissues from Lethal Embryos," *JEZ* 70 (1935): 197–204.

10. Beadle, "Recollections," *Annual Review of Biochemistry* 43 (1974): 1–13, on pp. 5–6; Beadle, "Foreword," in Sturtevant, *Genetics and Evolution* (San Francisco: Freeman, 1961), pp. iii–iv; Beadle to Schultz, 2 Jul 1970 (JS); Garland E. Allen, *Thomas Hunt Morgan: The Man and His Science* (Princeton, N.J.: Princeton University Press, 1978), p. 388.

11. Elof Carlson, "The Drosophila Group: The Transition from the Mendelian Unit to the Individual Gene," *JHB* 7 (1974): 31–48, and *Genes, Radiation, and Society: The Life and Work of H. J. Muller* (Ithaca, N.Y.: Cornell University Press, 1981), chap. 5; Jack Schultz, "Innovators and Controversies," *Science* 157 (1967): 296–301; Allen, *Thomas Hunt Morgan,* 188–208.

12. Schultz to Beadle, 31 Jul 1970 (JS).

13. Beadle, "Alfred Henry Sturtevant," American Philosophical Society, *Yearbook* (1970), 166–171, on 170.

14. Carlson, *Muller* (note 11), Chap. 5.

15. Schultz to Beadle, 31 Jul 1970 and Schultz to T. Caspersson, 31 Aug 1940, 18 Dec 1942 (JS).

16. Sturtevant to Stern, 6 Sep [1934] and Dobzhansky to Stern, 14 Jan 1935 (CS).

17. Morgan, "Genetics and the Physiology of Development," *AN* 60 (1926): 489–515; E. E. Just, "Phenomena of Embryogenesis and Their Significance for a Theory of Development and Heredity," *AN* 71 (1937): 97–112; E. W. Sinnott, "The Genetic Control of Developmental Relationships," ibid., 113–119.

18. Jan Sapp, *Beyond the Gene: Cytoplasmic Inheritance and the Struggle for Authority in Genetics* (New York: Oxford University Press, 1987), 3–4; Richard B. Goldschmidt, *Physiologische Theorie der Vererbung* (Berlin, 1927), and *Physiological Genetics* (New York: McGraw-Hill, 1938); Curt Stern, "Richard Benedict Goldschmidt," *BMNAS* 39 (1967): 141–192.

19. F. R. Lillie, "The Gene and the Ontogenic Process," *Science* 64 (1927): 361–368.

20. Lily E. Kay, "Selling Pure Science in Wartime: The Biochemical Genetics of G. W. Beadle," *JHB* 22 (1989): 73–101, on 77; Beadle, "Recollections" (note 10), 5.

21. Beadle, "Genes and Chemical Reactions in Neurospora," *Science* 129 (1959): 1715–1719, on 1716.
22. Sturtevant, *Genetics* (note 10), 320–326; F. J. Ayala, "Theodosius Dobzhansky," *BMNAS* 55 (1985): 163–228; Morgan, "Calvin Blackman Bridges," *BMNAS* 22 (1931): 31–48; T. F. Anderson, "Jack Schultz," *BMNAS* 44 (1975): 393–427.
23. Schultz's application for NRC fellowship, and reports, 1928–1929; Selig Hecht to Schultz, 13 Dec 1940; Schultz to Morgan, 3 Jan 1929 (all in JS).
24. Schultz, "Aspects of the Relation Between Genes and Development in Drosophila," *AN* 69 (1935): 30–54, on 30–31.
25. Hanson, "Diary," 4–5 Sep 1936 (RF, 1.1 205D 7 88).
26. Beadle, "Recollections" (note 10), 6.
27. Ephrussi to Schultz, 7 Nov 1934 (JS); Ephrussi to Dunn, 13 Dec 1934 (LCD); Ephrussi, "The Cytoplasm and Somatic Cell Variation," *Journal of Cellular and Comparative Physiology* 52 (1958), suppl. 1: 35–53, on 36.
28. Beadle, "Recollections" (note 10), 6–7.
29. Ibid.; Beadle, "Genes" (note 21), 1716; Beadle to Demerec, 5 June 1935 (MD).
30. Work on the nutritional requirements of "pure" tissues in culture, which Ephrussi planned to do in late 1934, may have been a preliminary to experiments with Drosophila disks. Ephrussi to Schultz, 7 Nov 1934 (JS). On Ephrussi's lab, see Burian et al., "Singular Fate" (note 9), 391–394.
31. Beadle to Demerec, 5 June 1935 (MD).
32. Ibid.
33. Ibid.; Beadle, *Genetics and Modern Biology* (Philadelphia: American Philosophical Society, 1963), 11–12; Beadle (note 21), 1716; Ephrussi and Beadle, "A Technique of Transplantation for *Drosophila,*" *AN* 70 (1936): 218–225; Ephrussi to Beadle, 9 Sep 1938 (GWB, 1 26).
34. Ephrussi and Beadle, "La transplantation des disque imaginaux chez la Drosophile," *CRAS* 201 (1935): 98–100, and "Transplantation in *Drosophila,*" *PNAS* 21 (1935): 642–646; Sturtevant, "Use of Mosaics" (note 8), 307.
35. Hanson, "Diary," 14–27 Aug 1935 (RF, 1.1 205D 7 87); Demerec to Beadle, 27 Sep 1935 (MD); A. F. Blakeslee to J. C. Merriam, 25 Jan 1936 (MD); Demerec, "Memorandum of Work of Dr. G. W. Beadle," 27 Nov 1935, in folder, "Gene Study" (MD).
36. Ephrussi and Beadle, "Technique of Transplantation," 224; Ephrussi to Demerec, 26 Apr 1937 (MD); Beadle, "Collection of Eggs," *Drosophila Information Service* 4 (1935): 64.
37. Beadle and Ephrussi, "The Differentiation of Eye Pigments in Drosophila as Studied by Transplantation," *Genetics* 21 (1936): 225–247, on 226–227.
38. Morgan, C. Bridges, and Sturtevant, "The Genetics of *Drosophila,*" *Bibliographia genetica* 2 (1925): 1–262, on 40–53.
39. Beadle to Demerec, 16 Jul and 13 Sep 1935 (MD): "One has to work in a country where Drosophila doesn't exist," he wrote, "to appreciate DIS [Drosophila Information Service] and the cooperation that goes with it."
40. Beadle and Ephrussi, "Differentiation of Eye Pigments," 230; Beadle, "Genes and Chemical Reactions" (note 21), 1716–1717.
41. Beadle and Ephrussi, "Differentiation of Eye Pigments" (note 37), 242–244; Ephrussi and Beadle, "Development of Eye Colors in *Drosophila*: Transplantation Experiments on the Interaction of Vermillion with Other Eye Colors," *Genetics* 22 (1937): 65–75, and "Development of Eye Colors in *Drosophila:*

Diffusible Substances and Their Interrelations," *Genetics* 22 (1937): 76–86, on 80–81.

42. Ephrussi and Beadle, "Development of Eye Colors in *Drosophila*," 85.

43. Schultz, "Aspects" (note 24), 42, 31–33, 41–43.

44. H. Roman, "Boris Ephrussi," *ARG* 14 (1980): 447–450; A. Meyer to Wilbur Tisdale, 2 Mar 1936, suppl. III, 2–4; Ephrussi, "Report, 1937–1938;" and Fauré-Fremiet memo, 13 Jul 1938, all in RF, 1.1 500D 12 127. Ephrussi to Dunn, 20 Nov 1935 (LCD); Demerec to Hanson, 7 Mar 1936 (MD).

45. Ephrussi report, 25 Apr 1939 (RF, 1.1 500D 12 127), 8; Ephrussi to Demerec, 4 Oct 1936 (MD); André Lwoff, "Recollections of Boris Ephrussi," *Somatic Cell Genetics* 5 (1979): 677–679.

46. Beadle to Demerec, 15 Feb and 13 Apr 1936; G. L. Streeter to Beadle, 8 Jan 1936; Demerec to Beadle, 17 Jan, 3 Apr 1936 (MD).

47. Joshua Lederberg, "Edward Lawrie Tatum," *ARG* 13 (1979): 1–5, and *BMNAS* 59 (1990): 357–386.

48. C. V. Taylor to Weaver, 5 Apr 1937, Hanson to R. L. Wilbur, 3 May 1937, and Hanson, "Diary," 6 May 1937 and 13–23 Apr 1938 (RF, 1.1 205D 8 105–106).

49. C. H. Danforth, "Charles Vincent Taylor, 1885–1946," *BMNAS* 25 (1949): 205–225; Hanson, "Diary," 7–8 Nov 1934, Taylor to Weaver, 1 Mar 1935 and reports, and W. Weaver, "Diary," 11 Mar 1936, 27 Jan 1937, 28 Jan 1939 (all in RF, 1.1 205D 8 103–107); Taylor to Weaver, 23 Aug 1938 (RF, 1.1 205D 10 135).

50. Ephrussi report, 25 Apr 1939 (RF, 1.1 500D 12 127), 7–8.

51. Burian et al., "Singular Fate" (note 9), 397–400; Meyer to Tisdale, 2 Mar 1936 (RF, 1.1 500D 12 127.7), 3–4, suppl. III, 2–4; Demerec to Ephrussi, 13 Nov 1936 (MD); Ephrussi and Beadle (note 34).

52. R. B. Howland, B. P. Sonnenblick, and E. A. Glancy, "Transplantation of Wing-thoracic Primordia in *Drosophila melanogaster*," *AN* 71 (1937): 158–166; C. W. Robertson, "The Metamorphosis of *Drosophila melanogaster*, Including an Accurately Timed Account of the Principal Morphological Changes," *Journal of Morphology* 59 (1936): 351–399; Glancy and Howland, "Transplantation of Mutant and Wild Type Bristle-bearing Tissues in *Drosophila melanogaster*," *BB* 75 (1938): 99–105.

53. Beadle and Ephrussi (note 37); Beadle, "The Development of Eye Colors in Drosophila as Studied by Transplantation," *AN* 71 (1937): 120–126.

54. Ephrussi, "Chemistry of 'Eye Color Hormones' of Drosophila," *QRB* 17 (1942): 327–338, on 327–328, 331–332.

55. Transplanted eyes and gonads were later used in the same way to alter the development of eye color in hosts.

56. Ephrussi, "Chemistry of 'Eye Color Hormones,'" 330–332.

57. Ephrussi to Demerec, 30 Nov 1936 (MD); Ephrussi to Beadle, 18 Oct 1937 (GWB, 1 26); Ephrussi report, 25 Apr 1939 (RF, 1.1 500D 12 127); Yvonne Khouvine, Ephrussi, and M. H. Harnly, "Extraction et solubilité des substances intervenant dans la pigmentation des yeux de *Drosophila melanogaster*," *CRAS* 203 (1936), 1542–1544.

58. Kenneth Thimann and Beadle, "Development of Eye Colors in Drosophila: Extraction of the Diffusible Substances Concerned," *PNAS* 23 (1937): 143–146.

59. Beadle and L. W. Law, "Influence on Eye Color of Feeding Diffusible

Substances to *Drosophila melanogaster,*" *PSEBM* 37 (1938): 621–623; Ephrussi to Beadle, 4 Aug 1937 (GWB, 1 26).

60. Ephrussi to Beadle, 7 Aug, 18 and 22 Oct 1937 (GWB, 1 26). Research had also revealed that Drosophila had the same chemistry of eye pigments as larger insects like Ephestia or Calliphora, from which active substances could be extracted more easily and in larger quantities—an essential prerequisite for chemical isolation.

61. Ephrussi to Beadle, 25 Mar, 8 May, 2, 13, 16, and 29 June, 4 Nov, 5 Dec 1938, 4 Mar, and 23 Oct 1939 (GWB, 1 26).

62. Ephrussi to Beadle, 9 Jan 1938 (misdated 1937), 8 May, 26 Dec 1937 (GWB, 1 26).

63. Edward Tatum and Beadle, "Development of Eye Colors in Drosophila: Some Properties of the Hormones Concerned," *JGP* 22 (1938): 239–253.

64. Ephrussi to Beadle, 2 June 1938, and 5 Dec 1938, 4 Mar 1939 (GWB, 1 26).

65. Ephrussi to Beadle, 18, 22 Oct, 8 Nov, 26 Dec 1937 (GWB, 1 26).

66. Ephrussi, "Chemistry" (note 54), 332–334; Khouvine, Ephrussi, and Simon Chevais, "Development of Eye Colors in Drosophila: Nature of the Diffusible Substances; Effects of Yeast, Peptones, and Starvation on their Production," *BB* 75 (1938): 425–446; Tatum and Beadle, "Effects of Diet on Eye-color Development in Drosophila melanogaster," *BB* 77 (1939): 415–422; Ephrussi to Beadle, 21, 25 Mar, and 8 June 1938 (GWB, 1 26).

67. Ephrussi to Beadle, 21 Jul 1939 (GWB, 1 26).

68. Ephrussi to Beadle, 8 May 1938 (GWB, 1 26).

69. Ephrussi to Beadle, 29 Nov, and 30 Dec 1937 (GWB, 1 26). Ephrussi suggested publishing the method in a "more or less neglected journal."

70. Ephrussi to Beadle, 15 Oct, 5 Dec 1938, 4 Mar, 17 Apr, and 21 Jul 1939 (GWB, 1 26); Harry Miller, "Diary," 1 Jun 1939 (RF, 1.1 500D 12 127).

71. Ephrussi, "Chemistry" (note 54); Tatum and Beadle, "Effect of Diet on Eye-color Development"; Tatum, "Development of Eye Colors in Drosophila: Bacterial Synthesis of v^+ Hormone," *PNAS* 25 (1939): 486–490; Ephrussi to Beadle, 17 Apr 1939 (GWB, 1 26).

72. Ephrussi, "Chemistry" (note 54); A. Butenandt, W. Weidel, and E. Becker, "Kynurenine als Augenpigmentbildung auslösendes Agens bei Insekte," *Die Naturwissenschaften* 28 (1940): 63; Tatum and Beadle, "Crystalline Drosophila Eye-color Hormone," *Science* 91 (1940): 458.

73. Ephrussi to Beadle, 19 Jul, 4 Aug, 26 Dec 1937, and 14 Feb 1938 (GWB, 1 26); Ephrussi, Khouvine, and Chevais, "Genetic Control of a Morphogenetic Substance in Drosophila melanogaster?" *Nature* 141 (1938), 204–205.

74. Ephrussi to Beadle, 17 Apr, 21 Jul, and 25 Jul 1941 (GWB, 1 26).

75. Beadle to Ephrussi, 17 Nov 1941 (GWB, 1 26).

76. Beadle and Tatum, "Experimental Control of Development and Differentiation," *AN* 75 (1941): 107–116; Sewall Wright, "The Physiology of the Gene," *PR* 21 (1941): 487–527; Sturtevant, "Physiological Aspects of Genetics," *Annual Reviews of Physiology* 3 (1941): 41–56.

77. Notes taken by Carleton Schwerdt indicate that Tatum lectured on the nutrition of fungi on 18 Feb 1941, though this is not necessarily when Beadle had his brainstorm (Joshua Lederberg, personal communication).

78. Beadle, "Biochemical Genetics, Some Recollections," in *Phage and the Origins of Molecular Biology,* ed. John Cairns, Gunther S. Stent, and James D.

Watson (Cold Spring Harbor, N.Y.: Cold Spring Harbor Laboratory of Quantitative Biology, 1966; expanded edition 1992), 23–32, on 29; Beadle, "Recollections" (note 10), 7–8; Beadle, "Genes and Chemical Reactions" (note 21), 1717; Kay, "Selling Pure Science" (note 20), 80–81.

79. Beadle (note 21), 1717; Lederberg, "Tatum" (note 47), 2; Beadle, "Chemical Genetics," in L. C. Dunn, ed. *Genetics in the 20th Century: Essays on the Progress of Genetics During Its First 50 Years* (New York: Macmillan, 1951), 221–239, on 224–225.

80. Kay (note 20), 80.

81. Miller, "Diary," 1 June 1939 (RF, 1.1 500D 12 127), 3; Ephrussi, "Chemistry" (note 54), 336; Ephrussi to Beadle, 23 Oct, 4 Dec 1939, 19 Feb, 2 May 1940, 18 Apr, and 25 Jul 1941 (GWB, 1 26).

82. Sturtevant, "Physiological Aspects" (note 76), 48.

83. Ephrussi, "Chemistry" (note 54), 337.

84. Beadle and Tatum, "Experimental Control" (note 76), 114.

85. Tatum, "Nutritional Requirements of *Drosophila melanogaster,*" *PNAS* 25 (1939): 490–497, and "Vitamin B Requirements of *Drosophila melanogaster,*" *PNAS* 27 (1941): 193–197. The eminent Harvard biochemist Yellapragada SubbaRow had also expressed interest in insect nutrition.

86. Tatum, H. G. Wood, W. H. Peterson, "Growth Factors for Bacteria. V. Vitamin B, a Growth Stimulant for Propionic Acid Bacteria," *Biochemical Journal* 30 (1936): 1898–1904.

87. Beadle to Weaver, 18 Jan 1940; Hanson to R. J. Williams, 8 Oct 1940; Hanson, "Diary," 30 Dec 1940 (RF, 1.1 205D 10 135).

88. Beadle (ref. 10), 7–8; Weaver, "Diary," 17 Feb 1941 (RF, 1.1 205D 10 135).

89. F. W. Went to Beadle, 9 Nov 1942, and Beadle to Went, 30 Nov 1942 (GWB 2 45–46); Beadle, "Genetics and Metabolism in Neurospora," *PR* 25 (1945): 643–663, on 644–645; C. L. Shear and B. O. Dodge, "Life Histories and Heterothallism of the Red Bread-mold Fungi of the *Monila sitophila* Group," *Journal of Agricultural Research* 34 (1927): 1019–1042; Beadle, "Genes and Chemical Reactions" (note 21), 1717–1718. Dodge taught at the University of Southern California, near Pasadena, from 1934.

90. Nils Fries, "Über die Bedeutung von Wuchsstoffen für das Wachstum verschiendener Pilze," *Symbolae Botanicae Upsalienses* 3 (1938): 1–188; Lederberg, "Tatum" (note 47), 2. Lindegren provided Beadle not only with Neurospora cultures but also with the craft knowledge of making mutants and getting spores to germinate: C. C. Lindegren to Beadle, 17 Mar, 15 Dec 1941, and 20 Jul 1942; and Beadle to Lindegren, 3 Feb 1943 (GWB, 1 41).

91. Beadle, "Genetics" (note 89), 644–650; Arnold W. Ravin, "Francis J. Ryan (1916–1963)," *Genetics* 84 (1976): 1–25, on 4–5; Joshua Lederberg, "Francis J. Ryan," in *University on the Heights,* ed. Wesley First (Garden City, N.Y.: Doubleday, 1969), 105–109; Adrian Srb, "Beadle and Neurospora, Some Recollections," *Neurospora Newsletter* 20 (c. 1974): 8–9; F. J. Ryan, Beadle, and Tatum, "The Tube Method of Measuring the Growth Rate of Neurospora," *AJB* 30 (1943): 784–799; Beadle and Tatum, "Genetic Control of Biochemical Reactions in Neurospora," *PNAS* 27 (1941): 499–506. Tatum and Beadle, "The Relation of Genetics to Growth Factors and Hormones," *Growth* 6, suppl. (1942): 27–37. The "race tube" technique depended on Neurospora's habit of maintaining sharply bounded colonies.

92. Beadle, "Genetics and Metabolism" (note 89), 646–650; Beadle and

Tatum, "Neurospora II: Methods of Producing and Detecting Mutations Concerned with Nutritional Requirements," *AJB* 32 (1945): 678–686; A. M. Srb and N. H. Horowitz, "The Ornithine Cycle in Neurospora and its Genetic Control," *JBC* 154 (1944): 129–139.

93. Beadle to Lindegren, 25 Jul 1941, and 2 Jul 1942 (GWB, 1 41); Beadle to Ephrussi, 1 Aug, and 17 Nov 1941 (GWB, 1 26).

94. Ephrussi to Beadle, 5 Jan 1942 (GWB, 1 26).

95. Beadle, "Genes and Chemical Reactions" (note 21), 1717–1718; Beadle, "Progress Report," 24 Sep 1942 (RF, 1.1 205D 10 142); Beadle, "Genetics and Metabolism" (note 89), 652.

96. Beadle to Hanson, 27 Jan 1943 (RF, 1.1 205D 10 143).

97. Hanson, "Diary," 15–18 Dec 1941; Beadle to Hanson, 28 Nov 1941; Beadle memo, 18 Dec 1941 (RF, 1.1 205D 10 141).

98. Weaver to Fosdick, 24 Dec 1941 (RF, 205D 1.1 10 141).

99. Kay, "Selling Pure Science" (note 20), 88–90.

100. Ryan to Beadle, n.d [1942?], 27 June 1942 (GWB, 2 23); Beadle to Hanson, 24 Feb 1942, and "Report," and Beadle to Hanson, 3 Apr 1943 (all in RF 1.1 205D 10 142–143); Beadle to Ephrussi, 1 Aug 1942 (GWB, 1 26); N. H. Horowitz, "George Wells Beadle," *Genetics* 124 (1990): 1–6; Horowitz to Beadle, 14 Feb 1942 (GWB, 1 33); Srb, "Beadle and Neurospora" (note 91); B. McClintock to Beadle, 14 Dec [1942], and A. Hollaender to Beadle, 17 Dec 1942 (both in GWB, 1 34).

101. Ephrussi to Beadle, 22 Aug 1941 (GWB, 1 26).

102. Beadle to Ephrussi, 17 Nov 1941 (GWB, 1 26).

103. Beadle to Ephrussi, 1 Aug 1942, and Ephrussi to Beadle, 25 Jul 1941, and 4 Aug 1942 (GWB, 1 26).

104. Beadle to Lindegren, 7 Jan, and 2 Jul 1942 (GWB, 1 41); Beadle, "Progress Report," 24 Sep 1942 (RF, 1.1 205D 10 142); Beadle to E. Brand, 11 Feb 1943 (GWB, 1 7).

105. Beadle to Hanson, 2 Nov 1944, Beadle to Miller, 24 Aug 1945, 6 Sep 1945 (RF, 1.1 205D 10 144–145).

106. Beadle to Hanson, 2 Dec 1944 (RF, 1.1 205D 10 144); McClintock to Beadle, 9 Jan, 12, 22 Aug, 27 Nov 1944, and 8 May 1945 (GWB, 2 5); McClintock, "Neurospora I: Preliminary Observations of the Chromosomes of *Neurospora crassa*," *AJB* 32 (1945): 671–678; Beadle and V. L. Coonradt, "Heterocaryosis in *Neurospora crassa*," *Genetics* 29 (1944): 291–308.

107. Tatum and Beadle (note 91), 30, 34; Beadle to Hanson, 17 Jan 1944 (RF, 1.1 205D 10 144); Beadle to Lindegren, 25 Jul 1941 (GWB, 1 41).

108. Beadle memo, 18 Dec 1941, and Beadle to Hanson, 24 Feb 1942 (RF, 1.1 205D 10 141–142).

109. Beadle, "Recollections" (note 10), 8–9.

110. Latour and Woolgar, *Laboratory Life* (note 2), chap. 5.

111. Kay, "Selling Pure Science" (note 20); Beadle to Hanson, 15 Apr, and 13 Jul 1942 (RF, 1.1 205D 10 142).

112. Lederberg, "Tatum" (note 47), 3–4.

113. Latour and Woolgar, *Laboratory Life* (note 2), chap. 5.

114. Galison "Bubble Chambers" (note 2); W. D. Hackmann, "The Relationship Between Concept and Instrument Design in Eighteenth-century Experimental Science," *Annals of Science* 36 (1979): 205–224; Joel D. Howell, "Early Perceptions of the Electrocardiogram: From Arrhythmia to Infarction," *BHM*

58 (1984): 83–98; Elzen, "Two Ultracentrifuges" (note 3); James Wright, Jr., "The Development of the Frozen Section Technique, the Evolution of Surgical Biopsy, and the Origins of Surgical Pathology," *BHM* 59 (1985): 295–326; Lenoir, "Models and Instruments" (note 3).

115. Beadle to Ryan, 4 Jan 1943 (GWB, 2 23); also, Beadle to Brand, 4 Jan 1942 (GWB, 1 7).

SOCIAL

The Quiet Revolution of the 1850s: Social and Empirical Sources of Scientific Theory

Alan J. Rocke

Introduction

History of science as an academic discipline grew to maturity in the twentieth century, with important affinities to the history of ideas and of philosophy. With increasing success, the hegemony of such an approach has been challenged in the last generation, in part as a consequence of the continued decline of positivist and realist currents in the philosophy of science. A central and now well-accepted element of postpositivist philosophy is the claim that theories are in principle seriously underdetermined by empirical evidence; this has been expanded by advocates of the "strong program in the sociology of scientific knowledge" to the position that scientific knowledge, like any and all other forms of human belief and doctrine, is entirely socially constructed, fully independent of empirical investigations.

In coining this phrase and defining the program, David Bloor was careful to specify that causes other than social might in particular situations be determinative. This appears to allow entrée to the influence of empirical evidence. In practice, however, many social constructivists tend to proceed as if no such influence were ever important. In his quest for Bloor's injunctions toward "symmetry" and "impartiality" Bruno Latour was moved to the position that "nothing extraordinary and nothing 'scientific'" ever happens in laboratories, and so we must "abolish the distinction between science and fiction"; similarly, Harry Collins believes that "the natural world has a small or nonexistent role in the construction of scientific knowledge." As Barry Barnes and Steven Shapin put it, "The intense concern of earlier generations with the special status of science and its allegedly distinctive characteristics has begun to ebb away."[1]

The strong programmers have been met by strong criticism, including charges of purveying "voodoo epistemology" and the "pseudoscience of science."[2] The social constructivists have also been known to use terms of opprobrium, referring to the "paralysis" of philosophical minds and the "bleatings of recusant epistemologists and methodologists."[3] Difficulties with the positions of many sociologists of scientific knowledge include a tendency to overshoot in their correctives, a tendency toward the construction of straw men, and their generally unconvincing responses to a variety of reflexivity problems. As far as the first of these issues is concerned, no one, I think, would deny that the overall historiographic changes in the profession over the last generation have been highly salutary, providing a more contextual, more realistic, and more fully historicist vision of the past. Nearly all historians of science admit the underdetermination thesis and regard social causation as a vital or even central element of historical explanation. But admitting underdetermination does not mean that the resulting gap must necessarily be filled *exclusively* by social causation; and even gross underdetermination does not mean *zero* determination.

Moreover, the assertion of a distinctive character to science need not be taken in the way that some positivist-whig historians once viewed it, as a sort of purely cerebral and bloodless—and always successful!—pursuit of ontological truth. Such a caricature is often the model attacked by strong programmers, but in fact modern historians and philosophers have long been constructing much more sophisticated and carefully nuanced conceptions of the nature of the scientific enterprise. Thomas Kuhn, one of the standard-bearers of the new philosophy of science, hastened to correct the widespread impression that he had opened the gates to irrationality and caprice. Even subjective and intuitive approaches to science, he said, eventually result in knowledge that is "systematic, time-tested, and in some sense corrigible." He took exception to the unitarian "Kuhnians" who viewed him as the great leveler, taking the "hard" sciences down to a softer level: the development of the natural sciences "may resemble that in other fields more closely than has often been supposed, [but] it is also strikingly different."[4]

The apparent general conformability of the world with our (admittedly socioculturally based) investigations of it does indeed suggest a certain epistemological asymmetry. I believe that there is in fact a degree of asymmetry here, and that science is a distinctive sort of enterprise; a central paradox then requires resolution, namely, how thoroughly socially grounded investigations can nonetheless exhibit those special characteristics—success, power, predictive heuristics, technological applications, and ultimately what appears to be gradual ontological progress—that we have come to associate with science and that are dismissed by social constructivists. This problem is still an open and difficult one.

But the position of the constructivists also contains paradoxes. For instance, a thoroughly symmetrized approach conceals an implicit scientism more radical than any they are attacking, for many of the strong programmers confidently assert that *all* nature, including social relations and social epistemology, is as transparent to human reason and empirical effort as the scientists' far more delimited territory. But, ironically and paradoxically, in order to be consistent they must at the same time relinquish all claims to have succeeded in erecting anything other than a socially determined castle of cards. Such circularities and reflexivities have often been discussed by both advocates and by opponents of the sociology of knowledge movement.[5] Many, including the present author, believe that the constructivists' attempts to show that such circularities are not vicious have not been successful.

If science has certain distinguishing traits that make the history of science special in some respects, chemistry may have some characteristics that distinguish it from the other sciences. The study of historical or contemporary physics leads naturally to the deep waters of epistemology and ontology, and to fundamental ideas about time, space, force, and matter; the study of biology, geology, and astronomy has led to searching examinations of the place of the human species with respect to nature and to God. By contrast, chemistry as the most central of the sciences also seems to be the most insulated from philosophical issues, or at least more or less *mediated* from its deeper implications. In scales of magnitude, chemists inhabit an intermediate world between microphysics and cosmology, and theirs has always been the discipline that has seen most ready technological application. Thus there is a sense in which I think it is correct to say that chemists themselves have been nudged by some of the intrinsic characteristics of their field toward a less ideological and more practical approach than for many of their scientific colleagues in other fields. (I would not want to put too fine a point on this suggestion, for I would be the last to deny chemistry its intellectual fascinations and philosophical implications—but they can require more teasing out than for biology and physics.)

These considerations may help to explain why no true philosophy of chemistry has ever been developed—or, more precisely, not to the same degree that there are mature and lively philosophies of biology and of physics.[6] They also may help to explain why the history of chemistry may well be the least popular of the major branches of history of science—as less obviously idea-laden, many find it less interesting than biology or physics! By the same token—and here we come to the concern of this paper—this sociocultural insulation of chemistry, and the effects of such insulation on its practitioners, may make it a less fruitful area of application for the social constructivists. Strong program theorists and associated historians tend to focus on relatively static controversies, on psi phenomena, on ethical or social dilemmas involving science, or on modern physical theories that still have rather shallow empirical bases. Many intrinsic characteristics of these categories make them relatively good subjects for their categories of analysis. The situation looks quite different, I want to argue, when one turns to a mature and empirically rich science of chemistry, especially when one examines cases of sudden and dramatic changes in chemists' views on the central issue of the nature of the composition of matter. Many such revolutions are known in the history of chemistry.

In this essay I wish to argue for the continued vitality of cognitive history of chemistry, and the incompleteness of the social constructivist view. In conjunction with many other recent writers, I urge a pluralistic approach, wherein the valuable insights and methods of advocates of social causation in science are allied with and complemented by a cognitive approach.[7] For this purpose I will use as a case study the remarkable series of changes that took place in chemical theory during the 1850s, especially in Germany.

My interest in this subject began with an investigation of the career of Hermann Kolbe (1818–1884), which initially appeared to require a predominantly social-historical approach.[8] I subsequently concluded and will argue, however, that cognitive and empirical factors were far more important in providing a satisfactory resolution of the historical problems connected with Kolbe, as also with Kolbe's contemporaries Adolphe Wurtz (1817–1884), August Kekulé (1829–1896), August Wilhelm Hofmann (1818–1892), and others. The argument needs to be hedged, however, and this is especially the case when attention is expanded from Germany to include France. Such hedging is less a concession than a positive affirmation of the need for methodological pluralism in the study of history of science.

A Precis of the "Quiet Revolution" in Chemistry

I use this phrase here to signify a series of changes during the 1850s that centered both on reforms of atomic weights and molecular formulas, and on the subdiscipline of organic chemistry. The essential elements of this extended event, prepared since the 1830s by the work of Justus Liebig (1803–1873), Jean-Baptiste Dumas (1800–1884), and

Auguste Laurent (1807–1853), and in the 1840s by Charles Gerhardt (1816–1856) as well, were the decline of Berzelian electrochemical-dualist theory, the development of "type" theories based on substitution reactions, the establishment of consistent ("two-volume") molecular magnitudes spanning organic as well as inorganic chemistry, the return from conventional equivalents to a modified version of Berzelian atomic weights, and, finally, the rise of the theory of the "atomicity of the elements," which comprised in part and led to in full what became known as valence and structure theory. In the following discussion, the molecular-magnitude and atomic-weight arguments will be considered as constituting the "Gerhardt-Laurent reform," which was prepared by the rise of type theory and followed by theories of valence and structure. The essential elements of structure theory were delineated by August Kekulé in 1857–58 and independently by A. S. Couper in 1858; they were given a particularly clear form by A. M. Butlerov from 1861. A largely equivalent and partly independent formulation of these ideas was due to Edward Frankland and Kolbe in 1857–60.

Is the word "revolution" appropriate to describe these changes? Others have merely referred to various new theories and conventions introduced during this period, culminating in the first international chemical conference, the Karlsruhe Congress of 1860. Within a few years after the Congress most European chemists had accepted most or all of the elements of the Laurent-Gerhardt agenda, and structure theory began to develop dramatically, especially in Germany. Its most dramatic deployment, and the event that marked its real coming of age, was Kekulé's theory of benzene structure, elaborated from 1865 on.

Considered only impressionistically, there are a number of justifications for emphasizing the magnitude of the changes. Many quantitative indicators of the very size and importance of the profession—number of chemists, number of papers published, total compounds known, technological applications and the explosive growth of chemical industries, and so on—suggest an inflection point shortly after the middle of the century.[9] In the years before 1850 the discipline was well developed and fully professionalized, but by any reasonable measure it was quite small and low in public profile.[10] The situation was dramatically different a generation later. To take one example, the Deutsche Chemische Gesellschaft experienced phenomenal growth immediately following its founding in 1867. Within a few years its *Berichte* ballooned in size, from a thin volume for 1868 to giant and unwieldy tomes that required repeated subdivision.

Also relevant are the centrality of organic chemistry in this revolution and the centrality of structure theory in organic chemistry. The early volumes of the *Berichte* are packed with structure-theoretical investigations, far outweighing all other kinds of papers put together. Structure theory clearly provided a virtually unlimited and highly fertile field for academic chemists; their numbers mushroomed and the German states began to compete with each other in building palatial laboratories, especially for organic chemists.[11] Standard heavy-industrial indicators such as coal, steel, and railroads excepted, the fine-chemicals industry took the leading role in Germany—and it became increasingly dependent on the scientific investigation of chemical substances, especially organic compounds studied via structure theory.

If structure theory was vital for the growth of chemistry in the second half of the century, the Gerhardt-Laurent reforms were clearly the prerequisite for structure theory. The reforms were introduced during a theoretically anarchic period. The investigations of substitution reactions by Dumas and other (mostly French) chemists had destroyed the basis of Berzelian theory, at least as applied to organic compounds, but for years there was nothing to put in its place other than Dumas's overly schematized types or Laurent's somewhat fanciful nuclei. Both at the time and retrospectively, commentators emphasized the confusion that prevailed during this period: water could be written HO, H_2O, or H_2O_2, and in urging reform Laurent and Kekulé each gleefully filled an entire page of text with currently defended formulas for acetic acid.[12] Four major systems of atomic weights and formulas, and many variations thereupon, were widely accepted in the 1840s and 1850s.

Dumas and Liebig, the inheritors of Berzelius's mantle as master theory-builders during the 1830s, became exhausted, embittered, and bewildered by the disputes, and both independently "retired" in 1840 from the theoretical dialectic of fundamental questions. Baptized by fire, chemists acquired a skeptical view of such issues and developed a deeply entrenched conventionalist approach toward atoms and molecules. Laurent and Gerhardt were almost alone in placing such questions center stage in the 1840s; but it would be a mistake to view them as clear-eyed modernists battling the entrenched conservatives. They too suffered many periods of uncertainty and confusion, dithered between positivist and hypothetico-deductivist commitments, and in their correspondence expressed nearly as much self-doubt as self-confidence.[13] Both in tactics and in logic they committed a number of serious gaffes.

The situation was only clarified when Alexander Williamson (1824–1904) developed the first strong experimental chemical evidence for the central element of the reform. Williamson's work (1850–54) was followed by similar investigations—to be described in the next section—by Hofmann, Gerhardt, Wurtz, William Odling, and Kekulé; by the time of Gerhardt's death in 1856 the reform was clearly winning the day, at least in Germany and England. This is not to deny a key role for the Karlsruhe conference itself and its canonical hero, Stanislao Cannizzaro. But to a degree that has hitherto been insufficiently appreciated,[14] the revolution had largely already been consummated by the time of the conference, at least in Germany. The relatively invisible quality of these changes has led me to choose the name "quiet revolution."

But however quiet it may have appeared to later observers, the participants themselves saw this period as revolutionary and looked to the 1850s as the critical decade. In 1852 Gerhardt viewed his work of that year as "terribly revolutionary," and on his deathbed four years later averred that he had advanced chemistry by fifty years.[15] Hofmann, alluding to the period from 1840 to 1865 from the perspective of the latter date, referred to the "convulsive struggles" leading to the "profound transformation" that had every right to be "termed a Revolution." His opinion of Gerhardt's 1852 paper at the time of its publication, as expressed privately to Cahours and to Liebig, is consistent with these phrases—and at that time Hofmann was by no means friendly toward Gerhardt.[16]

Much later, in a biography of Heinrich Will, Hofmann wrote of this period of rapid gains for the new views: "With some, this transition happened silently, they slipped as it were right into the new theory; in fact, there were some who the night before had been implacable opponents of Gerhardt and Laurent, and awoke the next morning completely converted." In his biography of Wurtz, Hofmann chose similar language to describe the period: "The time had arrived in which, one after another, the most ardent opponents—often overnight, and without providing any reasons whatever for their conversions—made their pronouncements."[17] The time Hofmann was referring to was several years before Karlsruhe, namely 1853, when Wurtz converted to the new chemistry, and 1854, when Will declared himself. Even the most diehard opponent of the revolution, Kolbe, referred to the "crisis" and "reformation" that was taking place around him, worrying as early as 1853 in letters to his friends that he might be left alone in the conservative camp.[18] By 1860 he was, or nearly so. Kuhn could wish for no more dramatic testimony of a gestalt-switch conversion experience than that of Lothar Meyer, who wrote that upon reading Cannizzaro's *Sunto* it was as if the "scales fell from my eyes, doubts disappeared, and a feeling of quiet certainty replaced them."[19]

What prompted all these conversions? There was considerable psychological and social-interest stake in the older chemistry, even though it had been weakened by attacks by the French. None of the respected and powerful leading chemists of the older generation, including Berzelius, Thenard, Dumas, Liebig, Wöhler, and Robert Bunsen, ever advocated the Gerhardt-Laurent reforms, and all except the first two lived long after 1860; by openly converting, a young chemist might even antagonize the old guard, who exerted powerful influences on careers. Tradition and received wisdom taught caution, as did the tragic fates of the radical reformers Laurent and Gerhardt. Furthermore, after having lived through several different reigning versions of atomic weight and formulas, there was an understandable hesitation to accept a change that would require once more committing to memory new formulas for all known chemical compounds, and at the same time relinquishing a convention (the Wollaston-Gmelin "equivalents") that nearly all considered perfectly empirical and theory-free. Indeed, the choice was no longer perceived, as it was in the 1820s and 1830s, as one between ontologically neutral conventional systems. Rather, since strong ontological claims were being made for the Gerhardt-Laurent reforms, conversion to the newest "fashion" would also perforce be seen as relinquishing a comfortable and socially reinforced positivist reserve.

Despite all these barriers, the reform quickly succeeded in Germany. I believe that the conversions, which, as I will argue here, occurred in larger numbers and earlier—hence more suddenly—than has been thought, can only have been prompted by the widely shared conviction that Williamson and others had demonstrated the actual truth of the new chemistry beyond reasonable doubt. The new proselytes were, if anything, acting against, rather than in accord with, perceived career and other social interests. I will even extend this argument to Kolbe himself, the last and staunchest German *opponent* of the reforms.

Williamson's Asymmetric Synthesis Argument and Its Influence

Williamson, a student of Liebig, converted to the French reform during his residence in Paris during the years 1846–49. In the summer of 1850, newly installed at University College London, he set out to demonstrate the truth of the new chemistry. It must be noted that until this time the arguments of Gerhardt and Laurent were largely schematic and esthetic, urging a thoughtfully rationalized rather than a positively and empirically verified system; there was little hard new evidence they could point to that unequivocally supported their ideas, and a few troubling anomalies persisted. They recognized these lacunae and agonized over them.[20]

Williamson devised a new synthesis of ordinary ether from alcohol, one of the most central of all organic-chemical reactions. The Williamson synthesis was not only an elegant, smooth, low-temperature and high-yield reaction, it also lent itself to innumerable modifications that could produce novel "tailored" ethers. On the basis of his synthesis of new asymmetric molecules such as ethyl methyl (as opposed to the ordinary diethyl) ether, Williamson devised a theoretical argument that was a good model of a Baconian experimentum crucis. By the terms of the older theoretical system, the reactants in the asymmetric reaction ought to have produced a mixture of two symmetrical ethers (that is, an equimolar mixture of diethyl ether and dimethyl ether). The fact that the product was a single homogeneous asymmetric ether could only be accounted for by assuming that chemists had been using inconsistent molecular magnitudes and that Gerhardt and Laurent's new view of the relation between alcohol and ether was correct. (Williamson's argument for the other element of their reform, a doubled atomic weight for oxygen, was not as logically tight, but it was reasonable and some even found it compelling.) In the following year Williamson used the same argument, but with a very messy reaction producing asymmetric ketones, to advocate Gerhardt and Laurent's formula for acetone—in the process removing one of their outstanding anomalies.[21]

The best evidence for the strong impact of this argument on the chemical community is its widespread deployment by others in the five years after 1850. In 1852 Gerhardt synthesized organic acid anhydrides by a reaction route analogous to Williamson's (and already predicted by Williamson the year before), and then applied Williamson's *logic* as well, by producing structurally *asymmetric* anhydrides that could be explained only by the new chemistry. Gerhardt regarded this as the most important paper of his life, and his colleagues agreed. His teachers Dumas and Liebig as well as his contemporary Hofmann (all hitherto inimical to Gerhardt) praised the work highly; in fact, this paper marked Gerhardt's breakthrough from ostracism to career success. In 1854 the asymmetric synthesis argument was again employed, by Williamson himself to the sulfoacids, by Odling in a wide-ranging and influential paper, and by the young Kekulé to thiacetic acid. Even Kolbe, an opponent, fashioned an asymmetric synthesis argument (1853) that he tried to turn against its creator, but a fallacious formula translation invalidated the reasoning, as Williamson quickly pointed out.[22] We will later discuss a final dramatic deployment, Wurtz's synthesis of "mixed" (that is, structurally asymmetric) radicals via the "Wurtz reaction" of 1855. Hofmann and Benjamin Brodie had tried to do precisely this right after Williamson's first paper appeared, but without success.[23]

The sober and cautious Hermann Kopp, seven years older than Williamson, recalled the latter's work as providing "brilliant confirmation" of the new chemistry, "demonstrating beyond question" the Gerhardtian constitutions, and quickly garnering numerous willing converts. Hofmann likewise described the asymmetric synthesis argument as "irresistibly convincing."[24] At the time of Williamson's first ether paper Hofmann was ripe for convincing, for during the past year he had discovered the secondary and tertiary amines; consequently he had on his own become a strong proponent of substitutionist type theories, relinquishing his earlier defense of dualist "copula" formulations.

Hofmann's friends Edward Frankland and Adolphe Wurtz were also becoming ardent substitutionists about this time, on the basis of the former's work on organometallic compounds and the latter's discovery of primary amines a few months before Hofmann's amine work. All of these contributions established the water, ammonia, and hydrogen types—collectively, the "newer type theory"—which stood as inorganic progenitors of large series of organic compounds that could actually or at least schematically be constructed by the substitution of radicals for hydrogen atoms. The evidence was so compelling that by 1856 even Kolbe became a type theorist. The newer type theory had a logic and empirical base independent of the Laurent-Gerhardt reform, but when in the course of the 1850s the reform was *allied* with type theory, valence and structure theory developed smoothly and naturally therefrom.

In France, Dumas reconciled with Gerhardt after 1852, and the elderly Baron Thenard was transformed into a warm advocate; Thenard helped Gerhardt to be appointed to two professorships at Strasbourg in 1855. Not only Gerhardt's students such as Chancel and Chiozza, but also Wurtz, Cahours, Pelouze, Malaguti and Quesneville adopted elements of his system. A few Russians and Americans, as well as such Britishers as Odling, Brodie, Williamson, and Henry Roscoe, became converts. By the summer of 1856—ironically, just before his untimely death—Gerhardt could look back on the struggles with satisfaction and forward to the future with confidence.[25]

In Germany the situation appeared even more favorable, Gerhardt's views gaining currency among many of the younger and mid-career leaders of the discipline. Leopold Gmelin had adopted Laurent's classification in 1848, and both Liebig and the émigré Hofmann had had high praise for the acid anhydride work. A new generation of German chemists—men such as Kekulé, Karl Weltzien, Adolf Strecker, Emil Erlenmeyer, Heinrich Limpricht, Ludwig Carius, Adolf Baeyer, and Leopold von Pebal—were declaring themselves as converts during the mid-1850s. Kolbe worried about how to respond to the new research, especially as he was writing the theoretical sections of his textbook in 1853, just as the tide appeared to be turning against him. To his scientifically literate close friend and publisher Eduard Vieweg, Kolbe wrote:

> I've had a hard time with it. You see, research by several chemists of the very greatest scientific importance, especially by Gerhardt and Williamson, has appeared in recent months, which threatens to destroy the entire existing structure of currently accepted doctrines in chemistry. Several weeks of my current vacation spent carefully studying these papers and their many implications were required before I came to a clear decision about them, during which time I could not even pick up my pen to write.[26]

He went on to report that he thought he had developed the "strongest proofs" against Williamson and Gerhardt, which he proceeded to include in his text as well as in an article sent to Liebig's *Annalen* and to the Chemical Society of London. But Williamson composed a devastating response, filled with effective substantive as well as rhetorical rebuttals, and there is no evidence that Kolbe's traditionalist defense was influential in the community.[27] Kolbe's letters to Vieweg in the late 1850s reflect his acknowledgment of increasing isolation, even though he remained confident that his views would prevail in the end—and for the next quarter century he predicted year after year that this would happen in the immediate future.[28]

Kolbe could take little comfort from the fact that the éminences gris of German chemistry—Liebig, Wöhler, and Bunsen—did not sign on, for he well knew that they were uninterested in the theoretical dialectic in general. "What opinion Liebig now has on the subject [of Kolbe's 1854 critique of the Williamson-Gerhardt theory] I do not know," he wrote Vieweg; "I suspect none at all, like Wöhler, since neither one seems to have much interest for such matters."[29] Wöhler, as Kolbe well knew, had never been inclined toward chemical theory. As mentioned earlier, in 1840 Liebig turned from theoretical matters to technical and applied chemistry, and quickly lost contact with what was happening in the field. Upon Wurtz's discovery of primary amines in 1849, Liebig needed Hofmann to remind him of his own 1840 prediction of such compounds.[30] In his correspondence during the 1850s and 1860s, he often confessed with embarrassment his ignorance of recent advances, but there was an element of defiance, too, for at the same time he averred that he did not even want to read the literature.[31]

The same was true for Kolbe's actual *Doktorvater,* Bunsen (1811–1899). There is probably no great chemist in history who was more averse to hypotheses and theoretical structures of all kinds than he.

After his great cacodyl investigation, which ended in 1841, he never returned to organic chemistry, which at this time was becoming intensely theoretical—and correspondingly intensely disputatious, a quality which also was anathema to the gentle Bunsen. Lothar Meyer reported Bunsen's "decided aversion toward the new business with formulas" in the 1850s—which applied both to the emerging structural school and the Kolbean school.[32] To Vieweg's suggestion that Kolbe enlist Bunsen's help in his efforts to spread his ideas, Kolbe responded that the idea was good in principle but impractical, as for years Bunsen was interested in different things than he, and thus incapable even if he were willing.[33]

In fact, it was out of the non- or even anti-theoretical laboratories of Wöhler, Bunsen, and Liebig that most of the theoreticians of the 1840s and 1850s, including structure theorists, emerged. In a sense this is not surprising, in that these three led by far the largest and best chemical institutes of Germany. Kolbe himself, theorist to the core, was a student of Wöhler and Bunsen. Limpricht, who was the first in Germany to write a textbook based on Gerhardt's system (*Grundriss der organischen Chemie,* 1855), also was a Wöhler student. Meyer, Pebal, and Carius were students of Bunsen. But the greatest number of reformers came from Liebig's lab. Gerhardt, Wurtz, Hofmann, Williamson, Kekulé, Kopp, Erlenmeyer, and Strecker all studied in Giessen during a period from the late 1830s to the late 1840s. Having converted to the reforms, many among the men just named went on to become structure theorists and then to establish schools of their own. Williamson taught George Carey Foster and Henry Roscoe, Kekulé taught Adolph Baeyer and influenced Butlerov and Meyer, Wurtz taught Couper and Friedel and also influenced Butlerov, and Hofmann taught William Henry Perkin and others.

One name I have so far omitted, a man whose influence has been much underestimated. Heinrich Will (1812–1890) was a Liebig student in the late 1830s, subsequently becoming Privatdozent and first assistant in Giessen. When Liebig was forced to open a branch laboratory for beginners in 1843, Will was placed in charge. As a consequence, he taught large numbers of students, many of whom achieved prominence in the next decade. Many of those who are counted (and counted themselves) as Liebig pupils actually were taught mostly by Will—examples include Hofmann, Strecker, Kekulé, and Erlenmeyer.[34]

What were Will's classes like? Although Will had a strongly empirical approach to his material, it is interesting to note that the topic of his disputation for habilitation (1844) was the thesis that there exist no compounds with an odd number of equivalents of carbon—a cardinal thesis of Gerhardt's chemistry.[35] A fellow student of Kekulé's during the years 1849–51 reported that Will's lectures opened many windows, and began to destroy the hegemony of the older radical theory in his students' minds.

> In contrast to Liebig, who often enjoyed saying of fairly obscure areas, "Gentlemen, this we know with absolute certainty," Will was fond of suggesting, even for uncontested issues regarding the constitution of organic compounds, that the matter could be considered from another point of view, from which were often discovered relationships that would otherwise have remained unremarked and unknown.[36]

Will succeeded Liebig as *ordentlicher* Professor when Liebig transferred to Munich in 1852, whereupon he became more open in his apostasy. Jacob Volhard, who took Will's classes in 1853, reported that Will provided a "clear, convincing, indeed enthusiastic portrayal" of Gerhardt's system, to the point that Volhard and his student colleagues learned to swear upon type theory and honor Gerhardt as the reformer of organic chemistry.[37] Then, four months after the first fascicle of Kolbe's textbook appeared, and a month after the appearance in German of Williamson's rebuttal to Kolbe's diatribe, Will published an extended commentary "On the Theory of the Constitution of Organic Compounds." Will particularly singled out Williamson's asymmetric synthesis argument as placing the issue beyond doubt, and dispassionately countered each of Kolbe's "proofs." Will, the respected model of the midcareer establishment German chemist and inheritor of Liebig's mantle, was publicly declaring himself for the reform.[38]

Will's colleague and fellow Liebig protégé Hermann Kopp (1817–1892) also gravitated to the new views, but not as forthrightly. Expressions he permitted himself in published papers and in his annual reports on the science left little doubt by 1854 that he, too, had become convinced—as Kolbe acidly noted in letters to a friend.[39] Kopp's defection, an intimate friend of Kolbe's, was a bad blow, as was that of another good friend from the Liebig camp, Strecker, whose text Kolbe had been using for his classes. Strecker had been successively student, assistant, and Privatdozent under Liebig and Will until his call to Christiania (Oslo) in 1851. In his papers since 1848 he had clearly been inclining toward some of Gerhardt's views; this trend was widely noticed in the early 1850s, and at the Karlsruhe Congress Strecker was eloquent in his support of the reformed atomic weights and molecular formulas.[40] In short, the whole Giessen school and elements of the Göttingen and Heidelberg schools could now be viewed as reformers. Even Gerhardt's old teacher and the publisher of the *Journal für praktische Chemie,* Otto Erdmann, appeared to align himself with the reform.[41]

All of this happened by the mid-1850s. In such an environment, the fact that the young Kekulé chose a defense of Gerhardt as the topic of his disputation for the *venia legendi* in Heidelberg in 1856 no longer seems as radical as it has been viewed. When Roscoe asked Pebal for his endorsement of the Karlsruhe program, Pebal complied but thought it was superfluous, as all *reasonable* chemists were already of the same opinion, and Kolbe would certainly never be converted anyway![42]

If I am right that the great majority of theoretically oriented German chemists had become Gerhardtians *before* Karlsruhe, then two apparent anomalies need to be resolved. Having succeeded in their program, why did the reformers perceive the necessity for the conference at all? And again, what has misled modern historians into placing the consummation of the reform *after* 1860?

As regards the first of these questions, it must be stressed that there continued to be widespread indifference as well as resistance in principle to discussion of such fundamental issues, especially among the older generation—a phenomenon which has been discussed above. Not all prominent chemists were theoretically oriented, including some, such as Liebig and Dumas, who had been passionate theorists in their youth. Furthermore, the reform had succeeded more fully in some countries than in others—Germany much more than France, for example—and although the impetus for the conference came from Kekulé and Weltzien, it was aggressively international from the start. Moreover, even among the reformist camp there was widespread confusion and lack of unity. The reform had a number of interlocking parts, and many accepted some pieces while rejecting others. Finally, such conventional issues as atomic weights, formulas, and notation were still in a frightfully confusing state and badly needed sorting out.

The last point also provides an explanation for why this revolution appeared to happen later than it did in fact. There has been a tendency to assume that every user of conventional equivalents ipso facto rejected the reforms, or even disbelieved in the existence of chemical atoms; many have concluded that the reforms were accepted late, since acceptance of atomic weights was not widespread until after 1860. However, such conclusions are flawed, since many full-fledged reformers used equivalents for a variety of purposes. Gerhardt himself only utilized the new atomic weights in the last part of the last volume of his monumental *Traité de chimie* (1853–56); Kekulé employed equivalents in selected contexts in 1856–58, despite his decisive conversion by 1854; and Hofmann began to write atomic weight formulas only in 1860, although there is no doubt of his membership in the reform group several years earlier. Conversely, Kolbe's own shift to modern atomic weights in 1868, prompted by peer pressure, cannot be taken as indicative of a change of heart at that time. In short, use of one convention or another by an author never provides sufficient evidence for his theoretical convictions.

To summarize: If one takes Gerhardt's first paper on atomic weights (1842) as the beginning of the reform movement, and places its conclusion after Karlsruhe, one arrives at the conventional view of a gradual period of transition extending over more than two decades. But I argue here that it was only Williamson's work (1850–54) that provided compelling evidence for reform, and that at least in Germany this evidence proved immediately and decisively persuasive.

At least in Germany—but what about the homeland of reform, France?

The Conversion of Wurtz

Adolphe Wurtz, the eldest son of a Lutheran pastor, grew up near Strasbourg and earned his medical degree at that university; his native language was Alsatian, but he spoke French and Hochdeutsch without accent. He studied at Giessen for one semester, and always considered Liebig one of his mentors. But in spring 1844 he journeyed to Paris seeking a chemical career, and quickly worked his way into Dumas's inner circle. In 1845 he became Dumas's *préparateur* at the Faculté de Médicine and *Chef des travaux chimiques* at the École Centrale. He was soon to inherit the master's mantle.[43]

Of course, fellow Alsatian Charles Gerhardt was also a presence in Wurtz's early years. Since Gerhardt and Wurtz were schoolfellows at the same Gymnasium, and since Wurtz translated Gerhardt's first book into German, and furthermore because Wurtz later became an ardent apostle of the new chemistry, it has been assumed that Wurtz and Gerhardt were good friends from childhood, and that Wurtz came to Paris already in the reformist camp. But the situation appears to be quite the opposite. Wurtz and Gerhardt apparently did not know each other in Strasbourg, and Wurtz's translation was merely a business proposition. Gerhardt was in Montpellier for the first four years after Wurtz arrived in Paris; even after Gerhardt's return to the capital Wurtz saw little of him. Wurtz scarcely knew Laurent. During Gerhardt's lifetime Wurtz never adopted the new atomic weights, and he maintained close relations with both Liebig and Dumas during periods in which both masters were furious with Gerhardt.[44]

The relationship between Wurtz and Gerhardt was never more than coolly correct. Gerhardt resented the fact that the younger Wurtz was favored by the powerful Dumas: Wurtz was made *agrégé* at the Faculté de Médicine in 1847, then took over Dumas's lectures on organic chemistry two years later, simultaneously holding his post at the École Centrale, all while Gerhardt was without any official position. (Gerhardt celebrated a brief moment in the sun as a result of the 1848 insurrections, but it passed with the onset of reaction.) In 1852 Wurtz married the daughter of a wealthy Paris banker, and in 1853 he succeeded both Dumas and Orfila at the Faculté, providing a secure career. "What luck that guy has," Gerhardt exclaimed to a correspondent.[45]

Wurtz appears to have been a quite sincere disciple of Dumas's chemistry during the late 1840s and early 1850s. Wurtz's early papers exhibit a distaste for dualism, a concern for merging organic and inorganic chemistry under unitary assumptions, and a predilection for Dumas's version of type theory—all perfectly conventional views at that time and place. But in 1853 he reached a decisive turning point in his theoretical commitments. Under the influence of Gerhardt's synthesis of compound amides and of Williamson's ether papers, Wurtz declared his acceptance of leading propositions of the reform. In his earliest surviving letter to Williamson, Wurtz described the "sensation" made in Paris by a summary of the ether research that Wurtz had solicited from Williamson for the January 1854 number of the *Annales de Chimie*.[46]

In 1855 Wurtz published two remarkable papers, both of which show the strong, and explicitly acknowledged, influence of Williamson. Inspired by Williamson's development of the multiple water type, Wurtz showed how a triple water model could well account for the reactions of glycerin. That summer he revealed what subsequently became known as the "Wurtz reaction," using it to make a theoretical argument—modeled after Williamson's, as he acknowledged—to provide what he regarded as "decisive" and "conclusive proof" that the interpretation of the new school regarding the isolated hydrocarbon "radicals," that is, that they were dimeric molecules rather than monomeric radicals, was superior to Kolbe's and Frankland's explanation. In a footnote he also proposed a speculation regarding atomic structure that could possibly explain the phenomenon of valence—a speculation which, I have argued elsewhere, may well have provided the proximate inspiration for Kekulé's structure theory.[47]

The coincidence in date of his attainment of financial and career security and his conversion to the reform would seem to provide prima facie support for a direct connection between the two events. Might he not in fact have been converted much earlier, refraining from declaring himself for fear of alienating Dumas and others during a period when he was still trying to achieve the difficult task of establishing a Parisian career? Once settled and secure, he could then safely reveal himself.

But this explanation does not fit a number of circumstances connected with the event. First, it was not until 1852 or even 1853 that Williamson's ether work began to be really noticed on the Continent; I have traced a general correlation between the recognition of Williamson's arguments by chemists and the conversions of many to the reform movement. It was only in the summer of 1853 that Kolbe became concerned about the threat to traditional chemistry. In Wurtz's case as well, this pattern seems to hold; why else would Wurtz regard it as necessary in late 1853 to solicit a French resumé of the three English ether papers of 1850–51 for the *Annales?* If this presumption is correct, his simultaneous professorial appointments and conversion experience appear to be coincidence rather than cause and effect—although, to be sure, any careerist inhibitions that might have impeded a change of theoretical loyalty had been lessened.

There is also more direct evidence for a genuine change of heart in 1853. In 1851 and 1852 Wurtz proposed a number of formulas in which he seemed to go out of his way to contradict the reform movement.[48] Had he been a closet reformer he surely would have chosen more theoretically noncommittal, indeed more conventional, formulas, in order maximally to preserve his perceived orthodoxy *as well as* his future flexibility. Furthermore, Wurtz engaged in a revealing public polemic with Gerhardt, trading "position papers" back and forth in the biweekly issues of the *Comptes rendus* during the summer of 1853. In this debate we can follow the gradual changes that Wurtz's opinions underwent, and can discern a number of points of difference that remained, such as Wurtz's retention of an important philosophical element of Dumas's type theory that he only later discarded. A student of Wurtz also described the numerous vigorous private discussions between Wurtz and Gerhardt that summer.[49]

Finally, in Wurtz's case his retention of equivalents for several years after 1853 seems to have indicated some residual doubts, not over the Williamsonian molecular magnitudes issue but rather over the atomic weight reform. In 1855 he declared himself a member of the movement more unequivocally than he had two years earlier, but he appears to have become fully convinced of the ontological truth of the new weights and formulas only as the result of his own work done during an immensely productive period of his life, 1856–58. He was probably also influenced by Kekulé's and Couper's structure theory papers of 1857–58. It was in the autumn of 1858 that Wurtz finally resolved to use only the new atomic weights in future papers. To summarize all of this, it seems clear that Wurtz's intellectual odyssey was guided to a greater degree by cognitive and empirical factors than by any particular combination of social interests.

From the beginning of 1859 Wurtz became not only an apostle but also a proselytizer for reform. He was involved with the early leadership of the Société Chimique and of a new journal, and he was, with Kekulé and Weltzien, a principal organizer of the Karlsruhe Congress. He wrote a heavily subtexted joint *éloge* for Gerhardt and Laurent (in 1862!—that is, long after their deaths); presented invited historical lectures to the Société Chimique, the Collège de France, and the Chemical Society of London; wrote a textbook; and finally published a full, formal history prefacing a new multivolume chemical dictionary. Every one of these forums was designed to defend and propagate the new chemistry in a country still dominated by older ideas.[50]

Unfortunately, none of his stratagems was notably successful. Before his death Gerhardt had made a number of partial converts in France and had gained the respect of his colleagues, but Wurtz was one of the few complete converts among established figures. Just as in Germany, older chemists such as Thenard, Dumas, and Antoine Balard remained unconvinced, even if more friendly; but unlike in Germany, important midcareer and younger chemists such as Victor Regnault, Henri Deville and Marcelin Berthelot also retained their skepticism. Wurtz was by no means totally alone—his brilliant student Charles Friedel, and the somewhat peripheral but perceptive and original Alfred Naquet are examples of fellow travelers—but for many years Wurtz justly felt isolated.

Wurtz revealed that feeling in a letter to his old teacher Liebig. At the last election to membership in the Académie des Sciences, Fremy had been chosen, and, Wurtz conceded, was not undeserving; but even Berthelot had received more votes than he. Balard had voted for Deville, and Dumas had not supported him. This was an unfortunate omen, indicating that "for my future nothing is secure."[51] In sum, he felt more appreciated in Germany than in his homeland. It is curious that this pessimistic letter was written in February 1858, just *before* Wurtz decided to shift to the new atomic weights and to carry out the extended public campaign for the new chemistry described above; it also suggests that he did not yet perceive himself as professionally secure. Nothing could have been better calculated to increase his feeling of insecurity and isolation than to begin such a high-profile campaign for views that he knew would be unpopular with important people. That he went right ahead with it suggests that he was convinced that the new chemistry was correct and would ultimately prevail. He must have been heartened by the obvious successes of the Gerhardt-Laurent reforms in Germany. Until the final victory in France, however, he was bound to create a professionally uncomfortable life for himself. He could not have known in 1858 that at his death twenty-six years later the reforms would still not have fully succeeded in his native France.

The Conversion of Kolbe

Up to this point we have used Kolbe merely as a foil to the reformers, and he deserves fuller discussion. Like Wurtz the eldest son of a rural Lutheran pastor and born less than a year later (to increase the similarities, both were native German speakers, sincere Lutherans, and ardent patriots, and died in the same year), Kolbe became a student of Friedrich Wöhler at Göttingen and Robert Bunsen at Marburg. An eighteen-month assistantship in London (where he befriended the younger Edward Frankland) and a four-year stint as editor for Vieweg in Brunswick preceded his call as successor to Bunsen in 1851. In 1865 he was named professor at Leipzig, where he lived until his death in 1884.[52]

Blessed with his father's great self-confidence and forthright manner, Kolbe also had prominent traits of traditionalism and xenophobia in his character. By all reports he was a brilliant teacher and had a benevolent and attractive manner with students, friends, and family. However, to strangers and foreigners he could be distant and suspicious, and those he perceived as inimical to him or his interests he never hesitated to attack. In his later years these attacks became ferocious, crude, and sometimes irrational, and did much damage to his reputation.

Kolbe's heroes were his teachers Wöhler and Bunsen, along with Wöhler's mentor Berzelius and his best friend Liebig. Having absorbed from his mentors a very negative view of the work of Dumas, Laurent, and Gerhardt during the stressful period of the late 1830s and early 1840s, he was the most resistant of all his confrères to the reform movement. He defended Berzelian electrochemical-dualist ideas in organic chemistry long after they had been relinquished by virtually everyone else. He only gradually came to see that the dualist copula theory could not stand in the face of the proliferation of substitution reactions that transformed the field in the 1840s and 1850s.

We have already noted that in the fall of 1853 Kolbe felt very threatened by Williamson's asymmetric synthesis argument and tried to respond to it, but he used a flawed argument that drew both Williamson's and Will's critiques and was without perceivable influence. Frankland's formidable attack on copulas in his valence-theory paper of 1852, and Wurtz's 1855 synthesis of "asymmetric" radicals such as butyl-amyl followed by an application of the Williamsonian argument to establish analogous dimeric constitutions for ethyl, butyl, amyl, and the other Kolbean "radicals," proved to be the final straws. In 1850 Hofmann had attempted to do what five years later Wurtz succeeded in doing, and had predicted a boiling point for butyl-amyl precisely matching

Wurtz's new compound; this and other physical predictions were based on the work of Kopp, who also noted the remarkable triumph of the reformers. Although Wurtz was a hated Frenchman, Frankland, Hofmann, and Kopp were perhaps Kolbe's three closest and most respected collegial friends. In an encyclopedia article completed in the first week of 1856, Kolbe fairly summarized Wurtz's new reaction and reasoning, according it a high significance. Without even attempting refutation, however, he still demurred: "It remains for the moment undecided and questionable whether these facts have sufficient probative force" to compel a doubling of the radical formulas. Residues of the copula theory can still be discerned in this article.[53]

However, a vigorous correspondence with Frankland during the year 1856 sufficed to remove Kolbe's last doubts and relinquish his grip on copulas. By the end of that year the two friends had jointly succeeded in formulating the "carbonic acid theory," a full-fledged type theory in which carbonic acid (formulated in equivalents with four oxygen "atoms") became the progenitor for many organic derivatives. This was the basis for Kolbe's later claim that he and Frankland, rather than Kekulé, deserved the credit for recognition of the tetravalence of carbon. By 1858 Frankland fully accepted the asymmetric synthesis argument for reformed molecular magnitudes, and two years later he even tried out a Williamsonian argument of his own. (Kolbe did not begin to write dimeric formulas for hydrocarbon radicals until 1864.)[54]

Kolbe understood the importance of these developments. On 4 January 1857 he wrote Vieweg, who had been growing impatient with Kolbe's delays, illnesses, and theoretical meanderings: "It is a joy to be able to say to you that I have now come to a clear understanding of the important theoretical questions with which I have so long been continuously occupied."[55] He subsequently referred to the period of "crisis" and "reformation" that organic chemistry was going through. In fact, Kolbe had capitulated: in his development of the carbonic acid theory, he had become a "typist" by any reasonable definition. In two crucial fascicles of his textbook published in early 1858 and early 1859, Kolbe eliminated all references, both verbal and substantive, to copulas; he now accepted the polybasic character of radicals and acids, a cardinal thesis of the reformed chemistry; and he freely used the concepts and the terms "substitution," "atomicity," and "types." He published a summary "confession of faith" in the form of a major paper for Liebig's *Annalen,* written in the summer of 1859 and published the following year.[56]

This paper constituted Kolbe's virtually final formulation of the theory of organic compounds; after more than a decade of wandering in the theoretical wastes, Kolbe felt he had arrived in the promised land of chemical truth. To the end of his life he regarded the ideas in the paper as necessary and sufficient to eventually expose to the light of day all the exotic creatures in Wöhler's tropical jungle of organic chemistry.

It must have been very discouraging for Kolbe to see that the response of his colleagues was to place him in the midst of what he termed "the whole mindless business"[57] of Gerhardtian type theory. In a review of Kolbe's major 1860 paper Wurtz wrote: "Monsieur Kolbe has so fully adopted the fundamental idea of types that not only does he want to multiply them, but even, with Gerhardt, to assume condensed types, as are represented by molecules of carbonic acid." The carbonic acid type, Wurtz affirmed, is nothing more than the water type with diatomic carbonic oxide functioning as the oxygen atom, just as Williamson had formulated it in 1851. For his part, Kolbe readily conceded that he was now pursuing types. His, however, were "real" and "natural" entities, whose direct genetic relationships with daughter compounds were clear and empirical. His opponents' types, by contrast, were merely schematic, formal, and conventional (that is, fictitious), whose relationships were unnatural and superficial. Wurtz, of course, argued that what Kolbe saw as "dead schematism" was in reality a more general and flexible realization of the same principle. "If, then, the ideas this chemist presents are novel, the innovation resides rather in the form than in the essence, and I believe that I have shown that even the form is not fortunate; in truth, he combats Gerhardt's types by counterfeiting them."[58]

Kolbe's friend Kopp saw the matter similarly, although he phrased his criticisms more kindly. "Kolbe still argues against relating organic compounds in general to the hydrogen, water, and ammonia types, as also against the recent assumption of mixed types. But he reveals himself as a de facto adherent of the 'type theory' by conceiving organic compounds as derivatives of inorganic compounds." Kopp also pointed out that Kolbe had used multiple types to formulate diacids, and suggested the influence of Gerhardt. Three years earlier Kopp had also referred to the Gerhardtian and Williamsonian elements in the earliest version of the carbonic acid theory.[59] Privately as well, friendly correspondents tried repeatedly to convince Kolbe that he no longer had any important differences with the type theorists, and later with the structural theorists. Even his son-in-law, student, and partisan defender Ernst von Meyer agreed that Kolbe's views were essentially those of structure theory.[60] When Frankland pushed too hard in this direction the year before Kolbe's death, Kolbe angrily cut off correspondence with his old friend.[61]

In fact there were some substantive differences between Kolbe's development of valence theory and that of the descendants of type theorists, the structuralists. In particular, the vestiges of dualism that Kolbe retained prevented him from countenancing chain formation or carbon skeletons, and this resulted in some operational distinctions in the application of his theory. The degree of overlap was striking, however, and Kolbe was able to use his faux types for over a decade to create a successful and impressively productive research school whose contributions easily rivaled, and largely dovetailed with, the structural schools of Kekulé, Erlenmeyer, Butlerov, Baeyer, Crum Brown, and others.

Structural Chemistry: A "German Science"?

Wurtz's history of chemistry, published in 1868, began: "Chemistry is a French science. It was founded by Lavoisier, of immortal memory."[62] This line created an international uproar whose aftershocks lasted into the present century. However much Wurtz tried subsequently to exonerate himself by having intended merely to indicate the *birthplace* of chemistry in Lavoisier's hands, other nationalities, and especially the Germans, were outraged at the implied presumption and arrogance. Emotions were naturally further inflamed as a result of the Franco-Prussian War, which France handily lost.

These were not good times for Kolbe, although his teaching program was phenomenally popular and he had acquired the *Journal für praktische Chemie* as a personal organ for his various campaigns. The emotions of the war, a new bout of ill health, the death of his father in 1870 and of his beloved wife in 1876, and his increasing isolation in the collegial community all took their toll. But Kolbe's intellectual isolation was largely self-imposed. To cite one example, having been elected, with Liebig, Wöhler, and Bunsen, a charter Honorary Member of the new Deutsche Chemische Gesellschaft, he angrily resigned in 1871 because the Society had not defended his critique of Wurtz's dictum when it met public foreign opposition.[63]

In the late 1870s Kolbe became obsessed with his crusades. The more popular and successful structural chemistry became in Germany, the more he despised it; the more disciples Kekulé and others of the modern school gathered, the more he was convinced they were all thieves and charlatans. Ironically, the French were far less oriented toward structure theory than the Germans; Kolbe noticed this fact with alarm, for to him it indicated a surprising source of French strength that was dangerous for the future health of the German chemical community. It seemed to him that history was repeating itself. A half-century earlier Liebig had to travel to France to learn scientific chemistry, since Germany was then dominated by *Naturphilosophie;* Germany was again in danger of being perverted by pseudoscience, and leaving France the leadership role once more. The irony was, as Kolbe well knew and loved to point out, that this same unscientific structural chemistry that had so completely conquered his homeland was a direct product of *French* chemistry—namely an outgrowth of the type theories of Dumas, Laurent, Gerhardt, and Wurtz.[64]

Here Kolbe had a point. Kekulé and other (predominantly German and German-educated) chemists had developed structural chemistry from that essentially French background. Indeed, Wurtz's isolation in France was a sort of a mirror image of Kolbe's in Germany, placing the contretemps over his chauvinist historical comment in even sharper relief. However, read with attention to the thematic orientation of the entire *Histoire,* and placed in context with Wurtz's other interpretive, historical, and polemical writings of the 1860s, the apparently frivolous chauvinism of his opening motto is subject to an additional interpretation.

Wurtz's history consisted of five long chapters, entitled "Lavoisier," "Dalton and Gay-Lussac," "Berzelius," "Laurent and Gerhardt," and "Theories of the Present Day." The principal theme was theories of composition, especially atomic theory and "atomicity" (valence and structure). Viewed as a piece of literature, the section on Lavoisier was only a dramatic introduction; the real climax of the piece was the classic chapter on Laurent and Gerhardt (heavily borrowed from his *éloge* of six years earlier), and the dénouement was the discussion of modern theories of structure.

I suggest that Wurtz's dictum carries a heavier thematic load than mere chauvinism. It was not so much Lavoisier and the first chemical revolution that Wurtz wanted to promote, as Lavoisier's countrymen Laurent and Gerhardt (not to mention Wurtz himself, aided by foreign Francophiles such as Williamson and Kekulé), who were the authors of the still incompletely consummated *second* revolution. The work was directed inward rather than outward, its intended audience Wurtz's countrymen, for they were the ones that needed to hear this particular message. What better way to persuade them to join the new movement than to appeal to their patriotism by arguing for the continued dominance of French chemistry in the international arena—for what Lavoisier had begun, Laurent and Gerhardt (and Wurtz) had completed. If I am right in this, we have here an example of nationalism put to rhetorical purposes, but for a cognitive, internalist goal and not for mere chauvinist puffery. But it was difficult for foreigners to get past that first fearsome line.

Wurtz's clever strategy, if such it was, failed to elicit the desired response in France; nor did a final historical manifesto on *La théorie atomique* published a decade later, with virtually the same theme, do any better. Ironically, by the late 1860s and 1870s Wurtz may have been fighting against an entrenched French prejudice that structure theory was German, hence foreign.

An examination of Kolbe's career reveals another kind of irony. No one had more contempt for the French or their theories in the late 1840s and early 1850s than Kolbe. However, the striking new reactions and brilliant arguments by Williamson, Gerhardt, Wurtz and Frankland during the early 1850s that convinced most of Kolbe's German colleagues to accept the French-English theories were by no means lost on Kolbe either. We have seen that by 1857 he had developed a theory of his own that was strikingly similar to the Williamson-Gerhardt newer type theory. Kolbe denied, with all the energy at his command, the arguments of friends as well as rivals that he had joined rather than defeated his enemies. But the critics were essentially right. Kolbe's pathological prejudices had failed to prevent him from understanding and being persuaded by the hated French ideas; they had only operated to prevent him from *believing* he had adopted them.

Concluding Remarks

In according a determinative role to the impact of new empirical evidence and ratiocination as explaining conversions to this revolution, and thus urging, by extension to the general case, a stronger role for purely cognitive factors in the process of scientific change than is granted by many social historians of science, I do not advocate a return to the bad old days of untrammeled internalism. Indeed, sociologists of science could—and, I hope, will—have a great deal of fun with even this case, despite its having been particularly suited for my thematic purposes. (If scientific theories are strongly underdetermined by empirical evidence, all the more so must be historical interpretations!)

For instance, by focusing on the favorable reception of the reforms in Germany, I have said little about why they made such slow gains in France. The differential reception in the two countries must necessarily have been due to social, cultural, and institutional factors, since the French had essentially equal access to the same papers of the early 1850s that so quickly converted their German neighbors. Indeed, as I have emphasized, they had even more direct access to the personal influence of the leaders of the movement, Laurent and Gerhardt, and subsequently Wurtz. Even Williamson, the key figure in the transformation of the evidence from circumstantial to decisive, was a Francophile. Why were the French so oblivious to this movement?

A full answer to this question has not yet emerged.[65] Clearly, the fact that Laurent and Gerhardt were social misfits (provincials, ardent republicans, materialists, and possibly both of Jewish extraction), and the circumstance that they almost defiantly refused to play the conventional social games, had the result that their true merits were much slower to be recognized than might otherwise have been the case. There might even have been an element of psychological backlash, in which the Laurent-Gerhardt reforms were regarded, in the time of political reaction after their deaths, as "tainted" or even dangerous—for look what a tragic fate the protagonists shared! Moreover, after the reforms caught on in Germany—which, as I have argued, happened even before the death of Gerhardt—they could be viewed, perversely, as German, hence foreign and suspect. Nor can it be mere coincidence that the first French chemist to base his textbook on the work of Laurent and Gerhardt was also the republican and scientifically peripheral Alfred Naquet. Additionally, it appears that the positivistic tone of French science, noticeable since before the beginning of the nineteenth century and continuing for more than a hundred years, worked to the disadvantage of the obviously theory-based movement described here. Why French science had this particular complexion is, again, incompletely understood, but one that must be due primarily to social and cultural forces.

Other hypotheses along this line are not hard to find. The decentralization that was so characteristic of Germany in the decades before 1871 has often been depicted as driving vigorous and ultimately beneficial rivalry between the various states, most notably in the economic sphere and between university administrations.[66] Competition was less prominent in France[67] or in Britain. This operated on many levels—local, regional, national and international, as well as in matters of prestige, prosperity, military and technical superiority, and so on. German as well as foreign observers saw an important cause of Prussian dominance over Austria and France in the superiority of German science, propelled by a competitive research ethos. Whether this sort of case can be sustained or not, there is little question that the research ideal had declined in France during the decades that it had increased so powerfully in Germany, and after 1871 the French began aggressively to address the problem.

Finally, it is necessary to point out, at the risk of belaboring the obvious, that history is a highly contingent process manufactured by individual actors, and that individuals can notably succeed or fail to exert powerful influence in their peer communities. For whatever reason, after the death of Gerhardt in 1856 France lacked a critical mass of reform-minded chemists who could make the case for the new ideas.[68] Wurtz was professionally lonely, just as Kolbe was isolated on the other side of the debate and on the other side of the Rhine. More generally, the rise of the German research ideal so prized and emulated was associated with only a handful of leading actors.

I am far from alone in urging a flexible approach as regards the classic (many would say outworn) dichotomy between cognitive and social history of science. Flexibility, pluralism, and an eclectic and empirical approach is often a recipe for success in science; I have argued that these attributes were precisely the necessary ones to accommodate the rise of structure theory, and conversely they were the ones that Kolbe notably lacked after 1870. The same attributes are also beneficial for the historian. Social and cultural forces are powerful, pervasive, and efficacious, and so is the strength of ideas and evidence pursued by conscientious scientists on the "agonistic field" in which one marshals resources. In the valuable perspectives provided by recent sociological studies, the power and vitality of scientific ideas and logic, the constant regulating appeal to the empirical world, and the contingent influence of individuals ought not be underestimated.

For their kind assistance and permission to publish from documents in their possession, I wish to thank the staffs of the library of the Royal Society of Chemistry, London, the Bayerische Staatsbibliothek, and the library of the Deutsches Museum, Munich; also Mr. and Mrs. Raven Frankland and Professor Colin Russell, Edward Frankland archive; William Brock, Harris Collection, now in D. M. S. Watson Library, University College, London; and Michael Langfeld, Ilse Dobslaw, and others at Vieweg Verlag archive, Wiesbaden. This research was generously supported by the National Endowment for the Humanities, grant RH-20801-88.

Notes

1. The strong program is defined and characterized in David Bloor, *Knowledge and Social Imagery* (London: Routledge & Kegan Paul, 1976), pp. 1–19. The quotes by Latour are found in K. D. Knorr-Cetina and Michael Mulkay, eds., *Science Observed: Perspectives on the Social Studies of Science* (London: Sage, 1983), p. 141, and S. Woolgar, ed., *Knowledge and Reflexivity: New Frontiers in the Sociology of Knowledge* (London: Sage, 1988), p. 166; see also Bruno Latour and Steve Woolgar, *Laboratory Life: The Social Construction of Scientific Facts* (London: Sage, 1979), and Latour's books *Science in Action: How to Follow Scientists and Engineers Through Society* (Cambridge, Mass.: Harvard University Press, 1987) and *The Pasteurization of France*, trans. Alan Sheridan and John Law (Cambridge, Mass.: Harvard University Press, 1988). Collins's sentence is in his "Stages in the Empirical Program of Relativism," *Social Studies of Science* 11

(1981): 3–10, on 3. Barnes and Shapin's statement is in their introduction to their edited volume *Natural Order: Historical Studies of Scientific Culture* (London: Sage, 1979), p. 9; see also Barry Barnes, *Interests and the Growth of Knowledge* (London: Routledge and Kegan Paul, 1977). For a useful collection of reprinted articles, see Barnes and David Edge, eds., *Science in Context: Readings in the Sociology of Science* (Cambridge, Mass.: MIT Press, 1982).

2. The first of these phrases is used in Paul A. Roth's elaborate critique *Meaning and Method in the Social Sciences: A Case for Methodological Pluralism* (Ithaca, N.Y.: Cornell University Press, 1987), pp. 152–225; the second is in Larry Laudan, "The Pseudo-Science of Science?" *Philosophy of the Social Sciences* 11 (1981): 173–198. Laudan's critique has been restated in his *Science and Relativism: Some Key Controversies in the Philosophy of Science* (Chicago: University of Chicago Press, 1990). Other prominent critics include Ronald Giere, *Explaining Science: A Cognitive Approach* (Chicago: University of Chicago Press, 1988); David L. Hull, "In Defense of Presentism," *History and Theory* 18 (1979): 1–15; Hull, *Science as a Process: An Evolutionary Account of the Social and Conceptual Development of Science* (Chicago: University of Chicago Press, 1988); and A. R. Hall, "On Whiggism," *History of Science* 21 (1983): 45–59.

3. The first of these quotes is from Bloor, *Knowledge and Social Imagery* (note 1), p. 45; the second is in John A. Schuster, "Constructing Contextual Webs," *Isis* 80 (1989): 493–496, on 494.

4. Thomas S. Kuhn, "Postscript—1969," in *Structure of Scientific Revolutions*, 2d ed. (Chicago: University of Chicago Press, 1970), pp. 174–210, on 175, 209.

5. E.g., Bloor, *Knowledge and Social Imagery* (note 1), pp. 13–14; Woolgar, ed., *Reflexivity* (note 1); Bloor review of the latter book, *Isis* 81 (1990): 155–156; Steve Woolgar, *Science: The Very Idea* (London: Tavistock Publications, 1988), pp. 43–44; Laudan, "Pseudo-Science of Science?" (note 2); and Hull, *Science as a Process* (note 2), pp. 3–6.

6. For some relevant references to recent treatments of philosophy of chemistry, see O. T. Benfey's review of W. J. Danaher's *Insight in Chemistry*, *Isis* 80 (1989): 159. In this context, one must distinguish the philosophy of science, which may contain specific examples from any number of particular sciences, from the philosophies of individual sciences. Conversations on this topic with Benfey, Roald Hoffmann, and Seymour Mauskopf have contributed substantially to my own ruminations.

7. Note, for instance, the historical and historiographical similarities between Martin Rudwick's *Great Devonian Controversy: The Shaping of Scientific Knowledge Among Gentlemanly Specialists* (Chicago: University of Chicago Press, 1985); David Oldroyd's *The Highlands Controversy: Constructing Geological Knowledge Through Fieldwork in Nineteenth Century Britain* (Chicago: University of Chicago Press, 1990); and James Secord's *Controversy in Victorian Geology: The Cambrian-Salurian Dispute* (Princeton, N.J.: Princeton University Press, 1986).

8. Rocke, "Kolbe Versus the 'Transcendental Chemists': The Emergence of Classical Organic Chemistry," *Ambix* 34 (1987): 156–168.

9. This is especially true for German chemistry. See Otto Krätz, "Der Chemiker in den Gründerjahren," in *Der Chemiker im Wandel der Zeiten,* E. Schmauderer, ed. (Weinheim: Verlag Chemie, 1973), pp. 259–284; Peter Borscheid, *Naturwissenschaft, Staat und Industrie in Baden (1848–1914)* (Stuttgart: Klett, 1976); and Jeffrey Johnson, "Academic Chemistry in Imperial Germany," *Isis* 76 (1985): 500–524.

10. When in 1848 Kekulé was considering a career in chemistry, his family

urged caution due to the poor prospects. There were then a total of only five chemical positions in the entire Grand Duchy of Hesse: three at Giessen, one at the Darmstadt trade school, and one in the mint (R. Anschütz, *August Kekulé* [Berlin: Verlag Chemie, 1929], 1:11–12).

11. Clearly, a market-economy model of academia, of the sort used by Joseph Ben-David and Avraham Zloczower, would be useful in helping to explain the rapid expansion of organic chemistry in Germany after 1850. As I argue in this paper, this social dynamic was only one factor, albeit an important one. For a concise description, analysis, and critique of the Zloczower thesis, see Steven Turner, Edward Kerwin, and David Woolwine, "Careers and Creativity in Nineteenth-Century Physiology: Zloczower Redux," *Isis* 75 (1984): 523–529.

12. Laurent, *Méthode de chimie* (Paris: Mallet & Bachelier, 1854), p. 28; Kekulé, *Lehrbuch der organischen Chemie* (Erlangen: Enke, 1859), 1: 58; L. Meyer, ed. notes in S. Cannizzaro, *Abriss eines Lehrganges der theoretischen Chemie* (Leipzig: Engelmann, 1891), pp. 53–58.

13. This is a continuing theme throughout the Gerhardt-Laurent correspondence: Marc Tiffeneau, ed., *Correspondance de Charles Gerhardt,* 2 vols. (Paris: Masson, 1918–25), vol. 1; see, e.g., Laurent to Gerhardt, 11 Dec 1846, 2 Feb and 4 May 1847, pp. 219, 222–225, and 232–233.

14. This applies both to my own treatment in *Chemical Atomism in the Nineteenth Century: From Dalton to Cannizzaro* (Columbus: Ohio State University Press, 1984), Chaps. 7–10, and to John Brooke, "Avogadro's Hypothesis and Its Fate," *History of Science* 19 (1981): 235–273, on 257, although both of us draw attention to the extended and multipartite character of the "Karlsruhe revolution."

15. Tiffeneau, *Correspondance,* 2: 124; Edouard Grimaux and Charles Gerhardt, Jr., *Charles Gerhardt: sa vie, son oeuvre, sa correspondance 1816–1856* (Paris: Masson, 1900), p. 290.

16. Hofmann, *Introduction to Modern Chemistry* (London: Walton & Maberley, 1865), p. v. In 1852 Hofmann wrote Cahours on the heels of Gerhardt's acid anhydride work, predicting "very grave modifications" to come in chemical theory: cited in Grimaux and Gerhardt, *Gerhardt,* p. 239. His words in a letter of 10 May 1852 to Liebig were similar: W. H. Brock, ed., *Justus von Liebig und August Wilhelm Hofmann in ihren Briefen (1841–1873)* (Weinheim: Verlag Chemie, 1984), p. 132.

17. Hofmann, "Heinrich Will: Ein Gedenkblatt," *Berichte der Deutschen Chemischen Gesellschaft* 23R (1890): 852–899, quotation from pp. 881–882; Hofmann, "Erinnerungen an Adolph Wurtz," in *Zur Erinnerung an vorangegangene Freunde,* 3 vols. (Brunswick: Vieweg, 1888), 3: 171–432, on 249.

18. Kolbe to Eduard Vieweg [16 Oct 1853], Vieweg Verlag archive, Wiesbaden, 311K (henceforth abbreviated VA), letter number 59; 23 Oct 1854, VA 87; 24 Mar 1857, VA 125 ("Es giebt eine kleine Reformation in der organischen Chemie"); 15 Oct 1859, VA 152; 24 Oct 1859, VA 153 (". . . die Krisis eingetreten ist"). Vieweg Archive has 524 letters from Kolbe to Eduard Vieweg, Heinrich Vieweg, and Franz Varrentrapp, from 1844 to 1884.

19. L. Meyer, in Cannizzaro, *Abriss* (note 12), p. 59.

20. See above, note 14. Hermann Kopp later drew attention to residual anomalies in the Gerhardt-Laurent system that impeded its acceptance in the 1840s, only some of which the reformers could convincingly rebut: *Entwickelung der Chemie in der neueren Zeit* (Munich: Oldenbourg, 1873), pp. 733–735.

21. On Williamson and his "asymmetric synthesis argument," see J. Harris and W. H. Brock, "From Giessen to Gower Street: Towards a Biography of Alexander W. Williamson," *Annals of Science* 31 (1974): 95–130, and Rocke, *Chemical Atomism* (note 14), Chaps. 8 and 9.

22. Described, with references, in Rocke, *Chemical Atomism* (note 14), pp. 215–229, 236–237, 251–257; also in Kopp, *Entwickelung,* pp. 738–767.

23. Wurtz, "Sur une nouvelle classe de radicaux organiques," *Annales de chimie et de physique* [3] 44(1855): 275–313; Hofmann, "Note on the Action of Heat upon Valeric Acid," *Journal of the Chemical Society* 3 (1850): 121–134; B. C. Brodie, "Observations on the Constitution of the Alcohol-Radicals, and on the Formation of Ethyl," ibid., pp. 405–411, on 411.

24. Kopp, *Entwickelung,* pp. 750–753, 763; Hofmann, "The Life-Work of Liebig," *Erinnerung* 1:195–305, on 273.

25. Grimaux and Gerhardt, *Gerhardt* (note 15), pp. 208–293 passim.

26. Kolbe to Vieweg, [16 October 1853], VA 59: "Es ist mir böse damit gegangen. In den letzten Monaten sind nämlich Arbeiten von dem allergrössten wissenschaftlichen Interesse von mehreren Chemikern, besonders von Gerhardt und Williamson erschienen, die das ganze bis jetzt bestehende Lehrgebäude der Chemie einzustürzen drohten. Ueber das gründliche Studium derselben, und der vielen dadurch berührten Verhältnisse sind mehrere dieser Ferienwochen vergangen ehe ich zu einem klaren Urtheil darüber gelangte, während welcher Zeit ich auch keinen Federstrich habe thun können."

27. Kolbe, "Critical Observations on Williamson's Theory of Water, Ethers, and Acids," *Journal of the Chemical Society* 7 (1854): 111–121; Williamson, "On Dr. Kolbe's Addition Formulae," ibid., pp. 122–139.

28. Kolbe to Vieweg, 23 October 1854, 31 August and 20 December 1857, 15 and 24 October 1859, 1 March and 9 April 1860, VA 87, 132, 134, 152, 153, 155, and 157.

29. Kolbe to Vieweg, 1 March 1853 [sic for 1854], VA 69: "Welche Ansicht Liebig gegenwärtig darüber hat, weiss ich nicht; ich glaube wie Wöhler gar keine, da beide sich für diese Dinge wenig zu interessiren scheinen."

30. Liebig to Hofmann, 23 April 1849, in Brock, ed., *Briefe,* p. 84.

31. E.g., Liebig to Wöhler, 14 April 1857, 27 February 1865, and March 1870, in *Aus Justus Liebig's und Friedrich Wöhler's Briefwechsel,* ed. A. W. Hofmann, 2 vols. (Braunschweig: Vieweg, 1888), 2: 42, 179, 280.

32. L. Meyer, "Leopold von Pebal," *Berichte der Deutschen Chemischen Gesellschaft* 20 (1887): 997–1015, on 1000. For Bunsen's aversion to organic chemistry and to chemical theory, see, e.g., Max Bodenstein, "Robert Wilhelm Bunsens Stellung zur organischen Chemie," *Die Naturwissenschaften* 24 (1936): 193–196.

33. Kolbe to Vieweg, 9 April 1860, VA 157.

34. Hofmann, "Heinrich Will" (note 17), pp. 852–899.

35. Ibid., p. 857.

36. R. Anschütz, *Kekulé* (note 10), 1: 16–17, from a letter to Anschütz, date not mentioned but probably shortly after the turn of the century, from Reinhold Hoffmann.

37. Jacob Volhard, *Justus von Liebig,* 2 vols. (Leipzig: Barth, 1909), 1: 351.

38. Will, "Zur Theorie der Constitution organischer Verbindungen," *Annalen der Chemie und Pharmacie* 91 (1854): 257–292.

39. Kopp, "Ueber die specifischen Volume flüssiger Verbindungen," *An-*

nalen der Chemie und Pharmacie 92 (1854): 1–32, on 24, 28; *Jahresbericht über die Fortschritte der Chemie* 7 (1854): 370–373; 10 (1857): 269–270; 13 (1860): 218–222; Kolbe to Vieweg, 1 July and 23 October 1854, VA 80 and 87. That Kopp as well as Will taught the reformed chemistry when Volhard was a student at Giessen (1852–55) is stated in D. Vorländer, "Jacob Volhard," *Berichte* 45 (1912): 1855–1902, on pp. 1860–1861.

40. On Strecker's conversion, see, e.g., his "Ueber einige Verbindungen der Milchsäure," *Annalen der Chemie und Pharmacie* 91 (1854): 352–367; Will, "Zur Theorie," p. 265; and Kolbe to Vieweg, 3 May and 1 Jun 1856, 20 Feb 1857, and 15 April 1860, VA 118, 119, 123, and 158. For his role at Karlsruhe, see Anschütz, *Kekulé* (note 10), 1: 680, 686–687.

41. On Erdmann's conversion, see Grimaux and Gerhardt, *Gerhardt* (note 15), p. 265; Erdmann had taken Gerhardt onto the editorial board of his journal in 1852. See also Kolbe to Vieweg, 1 July 1854, VA 80.

42. Pebal to Roscoe, 25 May 1860, H. E. Roscoe Collection, Royal Society of Chemistry, London.

43. The longest and best biography of Wurtz is Hofmann's *Zur Erinnerung* (note 17).

44. Grimaux and Gerhardt, *Gerhardt,* pp. 13–15, 76–77; Tiffeneau, *Correspondance* (note 13), 2: 302–313.

45. Gerhardt to Chancel, 1 Oct 1852, in Grimaux and Gerhardt, *Gerhardt* (note 15), p. 129.

46. Williamson, "Sur la théorie de l'étherification," *Annales de chimie et de physique* [3] 40(1854): 98–114; Wurtz to Williamson, 18 April 1854, Harris Collection, University College, London.

47. Wurtz, "Théorie des combinaisons glycériques," *Annales de chimie et de physique* [3] 43(1855): 492–496; Wurtz, "Radicaux" (note 23), esp. pp. 300–313; Rocke, "Subatomic Speculations and the Origin of Structure Theory," *Ambix* 30 (1983): 1–18.

48. Wurtz, "Sur un nouveau mode de formation de l'éther carbonique," *Comptes rendus hebdomadaires des séances de l'Académie des Sciences* 32 (1851): 595–596; Wurtz, "Sur l'alcool butylique," ibid., 35 (1852): 310–312.

49. Gerhardt and Chiozza, "Recherches sur les amides," *Comptes rendus hebdomadaires des séances de l'Académie des Sciences* 37 (1853): 86–90; Wurtz, "Sur les dédoublements des éthers cyaniques," ibid.: 180–183; Wurtz, "Sur la théorie des amides," ibid.: 246–250; Gerhardt, "Note sur la théorie des amides," ibid.: 281–284; Wurtz, "Nouvelles observations sur la théorie des amides," ibid.: 357–361; Auguste Scheurer-Kestner, "Charles Gerhardt, Laurent, et la chimie moderne," *Revue alsacienne* (August 1884), quoted in Grimaux and Gerhardt, *Gerhardt* (note 15), p. 198.

50. Wurtz, "Histoire générale des glycols," in *Leçons de chimie professées en 1860* (Paris: Hachette, 1861); "Éloge de Laurent et de Gerhardt," *Moniteur scientifique* 4 (1862): 482–513, also undated offprint, 32 pp., here used; "On Oxide of Ethylene, Considered as a Link Between Organic and Mineral Chemistry," *Journal of the Chemical Society* 15 (1862): 387–406; *Leçons de chimie professées en 1863* (Paris: Hachette, 1864); *Cours de philosophie chimique* (Paris: privately published, 1864); *Leçons élémentaires de chimie moderne* (Paris: Masson, 1867–68); "Histoire des doctrines chimiques depuis Lavoisier," in Wurtz, ed., *Dictionnaire de chimie pure et appliquée* (Paris: Hachette, 1868), 1: i–xciv.

51. Wurtz to Liebig, 3 February 1858: ". . . dass mir für die Zukunft nichts gesichert ist." Bayerische Staatsbibliothek, Liebigiana IIB, Wurtz file, letter 11.

52. The best biography is Ernst von Meyer, "Zur Erinnerung an Hermann Kolbe," *Journal für praktische Chemie* 138 (1884): 417–466. See also my biography, *The Quiet Revolution: Hermann Kolbe and the Science of Organic Chemistry* (Berkeley: University of California Press, 1993).

53. Kolbe, "Radicale; Radicaltheorie," *Handwörterbuch der reinen und angewandten Chemie* (Brunswick: Vieweg, 1856), 6: 802–807. The date of authorship is established by Kolbe to Vieweg, 5 January 1856, VA 114.

54. Kolbe, "Ueber die rationelle Zusammensetzung der fetten und aromatischen Säuren," *Annalen der Chemie und Pharmacie* 101 (1857): 257–265; Frankland, "On the Artificial Formation of Organic Compounds," *Proceedings of the Royal Institution,* 2 (1858): 538–544.

55. Kolbe to Vieweg, 4 January 1857, VA 122: ". . . es ist mir eine Freude Ihnen mittheilen zu können, dass ich jetzt über die wichtigen theoretischen Fragen, die mich bis lang unausgesetzt beschäftigten . . . mit mir im Klaren bin."

56. Kolbe, *Ausführliches Lehrbuch der organischen Chemie* (Brunswick: Vieweg, 1854–59), 1:567–575, 740–749; "Über den natürlichen Zusammenhang der organischen mit den unorganischen Verbindungen," *Annalen der Chemie und Pharmacie* 113 (1860): 293–332. At this time and subsequently Kolbe often referred to this article as his "Glaubensbekenntnis," e.g., in his letter to Vieweg of 24 October 1859, VA 153.

57. Kolbe to Vieweg, 1 March 1860, VA 155: ". . . die ganze geistlose Wirtschaft . . ."

58. Wurtz, *Répertoire de chimie pure* 2 (1860): 354–359; 3 (1861): 418–421.

59. Kopp, *Jahresbericht über die Fortschritte der Chemie* 13 (1860): 218–222; 10 (1857): 269–270.

60. Ernst von Meyer, *Lebenserinnerungen* (n.p., n.d., c. 1918), p. 40.

61. Frankland to Kolbe, 23 Sep 1883, Sondersammlungen, Deutsches Museum, Munich, MS. 3573; Kolbe to Frankland, 28 Oct 1883, Frankland archive, frame no. 01.02.1526.

62. Wurtz, "Histoire" (note 50), p. i.

63. For background to this event, see Rocke, "Pride and Prejudice in Chemistry: Kolbe, Hofmann, and German Anti-Semitism," in process of publication.

64. Kolbe to Volhard, 9 Jun 1876, Sondersammlungen, Deutsches Museum, MS. 3681.

65. But see, e.g., Terry Shinn, "Orthodoxy and Innovation in Science: The Atomist Controversy in French Chemistry," *Minerva* 18 (1980): 539–555; Mary Jo Nye, "Berthelot's Anti-Atomism: A 'Matter of Taste'?" *Annals of Science* 38 (1981): 585–590; Nye, *Science in the Provinces: Scientific Communities and Provincial Leadership in France, 1860–1930* (Berkeley: University of California Press, 1986); and Robert Fox, "Scientific Enterprise and the Patronage of Research in France, 1800–70," in G. L. E. Turner, ed., *The Patronage of Science in the Nineteenth Century* (Leiden: Noordhof, 1976), pp. 9–51.

66. Joseph Ben-David, *The Scientist's Role in Society,* 2d ed. (Chicago: University of Chicago Press, 1984); A. Zloczower, *Career Opportunities and the Growth of Scientific Discovery in Nineteenth-Century Germany* (New York: Arno, 1981); Steven Turner, Edward Kerwin, and David Woolwine, "Careers and Creativity in Nineteenth-Century Physiology: Zloczower Redux," *Isis* 75 (1984): 523–529; Peter Borscheid, *Naturwissenschaft, Staat und Industrie in Baden (1848–1914)* (Stuttgart: Klett, 1976); Jeffrey Johnson, "Academic Chemistry in Imperial Germany," *Isis* 76 (1985): 500–524; and Johnson, *The Kaiser's Chemists: Science*

and Modernization in Imperial Germany (Chapel Hill: University of North Carolina Press, 1990). Hull (*Science as a Process,* note 2) has depicted science as being characterized above all by a dynamic combination of cooperation and competition, in much the same way that natural selection is supposed to operate in the biological realm.

67. A French model of center-periphery competition has recently been explored in Nye, *Science in the Provinces* (note 65).

68. On this question see Fox, "Scientific Enterprise" (note 65).

Justus Liebig and the Construction of Organic Chemistry

Frederic L. Holmes

I

The great externalist-internalist debate of the 1970s now appears outdated, not because the questions raised during its prime have been resolved but because most of us grew tired of contesting the issue. It has been relatively easy to retire from confrontation, because rival claims could be settled merely by partitioning the history of science into two sets of subproblems. Those who were attracted to studying external situations have found ample scope to do so, while those who preferred what were lumped together as internalistic problems were also free to go their own way. A minority of eclectic historians have made efforts, generally applauded, to penetrate more than superficially across the perceived boundaries, but their exemplary studies have not been seen as resolving the issue. The debate has also been overshadowed in the 1980s by another arena for confrontation between intellectual and social approaches to science. In this debate it will be less easy to retire from the contest, because the field cannot be partitioned into separate problems. Rather, the two approaches collide over possession of the central problem of how scientific knowledge is acquired.

The polar positions in this debate are (1) that reliable scientific knowledge is arrived at through rational arguments and sound judgments based on available evidence, and (2) that scientific knowledge is constructed through social processes. Unlike the externalist-internalist debate, which took place largely among historians of science themselves (although it may have been modeled on similar debates that began earlier in other realms of history), the current debates over the construction of scientific knowledge arose in, and are to a large extent being waged from, disciplines external to the history of science. The most prominent proponents of rational or logical construction are philosophers of science, whereas the social constructionist arguments emanate largely from sociologists and a "science studies" group who define their methodologies as anthropological, ethnographical, or observer-participant. Recently, cognitive scientists have entered the fray as allies of the logical constructionists, viewing the human mind as able to discover knowledge through its capacity to process information.[1]

Despite the grounding of these positions in the disciplinary objectives and intellectual structures of other fields, the arguments invoked have inescapable relevance to the history of science. Philosophers, sociologists, and cognitive scientists have all bolstered their causes by recourse to historical examples, and a few of them have undertaken original historical case studies of impressive scope. Increasingly those of us whose primary pursuit is the historical development of scientific knowledge must be attentive to these positions and come to terms with their applicability to our own endeavor. Some historians have, in fact, already enlisted in the cause of one or another of the contesting forces.

Among the philosophers of science, Larry Laudan has taken the strongest stand on the question of whether belief in a scientific theory or concept is explainable in terms of arguments and evidence or in terms of the social circumstances in which the agent "found himself." "Where agents have sound reasons for their beliefs," Laudan contended in 1977, "those reasons are the most appropriate items to invoke in an explanation of the beliefs which these reasons warrant"; scientific beliefs should not be ascribed to social circumstances until after it has been shown that a rational intellectual account cannot be given. Contemporary cognitive sociology, according to Laudan, has failed "to explain any interesting scientific episodes."[2] Rejecting Laudan's view, which he assumed "to be prevalent among a number of philosophers and historians of science," Steven Shapin claimed in 1982 in an influential review of "History of Science and its Sociological Reconstructions" that there have been "many empirical successes of practical sociological approaches to scientific knowledge."[3] In 1986 Bruno Latour and Steve Woolgar playfully proposed "a ten-year moratorium on cognitive explanations of science." "We hereby promise," they wrote, "that if anything remains to be explained at the end of this period, we too will turn to the mind."[4]

Between such starkly opposed alternatives as those maintained by Laudan on the one hand and by Latour and Woolgar on the other, less staunchly committed participants in the debate naturally seek compromise positions. Typical of formulas that allow room for both an intellectual and a social approach is a cogent recent statement by David Gooding that "empirical access to Nature is both a cognitive and a social process. It is cognitive, in that experimenters learn through engaging a real (and often recalcitrant) natural world; it is social in so far as what they learn owes its significance and even its formulation to engagements with other (often recalcitrant) observers."[5]

Such formulations do not resolve the controversy, but do suggest a promising attitude from which to begin exploration of the further questions implied in these juxtapositions. It is only a good start to place an equal priority on the contribution of intellectual and social factors, to acknowledge that both nature and a social community of scientists are pertinent to the acquisition of scientific knowledge. If we are to reach a deeper understanding we must be able to weigh the respective contributions of intellectual and social factors, to delineate their interfaces, and to trace their interplay in far more intimate detail in a far greater variety of concrete cases, extended over longer spans of time than has been done up until now. Some progress toward this goal can be made by clearer definitions of the problem. An exemplary recent treatment by the philosopher of science Thomas Nickles provides an important pathway toward a reconciliation of the intellectual and the social constructionists.[6] A fuller resolution can only come, however, through large scale historical investigation.

If, as historians, we take up this challenge, I believe that we should guard against becoming partisan supporters of one or another of the sides already taken in the current debates. We should be aware that the positions that have been put forth are rooted in disciplinary considerations that are not our own, and that they rely on modes of analysis at which we are less experienced than the philosophers, sociologists, and cognitive scientists who deploy them. A perusal of the recent literature they have produced makes it evident that what are cast as logical arguments shade into disciplinary boundary disputes. They are in part contests about which of these fields is best qualified to explain the construction of scientific knowledge. (If this appears to be a sociological explanation for an intellectual question, it may be because I am less biased against social factors than I am sometimes taken by others to be.) Some historians of science have already identified themselves as social constructionists, whereas others defend the primacy of intellectual history. By doing so, we may be able to appear to influence this debate by bringing to bear historical evidence favoring one or another point of view; but in that way we will lose our best opportunity to draw from history insights concerning the central issues that even we ourselves may not anticipate.

To provide historical studies that will truly advance our grasp of these issues, rather than merely validate to some degree claims borrowed from other disciplines, is a more daunting task than it may appear on the surface. The engagement of the experimenter with nature, to which Gooding alludes in the statement previously quoted, is an individual activity, even if in the modern period we often mean an individual research group rather than a single person. This activity must be followed in fine detail if we are to understand how significant increments of scientific knowledge are gained. From my own experience reconstructing the scientific work of several individual scientists, I can assert with confidence that it is not enough to infer from published papers how scientists reach the conclusions they present therein. One must penetrate to the more intimate level of daily activity if one wishes to identify fully the complex interplay of intellectual factors, craft practices, and social complexes that direct the investigator along his or her pathway to new knowledge. Nor is it enough to reconstruct vignettes from such a pathway at critical or high points along it. To understand its dynamic one must have the patience to sustain the detailed trace over the years or decades that form the investigator's outlook and shape the discoveries that occur along the way.

Tracking individual scientists in this way helps us to understand the nature of creative scientific activity but is in itself only a first step toward an understanding of the construction of scientific knowledge. As the social constructionists are fond of repeating, no individual works in isolation. Not only are his or her contributions dependent on dense interactions with other members of a social community, but they are fragments of a broader picture, and they are shaped as much by their relationship to other investigations as by the sustained enterprise of the individual. If we wish to grasp the larger picture we must show how individual investigative pathways are linked into a network of interacting investigations that comprise the moving front of a shared set of research problems, or a scientific subfield. It is an open question whether we can assimilate this new level of complexity without losing the degree of resolution at the level of the individual pathways essential to understand how the experimenter engages nature.

Moreover, our picture of the network of interacting pathways must not be a mere snapshot of a field at a given time. The shapes of major pieces of scientific knowledge often emerge only over years, decades, or scientific generations, so that we must sustain our complicated multiscale narratives over prolonged periods. It may be that my criteria are too demanding for us to meet; but if that is so, then we ought to be modest in our claims about the degree to which we understand either the intellectual or the social forces that shape the construction of scientific knowledge.

II

To explore the feasibility of attaining such pictures is one of my objectives in a study in which I am currently engaged, that of the emergence of organic chemistry as a subfield of chemistry between 1820 and 1840. The story will be oriented around Justus Liebig (1803–1873), who came during the second half of this period to dominate that field (only to leave it shortly afterward); but it will also devote detailed attention to the work and influence of other leading figures such as Jöns Berzelius (1779–1848), Joseph Gay-Lussac (1778–1850), Jean-Baptiste Dumas (1800–1884), and Friedrich Wöhler (1800–1882), and to situate the contributions of many other figures, some of them relatively minor. I started out on this topic more than a decade ago, stimulated by the existence of a dense network of surviving correspondence among the leading and lesser participants. Their letters to one another provided a rich view of their scientific exchanges and concerns, their collaboration and competition, their personal bonds and conflicts, and their respective outlooks on the rapidly changing enterprise in which they were all engaged. To a large extent these letters constituted the medium of informal communication that we have come to perceive as so central to the cognitive and the social development of science. I intended to follow the experimental and theoretical progress of the field mainly at the level of detail displayed in the numerous research papers that the participants in this field produced, using the correspondence in part to supplement the published record, but more particularly to illuminate the web of personal activities and interactions underlying the scientific output.

In 1975 I wrote on this subject a manuscript that has not been published, but from which I abstracted an article on Liebig for the *Dictionary of Scientific Biography.* At the time I did not define the activities I described as elements in the social construction of scientific knowledge, but much of what I found would probably be so regarded in light of the literature I have discussed in the previous section. Shortly afterward I changed directions, embarking on the study of Hans Krebs and then of Lavoisier's work in biological chemistry, which fixed my attention for a long time on the fine structure of the investigative pathways of individual scientists.

Returning now to my earlier topic after so long a lapse, I will follow the general plan I had in mind then; but my perspective on the problem has, I believe, been sharpened over the intervening years in three dimensions: (1) My investigations of the research pathways of individuals based on laboratory notebooks has made it clearer to me that the reconstructions of the work of Liebig and his contemporaries that

I had based largely on their published papers reached mainly the public face of their endeavors, and that I had attained only attenuated glimpses of their private experimental engagements with nature. (2) The debates about the construction of scientific knowledge that have taken place since 1975 provide more structured viewpoints to test within the context of this particular case. (3) Recent scholarship on the history of scientific education has sensitized me to the connections between Liebig's teaching program and his research enterprise.[7] I recently published a paper on the latter topic and will confine myself here to the first two issues.

Justus Liebig is a marvelous subject for some of the favored themes in the recent literature on the social construction of scientific knowledge. An ambitious, passionate man, more entrepreneurial than intellectual in temperament, seeking authority and influence as avidly as he pursued experimental achievement, Liebig divided the inhabitants of his scientific world into friends and enemies, making little distinction between a challenge to his views and an attack on his person. Treating the international chemical community of his time as a set of shifting alliances, he cultivated the distinguished senior scientists from whom he hoped for support; sought collaborations with those of his contemporaries whom he admired; nurtured generously those younger chemists whom he perceived as actual or potential recruits to his cause; and employed every weapon at his disposal to undermine those whom he perceived as his foes or whose scientific performances he deemed inadequate.

Leader of a research school whose scale was unprecedented in its time, Liebig could bring to bear in support of the scientific positions he took, not only his personal experimental skills and accumulated knowledge but also the organized efforts of those in his laboratory who were learning to do chemical research in his manner. That advantage made it sometimes difficult for other chemists who might differ with him to compete in contests to resolve the questions in dispute. In these respects much of Liebig's behavior fits well the belief of Bruno Latour and Steve Woolgar that in science

> the set of statements considered too costly to modify constitute what is referred to as reality. Scientific activity is not "about nature," it is a fierce fight to *construct* reality. The *laboratory* is the workplace and the set of productive forces which makes construction possible.

"The cost of challenging" statements that have become reified through the processes that Latour and Woolgar describe "is impossibly high."[8] With his powerful laboratory, with the authority he acquired as editor of a journal he made preeminent in the field, and with his extensive network of contacts, Liebig commanded sufficient resources to raise the cost of opposing him to intimidating heights.

The degree to which Liebig saw the chemistry of his time as a fight to construct reality, and the fierceness with which he fought, are captured vividly in a letter that he wrote to his closest friend and ally, Friedrich Wöhler, in February, 1833 concerning the French pharmacist Pierre-Jean Robiquet. (For reasons that are not difficult to surmise, this letter was omitted from the Liebig-Wöhler correspondence published in the late nineteenth century.) I shall quote it at length for the picture it reveals of the manner in which Liebig wielded the power he believed he possessed to punish or reward.

> I have recently avenged myself in a splendid way on Robiquet. You know how fully he opens his ass's mouth at every opportunity in his articles, and especially in the *Journal de pharmacie,* where he regularly calls my analyses into question. For years it has been my most ardent wish that he might finally publish some organic analysis of his own. This has now happened, he has published a large work on meconic acid, a parameconic and pyromeconic acid and has croaked like a frog and strutted like a rooster over it. I have seen his article already in the galley proofs [a typical example of the way in which Liebig used his connections to obtain privileged advance information on the work of his competitors] and have isolated his meconic acid and have analysed it with every possible precaution. He has made an error of no less than 5 *atoms* of hydrogen, and his para [meconic acid] has gone the way of all easily shattered glass. . . .
>
> Yesterday I wrote him the following: "My dear Mr. Robiquet. In your last work on oxamid you claimed for yourself the existence of the benzoic acid radical, you have with incomparable frivolity questioned the results of works that have not been shaken out of the French arsenal, you have on every occasion loosed the hounds and apes of the *Journal de pharmacie* on German works, you have set yourself up as the arbitrator of matters that are beyond your reach, and in your latest work you have delivered the final proof that you are an ignorant and arrogant ass. Here are the analyses of meconic and parameconic acid. If I wished to I could flay you with them and expose you to ridicule in Paris, but only a French apothecary would do that. I make a gift to you of my analyses, do with them what you will; but if you ever again set your sights on fields for which you have no weapons, the devil himself will blast you from behind with his wintry winds."
>
> You see how I have triumphed, but we can only enjoy it between ourselves. Another reason not to destroy him is that he is the only person who can to some extent hold Dumas in check, and a fighter against the ill-intentioned spirit of Dumas is not to be scorned. He [Robiquet] has the loyalty of numberless apothecaries behind him, he must be retained.[9]

Liebig's shifting alliances, and the personal and national rivalries that motivated them, exerted a strong influence on the course of the most prominent controversy in organic chemistry during the 1830s, that over the various radical theories proposed by the leading chemists in the field. In 1828 Dumas and Polydore Boullay (1806–1835) had interpreted the reactions of alcohols with several acids, forming what they called ethers (all but one of which are now called esters), in terms of a persistent subgroup, C_2H_2, which they called "hydrogen bi-carbon" and envisioned as being exchanged as a unit in the reactions. Liebig was skeptical of Dumas's attempt to define such groups of atoms, but sufficiently influenced by the example so that he too began in 1831 to propose similar units to explain, for example, the relation between camphor and camphoric acid. In 1832 Liebig and Wöhler defined a benzoyl radical on the basis of their impressive study of benzoic acid and a series of related compounds that they produced from it. The benzoyl radical appeared so solidly grounded in their analytical results that Berzelius hailed the investigation as the most important that had ever taken place in organic chemistry.[10]

Meanwhile Dumas was defending his hydrogen bi-carbon radical against various criticisms, and even strengthened the case for it by finding further reactions that he could readily assimilate within its framework. Berzelius acknowledged that representing the composition of the compounds in this manner allowed one to place so many phenomena in a simple and concrete order that the idea was worthy of the attention it received, but he was not persuaded that the compounds were necessarily constituted in that way. Noting that the radical was a multiple of the formula for olefiant gas, he suggested in 1832 that it be named "aetherin." A year later, in response to new studies of phosphovinic, sulfovinic, and aethionic acid, substances recently identified as involved in the formation of ethers from alcohol, Berzelius formulated a different theory treating ether as $C^4H^{10} + O$ and alcohol as $C^2H^6 + O$: that is, as oxides of different radicals. In Dumas's theory the same radical (Berzelius's "aetherin") underlay both compounds. Liebig quickly backed most aspects of Berzelius's theory, but a little later he suggested a modification in which alcohol would be viewed as the hydrate of ether. Naming his own proposed radical "ethyl" ($E = C^4H^{10}$), Liebig represented ether as $E + O$, alcohol as $EO + H_2O$, and sulfovinic acid as $(EO + H_2O) + 2SO_3$.

During the next two years Dumas defended his views vigorously, whereas Liebig devoted considerable effort to gathering evidence intended to destroy the hydrogen bi-carbon theory. In March 1835 he wrote Berzelius that he had shown the foundation of Dumas's theory to be "completely false." Assuming that he and Berzelius stood together in support of the ethyl radical, Liebig was greatly disappointed when he realized that Berzelius rejected the modification he had made in Berzelius's own theory in order to preserve a common radical for alcohol and ether.

In 1837 Liebig tried to settle his differences with Dumas. During a visit to Paris he reached with the French chemist what he took to be full agreement on the theoretical issues in organic chemistry. Shortly afterward Dumas composed in both of their names a *Note on the present state of organic chemistry* announcing their new agreement and claiming that the laws of combination and reactions in organic chemistry were the same as those in inorganic chemistry, but that the role played by the elements in the latter was represented by combinations of elements, or radicals. "All of the art of the chemist" Dumas declared for both of them, "consists in manipulating [these compound radicals] while avoiding their destruction." The joint note was in fact so general that it avoided, instead of resolving, the questions upon which their previous disputes had centered. Their collaborative efforts soon foundered and they again went their separate ways.

In 1839 Liebig proposed another new radical theory, based on an acetyl radical (Ac = C_4H_6) with which, he asserted, "one can easily perceive that the opposing views [that is, the hydrogen bi-carbon or aetherin theory, Berzelius's radicals, and the ethyl theory] were basically the same." The aetherin and ethyl groups now appeared not as fundamental radicals but as combinations of acetyl with hydrogen.

In the sense that the acetyl theory of 1839 appeared to reconcile the hydrogen bi-carbon theory of 1828 and the several theories proposed in 1833, the debates occurring during these years can be viewed as a progressive development leading to successively better resolutions of the problems at issue. They are presented in that manner, for example, in Aaron Ihde's *The Development of Modern Chemistry* under the heading "The Growth of the Radical Theory."[11] This was, moreover, not merely an intellectual dialectic, for the points in dispute turned often on the accuracy of elementary analyses, the purity of critical substances analyzed, and the identification of new compounds and reactions. The experimental skills of the contestants were as critical to the outcome as was their conceptual acuity. The intensity of the contest, however, the tenacity with which each sought to maintain his position and to defeat the positions he opposed, cannot be fully explained on either cognitive or experimental grounds.

The scientific disagreements between Liebig and Dumas were in fact not fundamental. In principle both agreed that organic chemistry was changing so rapidly that no theory yet available was general enough to encompass all the data to which it could be subjected. It was important to formulate broad, though partial and provisional, theories to bring order into what would otherwise burgeon into a chaotic mass of analytical results. Both men repeatedly espoused the tenet that theories of organic composition were useful guides to research even though they might ultimately be proved wrong, and that debates were beneficial because competition induced the defenders of divergent views to discover and analyze more compounds in the endeavor to support them. In this assessment they were correct.

Why then did Liebig seek so stubbornly to establish his own theories and to discredit those of Dumas? The most compelling explanation is that Liebig saw himself as locked in a titanic struggle with Dumas to direct the development of the field. He convinced himself that Dumas was using his increasingly dominant position in Paris to impose his views on other French chemists, and he thought of himself as standing, united with Berzelius, in opposition to Dumas's imperial ambitions. He envisioned himself as rescuing French chemistry from the "false route" onto which Dumas's penchant for creating frivolous laws and hypotheses had diverted it. At the same time he was trying to force the chemists in Paris, which had once seemed to him the center of the scientific world, to acknowledge him and his followers as a dominant force in organic chemistry.

Similarly, Liebig's decision in 1837 to try to join forces with Dumas was less an intellectual rapprochement than a calculated political move. During the previous three years he had nurtured a friendship with the French chemist Jules Pelouze (1807–1867) in such a manner that it became also an alliance directed against Dumas. Pelouze became in effect the Parisian representative of the Giessen school. Just at the height of the campaign against Dumas in which Liebig had enlisted Pelouze, Pelouze became a candidate for a seat in l'Académie des Sciences. When Dumas informed Pelouze that Pelouze's association with Liebig's attacks upon him made it impossible for him to support Pelouze's nomination, Liebig suddenly became fearful that his actions might cost his friend dearly. It was that concern that prompted him, in May 1837, to write Dumas offering a reconciliation. Dumas accepted the offer with enthusiasm, but their understanding lasted for less than a year before a priority dispute between Pelouze and Dumas forced Liebig to choose between them. Even though Liebig regarded Pelouze's complaint as unjustified, his ingrained suspicion of Dumas caused him finally to abandon his old rival in order to preserve his more deeply rooted connection with Pelouze.

In 1836 Liebig had regarded the defense of his ethyl radical, in particular its application to the theory of the formation of ethers against the ether theory based on Dumas's hydrogen bi-carbon radical, as of paramount importance, because he thought that the most general questions concerning the nature of organic compounds revolved around this crucial case. By 1839, when he acknowledged with his acetyl theory that both his earlier views and those of Dumas "have the same foundation," he was prepared to be more accommodating, in part because he had lost much of his earlier confidence that there was any satisfactory single way to view organic composition. Although his change of attitude resulted partly from the experience of two more years of rethinking the unresolved problems in the field, it also reflected further shifts in his personal relationships within the field. In 1836 he had envisioned himself standing side by side with Berzelius against the overwhelming ambitions of Dumas and his followers. By 1839 his failure to persuade Berzelius to modify long held views in order to support his own efforts to revise the theory of acids had led Liebig to perceive Berzelius as an increasingly rigid, spent force in organic chemistry. The theoretical flexibility of Dumas that had formerly seemed to Liebig a mark of his superficiality now appeared to him better adapted to the fluid state of organic chemistry than did Berzelius's firm adherence to fixed principles. As Berzelius set himself also against the substitution theory of Dumas, Liebig thought that the French chemist was bound to prevail. Despite his continued personal distrust of Dumas and his repeated efforts to retain his ties with Berzelius across the strains created by their growing disagreements, Liebig found himself drawn toward an intellectual alliance with his formidable rival while distancing himself from his former idol.

We could easily multiply these examples of Liebig and his contemporaries competing with one another to shape the direction of their field, to impose their own views upon the field, and to remove from the field the influence of rival views. In the process some of them formed combinations directed against others, and they sometimes regrouped in other combinations when such shifts suited their competitive goals. The behavior of these historically distinguished leaders of the formative period in the emergence of one of the major fields of modern science appears to illustrate admirably Latour and Woolgar's concept of scientists as contenders in an agonistic field filled with social conflicts, disputes, alliances, and maneuvers. A committed social constructionist might select from these events ample material to portray scientific judgments as contingent upon social and professional status, vested interests, individual and national rivalries, and other factors beyond the realm of logical reasoning or the disinterested evaluation of evidence. Much of what Liebig, in particular, did was, as Latour and Woolgar assert for science in general, directed not at "nature" but at ensuring that his statements of scientific fact would, in the eyes of others in the field, constitute reality.[12] Such actions form, however, only one thread in a multitextured story.

Even as Liebig and Dumas sought power and influence to lead the development of a growing scientific field, and as Berzelius struggled to retain the influence he had once held, these three, as well as their less eminent colleagues, were all being swept along in a moving investigative stream that none of them could control. Each of them had to strive not only to direct but to adapt to changes that he could neither arrange nor anticipate. Liebig expressed the feeling of being caught up in currents moving of their own gathering momentum when he wrote Pelouze in 1834 that "with the impetus that organic chemistry has acquired, one truly does not know where it will stop, and where we shall be led. One becomes dizzy from so many discoveries." In another letter he wrote that "it is a true pleasure to work now in the most elegant part of chemistry, in organic chemistry, but it requires furious efforts to keep up and not to lose one's breath."[13] These were expressions of a sense not only that the field was rapidly growing and competitive but also that even someone in as powerful a position within the field as he was by then acquiring could not tell, let alone control, where it was all heading.

By 1839 Liebig believed that organic chemistry was moving in directions that not only escaped his control but were displeasing to him. So strong did his distaste for the latest trends become that in 1840, at the peak of his personal stature in the field, he abandoned it precipitously and moved into other areas. Nor were the others who had led the field through the previous decade content with its state. Berzelius watched helplessly while a younger generation lost confidence in his theories of composition that had earlier appeared to provide secure foundations for the field. Dumas still retained some momentum with his substitution theory; but neither he nor Liebig nor Berzelius was able to suppress the influence of the imaginative, far-reaching theories of organic composition and classification put forth by an upstart former assistant to Dumas named Auguste Laurent, even though each of them—for different reasons—regarded Laurent's views as disruptive to the stability of organic chemistry, and even though Laurent held none of the resources of authority that Liebig, Dumas, and Berzelius could still command.

One might well assert that the inability of the most powerful chemists in this field to channel its movement in the directions they preferred is a weak argument against a social constructionist interpretation of science, since powerful political leaders also frequently fail to achieve their agendas and can sometimes even be toppled by the people. It is equally cogent to suggest, however, that the progress of the field eluded their authority because they were collectively, as Hans Krebs liked to put it, "feeling their way into the unknown."[14] Social constructionists who do not subscribe to Latour and Woolgar's extreme view that "scientific activity is not 'about nature' " often describe nature as "underdetermined."[15] According to Shapin, "Reality seems capable of sustaining more than one account of it, depending upon the goals of those who engage in it"[16]; but the present case suggests that those who engage nature are also forced to adjust their goals in the face of realities they cannot themselves construct. The current debates about the construction of scientific knowledge leave room for the shaping effects both of "reality" and of social processes. The relative weights of these effects and the way they interact is something that we, as historians, should also not anticipate in advance of the historical cases we attempt to reconstruct.

Social constructionist models of scientific activity are naturally built around the public or semipublic faces of science: the formal and informal communication of information, organizations, open controversies, and the processes through which disagreements may be resolved. They deal mainly with what Gooding calls the engagements of investigators with other observers. To attain a balanced account of the construction of scientific knowledge, we must be equally attentive to the more private engagement of the experimenter with nature. In the case of the encounter between Liebig and Robiquet related in Liebig's letter to Wöhler, it really matters whether Robiquet did make an error of 5 atoms of hydrogen as Liebig claimed, or whether Liebig was simply arrogating the power to adjudicate what (as Harry Collins would state it)[17] should count as a good experiment. How can historians decide such matters? It is not good enough to look up in a modern textbook what is the accepted correct answer.[18] We must assess the situation in terms of the knowledge and practices of the 1830s. It may well be that by inspecting the analytical data that Robiquet and Liebig respectively presented on meconic acid and the other two related compounds, a historian deeply familiar with the chemistry of that period can ascertain which of them had attained, within the context of the time, the more accurate results. If not, then we must resist the temptation to interpret this episode either as a contest about whose statements are to be accepted as reality or as an instance in which nature yielded a better account to the more skilled experimenter.

I mentioned earlier that the published record of a scientist's research papers does not offer a direct view of the experimenter's engagement with nature. Geoffrey Cantor has recently viewed such texts as part of the rhetoric of scientific discourse and therefore as belonging to the public face of science.[19] I have viewed scientific writing rather as an integral phase in the process of creative investigation itself.[20] Nevertheless, there is a large gap between what such texts say and what Holton calls "science in the making as experienced in the scientist's own personal struggle."[21] My own reconstructions of investigative pathways from laboratory notebooks reveal critical differences between the course of the personal endeavor as it occurs in the laboratory and the representation of completed parcels of the work as it is reorganized to enter the public discourse. If we are to attain balanced pictures of the social and the investigative dimensions of the construction of scientific knowledge, then we must probe as deeply as our documents permit us behind the public arenas of science to the semiprivate worlds in which individual investigators, or investigative groups, confront most directly the objects of their study.

In the case of the participants in the events that I have discussed here, in particular for Justus Liebig, there do not appear to have survived laboratory notebooks of sufficient detail and coverage to reconstruct an unbroken research trail of the sort that I have been able to do with large segments of the work of Claude Bernard, Antoine Lavoisier, and Hans Krebs. Liebig's intense correspondence, however, especially his letters to Berzelius and to Wöhler, include reports of work in progress that are extensive enough to offer significant vistas onto his engagements with nature. It remains to be seen whether these documents, together with a group of rather unsystematic notebooks—recently added to the Liebigiana collection in the Bayerische Staatsbibliothek—in which he did jot down some of his experimental data, will suffice to allow penetration through the filters of his published research papers and the construction of a coherent, sustained account of his investigative progress. His letters to Wöhler do provide numerous vignettes that impart at least vivid impressions of Liebig's subjective feelings about the personal struggle of private science. My favorite so far, because it brings together the central themes with which I have wrestled in this paper, is the following passage, written in 1834:

> I become angry every day about the work I have [just] ended. It will excite neither attention nor interest, and will not be read, which is the same thing; and yet it is one of the best investigations that I have ever done. One sees that our satisfaction is dependent upon whether the objects we find, regardless of how much energy we spend on them, are incorporated now or in the future as a wheel or some kind of lever in the march [or machine?] of science. If this does not happen, then our work is worth nothing. One should draw from this the lesson to involve oneself only with matters that one can immediately apply. Still, I believed that I had the radical of uric acid . . . in my hands; but the devil has taken it back again. But you have now no feeling for chemical anxieties and for the sorrow of chemical disappointment.[22]

The final sentence in this passage alludes to Liebig's belief that Wöhler was too absorbed in his recent engagement to be married to concern himself with chemistry. The rest of the paragraph reveals beautifully the tensions felt by this passionate, moody, impatient man in the conflicting demands of the private and the social dimensions of scientific discovery. A discovery is worth making only if it is recognized as important by the scientific community. The text is an ideal case for the view recently argued by Augustine Brannigan that scientific discovery is a social process.[23] In the wake of the recent triumph that he and Wöhler had enjoyed in the reception of their discovery of the benzoyl radical, Liebig could be quite confident that the discovery of the uric acid radical would attract similar acclaim in the scientific world. For all his will and his experimental skill, however, he could not produce the result that would attain the recognition he desired. Nature appeared to Liebig in the guise of the devil, snatching from him the fruits of his labor.

Whether nature appears to us as the devil or through some more benevolent metaphor, we cannot, in our historical accounts of the construction of scientific knowledge, eliminate it or relegate it to a background phenomenon. We must confront the struggle of the scientist to wrestle from nature that which she or he may hope to present to the scientific world at large as fully and as seriously as we examine the social processes that surround, pervade, and motivate the investigative enterprise.

Notes

1. Peter Slezak, "Scientific Discovery by Computer as Empirical Refutation of the Strong Programme," *Social Studies of Science* 19 (1989): 563–600.

2. Larry Laudan, *Progress and Its Problems: Towards a Theory of Scientific Growth* (Berkeley: University of California Press, 1977), pp. 197–222, on 210, 219.

3. Steven Shapin, "History of Science and Its Sociological Reconstructions," *History of Science* 20 (1982): 157–211, on 158, 194.

4. Bruno Latour and Steve Woolgar, *Laboratory Life: The Construction of Scientific Facts,* 2d ed. (Princeton, N.J.: Princeton University Press, 1986), p. 280.

5. David Gooding, "'Magnetic Curves' and the Magnetic Field: Experimentation and Representation in the History of a Theory," in *The Uses of Experiment: Studies in the Natural Sciences,* ed. David Gooding, Trevor Pinch, and Simon Schaffer (Cambridge: Cambridge University Press, 1989), p. 192.

6. Thomas Nickles, "Justification and Experiment," ibid., pp. 299–333.

7. See in particular, Joseph S. Fruton, "Contrasts in Scientific Style: Emil Fischer and Franz Hofmeister: Their Research Groups and Their Theory of Protein Structure," *Proceedings of the American Philosophical Society* 129 (1985): 313–370; Joseph S. Fruton, "The Liebig Research Group—A Reappraisal," ibid., 132 (1988): 1–66; and Kathryn M. Olesko, *Physics as a Calling: Discipline and Practice in the Königsberg Seminar for Physics* (Ithaca, N.Y.: Cornell University Press, 1991).

8. Latour and Woolgar, *Laboratory Life,* p. 243.

9. Justus Liebig to Friedrich Wöhler, 20 Feb 1833, Liebigiana II A (58), Handschriftenabteilung, Bayerische Staatsbibliothek.

10. For a more extensive discussion of this and the following paragraphs, see Frederic L. Holmes, "Liebig, Justus von," in *Dictionary of Scientific Biography,* ed. Charles C. Gillispie (New York: Charles Scribner's Sons, 1970–80) 8: 333–338.
11. Aaron J. Ihde, *The Development of Modern Chemistry* (New York: Harper & Row, 1964), pp. 184–189.
12. Latour and Woolgar, *Laboratory Life,* pp. 105–149, 237. See also Bruno Latour, *Science in Action: How to Follow Scientists and Engineers Through Society* (Cambridge, Mass.: Harvard University Press, 1987), pp. 1–102.
13. Liebig to Jules Pelouze, April 16, 1834; June 14, 1834, Dos. Dumas, Archives of the Académie des Sciences, Paris.
14. Hans Krebs to F. L. Holmes, personal conversation.
15. According to Slezak, proponents of the "strong program" of social construction derive the term "under-determined" from the philosophical views of Willard Quine. See Slezak, "Scientific Discovery" (note 1), p. 587.
16. Shapin, "History of Science" (note 3), p. 194.
17. See H. M. Collins, *Changing Order: Replication and Induction in Scientific Practice* (London: Sage Publications, 1985), p. 89.
18. Shapin, "History of Science," p. 197, admonishes that to "lay a bet" on a modern textbook account is to "succumb to the very . . . 'presentism' that historians have so generally agreed to abominate."
19. Geoffrey Cantor, "The Rhetoric of Experiment," in Gooding et al., *The Uses of Experiment,* pp. 162–165.
20. Frederic L. Holmes, "Scientific Writing and Scientific Discovery," *Isis* 76 (1986): 220–235.
21. Gerald Holton, *The Scientific Imagination: Case Studies* (Cambridge: Cambridge University Press, 1978), p. 5.
22. Liebig to Wöhler, 8 Mar 1834, Liebigiana II A (69) Handschriftenabteilung, Bayerische Staatsbibliothek.
23. Augustine Brannigan, *The Social Basis of Scientific Discoveries* (Cambridge: Cambridge University Press, 1981).

II
Production

The Evolution of the Chemical Industry: A Technological Perspective

John Kenly Smith

In 1915 Arthur D. Little suggested structuring the emerging discipline of chemical engineering around the concept of unit operations. Previously the subject had been taught as if it were a branch of natural history. Production processes for chemicals were studied in detail and noted for their unique and unusual features. What the professors of chemical engineering had overlooked, Little pointed out, was that all chemical processes consisted of different combinations of a relatively small number of operations, such as distilling, cooling, heating, crystallizing, drying, and so on. By studying these "unit operations," a chemical engineer could apply his knowledge to the manufacture of any chemical, not just to the particular ones with which he was familiar. Little's observation put the shaky discipline of chemical engineering on a firm foundation.[1]

The writing of the history of the chemical industry is at the stage that chemical engineering was in when Little proposed his organizing concept. Most histories are organized around products or firms. The best ones have focused on the evolution of specific chemical technologies and the economic aspects of chemicals and chemical production. What is lacking is a larger framework within which to put the industry and its products. This lack of a framework has made chemicals intimidating for most professional historians, who subsequently look elsewhere for research topics. It is therefore not surprising that most of the work on the history of industrial chemical technology is being done by chemists and engineers. Their products are useful and insightful but usually not easily accessible to non-technically trained people. One encouraging development is the increasing interest by chemical companies in their histories. With greater access to corporate-held materials, there is hope that historians can develop a deeper understanding of the history of the industry. Because the boundaries of the chemical industry are somewhat arbitrary, however, perhaps the goal of producing a conceptually satisfactory history may never be realized. In this paper I will suggest some ways in which to comprehend the industry and point out areas in which we need to know more.

The central aspect of chemical technology has been economies of *scale*. A chemical plant cannot be simply a larger version of laboratory apparatus.[2] The reasons for this are many and involve changing property ratios and limitations of structural materials. For example, temperature and pressure conditions that can be easily contained in small vessels can become critical factors in larger scale operations. Another limiting condition is the cost of raw materials. Until the eighteenth century, chemical producers had only a few widely available substances, such as salt, around which to build chemical processes. At this time there was no chemical industry, I would argue, because of the small scale and high cost of manufacture. The modern chemical industry began in the second half of the eighteenth century with the lead chamber process for making sulfuric acid.[3] Relatively cheap sulfuric acid made possible other chemical processes, such as the Le Blanc process for making alkali.

Another important concept for understanding the chemical industry is *scope*, a term I borrow from Alfred D. Chandler, Jr.[4] In the context of the chemical industry, scope refers to the fact that as the number of industrially available chemicals increases, the number of potential products that can be produced increases exponentially. In other words, the more chemicals you make, the more you then can make. The dyestuffs industry is perhaps the best example of the importance of scope. Because dye molecules are complex, they are synthesized through multistep processes employing numerous intermediate compounds. The manufacture of one dye requires an enormous investment that a single product could not possibly justify. Therefore a dye manufacturer has to make hundreds or thousands of dyes to provide a large enough business to pay an adequate return on the enormous investment in the infrastructure of production facilities for intermediates.[5]

The scope of chemical production determines what can be produced, not only for one firm but for the entire industry. In the past, chemical companies have been among each other's best customers because different companies specialized in different chemicals. In the pre-World War II era, because such companies did not compete directly, the base of the entire industry was broadened by making a larger number of chemicals available for use as intermediates.[6]

To achieve either scale or scope in production requires a manufacturing process, which has been the heart of chemical technology and the chemical industry. Yet our knowledge of the history of process technology is rudimentary at best. A major exception to this is John Graham Smith's *The Origins and Development of the Heavy Chemical Industry in France,* which details the growing technical sophistication of sulfuric acid plants as they evolved during the nineteenth century. Sulfuric acid production involved chemical reactions (one of which involved homogeneous catalysis), absorption, and distillation. A late-nineteenth-century plant operated along modern principles even though they had not been explicitly articulated.

Central to chemical processes are, of course, chemical reactions. The history of the development of techniques for carrying out reactions on a large scale has only been explored for a few specific cases. In large-scale equipment "local" conditions affect the result of the reaction. Catalysts are usually necessary to make a reaction go. But the history of this very important and nonscientific technology has not received adequate attention,[7] even though the catalyzed chemical reaction has become the heart of the modern chemical process. For example, it has been incorporated in continuous flow processes. Engineering periodicals and textbooks of the 1930s refer to the transition from batch to continuous processes as an important development.[8] This raises questions about where these changes originated and how widespread was their adoption. There are also collateral considerations, such as the importance of instrumentation and process control for continuous processing. One would suspect that the emergent discipline of chemical engineering might be the driving force for process innovation; however, this remains to be proven. In the 1920s chemical engineers—mainly at MIT—began to develop quantitative tools for analyzing entire processes and the operation of individual units that when connected together make a process. Exactly how chemical engineering science and actual process engineering are related needs to be examined more closely. I suspect that the MIT-Exxon connections might be a good place to start.[9]

After process engineering became well established, there remains the question of how it spread to other, related industries such as food processing and brewing. Was it carried into these industries by newly minted engineers, or was some other mechanism involved? Overall, process engineering has been an important technological development that has affected many industries, yet much work remains to be done before we understand its history.

Although technology has been one side of the chemical industry, the nature and extent of markets has also shaped its development. Much of the demand for chemicals derives from industries that are closer to the ultimate consumer. For its first century, chemical producers sold the bulk of their output to the textile-processing industry. Today, virtually all modern industries consume large quantities of chemicals. For much of the twentieth century, chemical manufacturers grew twice as fast as overall industrial output.[10] This high rate of growth has been accomplished in part by companies developing new materials that have replaced natural ones. This has occurred at several different levels. One approach has been to replace a naturally derived material with a synthesized version. Before the LeBlanc process, alkali was made by burning kelp or wood to create alkaline residues. Since the early nineteenth century, synthetic producers have promoted their products on the basis of consistent quality and availability, as an advantage over natural products, which are influenced by weather and other variables.

In the second half of the nineteenth century a few chemical manufacturers began to make more complex materials that did not directly duplicate natural ones but could perform roughly the same function. The textile industry is the classic case of chemical producers revolutionizing a basic industry through the development of synthetics. Similarly in plastics, hundreds of small fabricators became virtually satellite operations of the chemical companies, depending on them for raw materials, fabricating techniques, product development, and advertising. In instances such as these, chemical firms created growth opportunities for themselves. Their being manufacturers of unfinished goods did not relegate them to waiting for increased demand to trickle down from their customers. After World War II the rapid expansion of polymer technology gave the chemical industry the power to transform a number of downstream industries.[11] Research and development gave the chemical companies the tools to accomplish this task.

In its beginning stages virtually all chemical innovation begins as a process of substitution. There is no point in developing a material for a function that does not exist. Once established in the marketplace, a new material may find novel applications based on its peculiar properties. The nature of chemical technology affects the research and development process in several ways. First, the relationship between the structure of a given chemical and its physical properties was pretty much a mystery until recent times. This explains the high degree of serendipity in many chemical discoveries. The scenario is usually as follows. A chemist identifies a need and sets out to discover a chemical to fulfill it. Whatever is found, it rarely satisfies the initial need. This is the critical point in the process. The chemist can either give up or try to develop other uses for the substance. After potential uses have been identified, questions of scale and scope take over. Can the material be made at a cost less than what the customer is willing to pay for it? Because the cost of the material is a strong function of the scale of manufacture, the size of a profitable plant can be calculated based on the expected price. If a profitable plant has to be a very large one, there will be considerable risk involved in the project because the market might reject the product or the first-generation technology might quickly become obsolete. A preferred strategy is to identify premium price applications for the new material so that a smaller plant can be built initially. Later, as experience with the manufacturing and application of the material increases, larger plants can be built and costs reduced. This opens new markets.[12] In general, the path from the laboratory to the marketplace has never been a straightforward one. The process has required considerable judgment at every step.

Looming above the question of research and innovation is the relationship between chemistry and the chemical industry. At this point, about the best that we can hope to do is to avoid the simplistic directional arrow that points from science to technology. A better metaphor is that of a feedback loop. Technology can evolve independent of science, but scientific investigations can provide new insights that open new avenues for technological development. The examples of Du Pont polymer research and the development of the synthetic dyestuffs industry show the interactive nature of the science-technology relationship and that the long-term continuity might come from technology, not from science. In other words, technology poses the questions that science addresses.[13]

In the remainder of this essay I attempt to apply the considerations discussed so far to the historical development of the chemical industry.

Although people have manufactured chemicals and used chemical processes since the beginning of civilization, the creation of a chemical industry awaited the Industrial Revolution. Before this era there was inadequate technology for large-scale production of chemicals and insufficient markets to justify investment in large plants. The demand for alkalies for textile processing led to the establishment of the Le Blanc soda process in Great Britain in the 1810s. This chemically primitive process, developed in the 1790s in France, reacted salt with sulfuric acid to produce sodium sulfate and hydrochloric acid gas—which escaped, laying waste to the surrounding countryside. The second step entailed mixing sodium sulfate with limestone and heating them to produce sodium carbonate (soda) and an equal quantity of worthless calcium sulfide. Its most expensive ingredient, sulfur, was completely wasted. The reason that the Le Blanc process was successful was that technology had been developed to manufacture sulfuric acid on a significant scale.[14]

The origins of modern process technology lie in the lead-chamber sulfuric acid process. The principal steps in this process were burning sulfur in air to form sulfur dioxide, and the reaction of this gas with air to form sulfur trioxide, which absorbs in water to form sulfuric acid. The major breakthroughs that led to this process were the use of large lead-lined chambers for the slow reaction of sulfur dioxide and air, and the discovery that small quantities of nitric oxide catalyzed this reaction. By 1850 French and British sulfuric acid plants had reached a high level of sophistication and incorporated most of the design principles of modern engineering.[15] This was the result of decades of incremental improvements, because the conceptual basis for these processes would not be elucidated for decades to come. These techniques did not diffuse into other processes, because no other chemicals were produced on such a large scale. Further development of the chemical industry in the nineteenth century came from directions other than bulk chemicals.

A separate industry grew up in the second half of the century around organic chemicals that had been found to be valuable dyestuffs. In 1856 William H. Perkin, a young English chemist, discovered a mauve dye made from aniline, one of the many compounds that could be produced from heating coal to make coke or gas for lighting streets. Perkin had hoped to synthesize quinine and help the empire expand into the tropics. His discovery launched the synthetic dye industry, which soon had the capability to produce a rainbow of colors in innumerable shades. The locus of this industry shifted to Germany, which had more technically trained people and exhibited a willingness to organize research on systematic lines. Dye research consisted of the synthesis and evaluation of literally thousands of compounds. There were no general principles from which one could predict whether a specific chemical would make a good dye. Perkin and other English chemists did not build the organizations that could manage the resources needed to do this kind of work, so the world's leading textile-producing country lost its dye industry.[16]

Synthetic dyestuffs contributed to the development of the chemical industry primarily through the synthesis techniques that chemists worked out to make a wide variety of compounds. From a manufacturing standpoint, dyestuffs relied more on scope than on scale. Dyestuffs and their intermediates were manufactured in batches, using technology that was more similar to laboratory equipment than to large, continuously operating plants. The sulfuric acid plants relied on scale; the dye industry depended primarily on scope.

A critically important development for the chemical industry in the long run was the nitration of the natural organic compounds glycerin and cellulose by European chemists in the mid-nineteenth century. Dynamite, nitroglycerin absorbed in diatomaceous earth, was invented by Alfred Nobel in the 1860s and soon became an important industrial product worldwide.[17] Du Pont became a chemical manufacturer in 1880 when it began to make dynamite. The new product required precise control of the chemical reaction of glycerin with a mixture of nitric and sulfuric acids, and the delicate separation of nitroglycerin from the acids. Dynamite replaced black powder in mining and construction, and was a typical chemical-based substitution.[18] Chemically similar nitrocellulose would eventually have a revolutionary impact on the industry.

In 1870 Albany tinkerer John Wesley Hyatt formed a solid plastic from a mixture of nitrocellulose and camphor. According to tradition, Hyatt was looking for a substitute for ivory billiard balls. In the event the new material was not hard enough for billiard balls, so Hyatt had to find other uses for the plastic he called *celluloid*. Proceeding cautiously, Hyatt developed his celluloid to look like ivory and tortoise shell. From this material he made brushes, hair combs, and numerous other articles. Celluloid was the first chemical product that could be molded by heat and pressure into a number of different shapes. It briefly enjoyed some popularity as a fiber. In the 1890s there were stories of men accidentally igniting women's ball gowns with their cigars. Because nitrocellulose burns with almost explosive intensity, the market quickly shifted over to rayon, a regenerated cellulose fiber, when it became available. Celluloid found more enduring applications in photography. It made roll film possible and increased the popularity of George Eastman's Kodak cameras. After 1900, celluloid became a household word when it was used as the base for the first moving picture film.[19] Following the example of celluloid, chemists began to look for other plastic materials that could be processed into a wide variety of goods.[20] It would take decades, however, to develop the tools needed to create new substances. The importance of celluloid as a model for chemical innovation is apparent in retrospect, but at the time other developments received much more attention.

The most dynamic sector of the chemical industry in the late nineteenth century was probably electrochemicals, which began to be manufactured on a large scale after the invention of the dynamo in 1870. Much of the early activity centered on finding a low-cost method of reducing aluminum ore to metal. By the time Charles Martin Hall in the U.S. and Paul-Louis-Toussaint Héroult in France contemporaneously discovered such a process in 1886, others had begun to see electrochemicals as a means to manufacture many chemicals and metals.[21] Three of the major American chemical companies, Dow, Union Carbide, and American Cyanamid, had their origins in electrochemical technology. In the long run this sector made available chemicals that could not be made economically by standard process technology. These chemicals included chlorine gas, sodium metal, and acetylene.[22] This latter compound was an important organic intermediate before the advent of petrochemicals. Electrochemicals broadened the scope of chemical manufacture considerably.

At the turn of the twentieth century, there existed not one chemical industry but a number of loosely related industries. But there were some signs of technological convergence. Process technology had become much more sophisticated because of the advancement of chemical science, the increasing use of catalysts, and the accumulation of skills by practicing chemical engineers. The development of physical chemistry and thermodynamics gave chemists concepts to understand the nature of chemical reactions. These relationships showed that many reactions that chemists had failed to make go were theoretically possible. In the 1890s, Wilhelm Ostwald in Germany explained that catalysts were agents that made recalcitrant reactions occur.[23] Through the use of these new techniques, the number of chemicals produced industrially increased rapidly after 1900, and engineers in Europe and the United States began to specialize in chemical technology. But it would be German chemists and engineers who would bring all the pieces together.

The development of the Haber-Bosch ammonia process between 1906 and 1912 was a technological tour de force that became a model for future chemical processes. The search for methods to fix nitrogen had begun in the late 1890s because the world's supply of fixed nitrogen for fertilizer and explosives came from the deserts of Chile. Of course, air was 80 percent nitrogen, but no one had found a way to break the strong bonds between the two nitrogen atoms, which rendered the molecule inert under ordinary conditions. The need to "fix" nitrogen synthetically was emphasized in 1898 by Sir Williams Crookes, president of the British Association for the Advancement of Science, who predicted the exhaustion of the Chilean nitrate beds in twenty or thirty years, leading to food shortages and widespread starvation. If chemists were not motivated by humanitarian concerns, nationalism provided another incentive. As the European arms race accelerated, no government could tolerate being dependent upon a strategic material obtainable only from a source thousands of miles away. In Germany, Fritz Haber, a professor at the Karlsruhe Engineering College, discovered a method to synthesize ammonia from nitrogen and hydrogen at 500 degrees Celsius and 200 atmospheres pressure using osmium and uranium catalysts. The process was inelegant and impractical, but it was a way to fix nitrogen. In 1909 Haber licensed his process to the BASF Company of Ludwigshafen. During the next few years a team of engineers led by Carl Bosch developed a commercially viable process. To come up with a cheaper and more efficient catalyst, BASF chemists tested 2,500 substances and ran 6,500 experiments. Even with the resulting improvement, however, the extreme temperature and pressure the process required could not be modified. Building large scale equipment to operate under such harsh conditions had never been attempted before, and some of the initial reactors were crafted out of Krupp cannons. In the course of developing the process, Bosch empirically discovered the embrittling effect of hydrogen on steel, which was vividly demonstrated by the alarming tendency of the reactors to blow up at frequent intervals. Fortunately for the German military, which had planned only for a short war and had not stockpiled large quantities of Chilean nitrates, the new ammonia process had been commercialized by 1914.[24]

While the ammonia process was the most dramatic example of the development of chemical technology, the first decades of the twentieth century witnessed rapid growth in the number of industrially produced chemicals. Not only did established companies broaden their product lines, but many entrepreneurs in the United States and Europe started small chemical companies. Most of the technology originated in Europe. This proliferation of chemical companies has not received adequate attention from historians.[25] The proliferation of chemicals is slightly better documented. Du Pont, for example, had been a gunpowder manufacturer, which required little chemical knowledge or technology. The shift to dynamite and nitrocellulose explosives demanded a higher level of chemical sophistication. After 1900 Du Pont began to make new explosives, such as TNT, which were purely chemical products. Thus, Du Pont became a chemical manufacturer through transformation of the traditional explosives industry.[26] At the same time, chemical companies began to develop products for other chemical and material processing industries.

The case of Rohm and Haas is probably typical of many small American chemical companies in the early twentieth century. Around 1906, German chemist Otto Rohm began to study leather tanning, looking for a way to make his fortune. A critical, yet particularly noxious, step in the process was the bating or softening of the hides with fermented dog dung. Because of the varying strength of this reagent, there was considerable variation in the quality of the product. Rohm discovered that enzymes extracted from animal pancreases did a better job. Soon he began selling this product to leather producers. An associate, Otto Haas, set up an American subsidiary several years later. In subsequent years, the firm of Rohm and Haas diversified into a wide range of chemicals for use in leather and textile processing. By supplying chemicals to these industries, chemical companies, such as Rohm and Haas, began to exert influence over their customers' businesses, a trend that would accelerate in the 1920s and 1930s.[27] Before this could happen, the American industry needed to increase its capabilities and resources.

World War I gave the young American chemical industry an enormous boost by eliminating German competitors in organics and by creating a high level of demand for a wide range of chemicals. The wartime experience increased the financial resources and the organizational and technological capabilities of the larger chemical companies. Overall, the increase in demand for chemical products was probably a more important factor in the growth of the industry than the cessation of trade with Germany.[28]

In the postwar era, chemical companies began to diversify their operations by moving into related technologies, product lines, or both. Du Pont actually started diversifying before the war, when government competition and harassment had threatened the company's nitrocellulose-based smokeless powder business. In this era, dynamite accounted for 80 percent of Du Pont's business, so the decision to diversify was not a matter of life or death for the company. Instead it reflected a determination to use know-how that the company had developed during fifteen years of smokeless powder research and development. To utilize its capabilities, Du Pont purchased the Fabrikoid Company (1910), maker of nitrocellulose-coated fabrics used as a leather substitute, and the Arlington Company (1915), the largest maker of celluloid plastics. Starting from its base in nitrocellulose technology, Du Pont moved into two entirely new product lines. Although these acquisitions were not great financial successes, they pushed Du Pont into making *materials,* not just chemicals. These first moves set the course of the company's evolution for many decades to come.[29]

The experience of Union Carbide was similar to that of Du Pont. The original firm had been organized around electrochemical technology used to make calcium carbide, metal alloys, and graphite. This latter material was used extensively in arc and early incandescent lighting systems. When the tungsten filament replaced graphite in the 1910s, Carbide had to look for new markets. In 1917, Union Carbide merged with the Prest-O-Light and Linde Air Products companies in order to promote oxyacetylene welding. Prest-O-Light had been formed in 1896 to make acetylene lamps for automobiles. The gas was generated by dripping water on calcium carbide. A few years later the company obtained the rights to a French patent involving the safe storage of acetylene in cylinders. Because acetylene decomposes readily, liberating enough energy to burst cylinders, it had not been used for lamps. The third company, Linde, made industrial gases by liquefying and distilling air. The process was developed in Germany at the turn of the century and is still in use today. Linde contributed oxygen to the oxyacetylene welding package and at the same time put Union Carbide in the gas business.[30]

A third major company, Allied Chemical and Dye Corporation, was formed in 1920 through the merger of five companies that produced heavy chemicals, coal tar organics such as benzene, and dyestuffs. American Cyanamid, which saw its nitrogen-fixing process superseded by the Haber-Bosch process, also diversified by acquisition but ran a distant fourth in size.[31] In 1929 Du Pont, Union Carbide, and Allied were the only three chemical companies among the top 100 U.S. corporations in assets. These three would hold their positions until Monsanto and Dow surpassed Allied in the 1950s. With total assets of $500 million, Du Pont ranked ninth overall and had 60 percent more assets than its competitors. Du Pont gained this advantage by pursuing a higher degree of diversification than Union Carbide or Allied.[32]

Beginning in 1916 and ending about 1930, Du Pont established itself in ten different businesses many of which had nothing to do with the company's technological base. The company undertook this ambitious program because it had large financial resources and extreme confidence in its own technical capabilities. Du Pont management believed that its chemists and engineers could transform traditional industries and create new ones. To accomplish this goal, Du Pont specialized in making strategic acquisitions of companies and nascent technologies. Then it went to work improving the technology and developing markets for the new products. Commercialization was Du Pont's strength. In this era, company management believed that invention was too unpredictable to be counted on as the source of new businesses.[33]

Du Pont could manage highly diversified product lines because of the new organizational structure it had adopted in 1921. At that time the company had divided its businesses into five operating departments. In the next decade Du Pont added six additional departments; three were built around licensed technology and three were acquired companies. The departmental structure that Du Pont had in place in 1930 survived virtually intact until 1970, with only three changes occurring during that period: a merger of two departments, and the formation of two new departments split off from old ones. Paralleling this organizational stability was a stability of product lines. Du Pont entered no major new businesses between 1930 and 1970, although many product lines were transformed in that interval.[34]

While Du Pont was putting together its chemicals empire, I. G. Farben in Germany and Imperial Chemical Industries, Ltd. (ICI) in Great Britain formed similar aggregations. In many aspects these companies were conglomerates because many of their component businesses had little in common with each other. For example, Du Pont's fields included heavy chemicals, paint, rayon fibers, and photographic film. It is clear that these chemical giants were not formed around a central technology. Their diversity was their strength. They produced more chemicals than other companies and consequently had the ability to combine these chemicals into a far broader range of products. Lacking the resources of the big three, other companies frequently joined together to produce new chemicals. Du Pont, I. G. Farben, and ICI considered entering any business having manufacturing operations that could be characterized as chemical processing, even though this encompassed a wide range of operations.[35] They had faith that over time a technological convergence would bring their businesses together. This was one of the goals of the research laboratory.

In the highly diversified firms, a central research laboratory had a difficult time picking research areas of broad interest. For this reason, ICI lacked one until 1939, and until the late 1920s Du Pont's central laboratory struggled to find a role for itself. Du Pont's central research director, Charles M. A. Stine, tried to identify the company's basic technologies. He picked out chemical engineering, catalysts, polymers, and a few other areas as fields in which Du Pont had big stakes and that had been neglected by academia. After receiving approval from the Executive Committee, Stine hired researchers to investigate these fields. In chemical engineering, Du Pont became one of the leading research institutions, academic or otherwise. Du Pont failed to bring catalysts out of the realm of art and into that of science. With polymers, Du Pont researchers made important contributions to the science and technology that transformed the chemical industry in subsequent decades.[36]

Polymers were at the cutting edge of science in the 1920s. Previously, considerable scientific authority denied their existence. However, a number of German chemists, particularly Hermann Staudinger, began to argue that long-chain molecules existed. This view received support from measurements of natural molecules such as cellulose and hemoglobin, which showed that they had enormous molecular weights. Academic chemists in America ignored this field, in part because it fell into the cracks between standard chemical disciplines. Ignoring the debate over the existence of polymers, industrial chemists had discovered methods to synthesize them for use as plastics and resins. In 1909 independent inventor Leo Baekeland discovered Bakelite, as he would call it, by heating phenol and formaldehyde under pressure until a solid resin had formed. He was not sure what had happened at the molecular level, but he had a product he could sell as an electrical insulator and for other uses. A few other similar products were developed by analogous reactions in following years.[37]

Du Pont's interest in polymers was primarily in cellulose, which formed the basis for the company's first major discovery, Duco lacquers, and its two fastest growing products, rayon fibers and cellophane films. To investigate the phenomenon of polymerization, Stine hired a thirty-one-year-old Harvard chemistry instructor, Wallace H. Carothers, and encouraged him to undertake a polymer research program. In his nine years at Du Pont, not only did Carothers help establish a firmer base for polymer science, but along the way his associates discovered neoprene synthetic rubber (1930) and nylon (1934).[38]

With new polymerization techniques developed by Du Pont and I. G. Farben, chemists in the 1930s attempted to polymerize all kinds of compounds. Chemists had been making polymers accidentally for years, but the appearance of a sticky liquid or solid was an inexplicable result and signaled a failed experiment to be relegated to the trash bin. Now chemists had the tools to interpret these results. Chemists at ICI and Du Pont discovered polyethylene and Teflon respectively, while doing experiments designed for other purposes. Systematic work also yielded results in the form of new synthetic rubbers, acrylic resins, and polystyrene. But these materials progressed slowly because companies lacked raw materials, large-scale polymerization techniques, and knowledge about potential markets. Only highly diversified companies had the capabilities to commercialize these new polymers.[39]

With the possible exception of I. G. Farben, no company other than Du Pont could have commercialized nylon in the 1930s. In retrospect, it appears remarkable how well Du Pont's expertise matched up with the needs of the nylon development program. The intermediate compounds were laboratory curiosities. Du Pont's long and trying venture into high-pressure technology finally paid off when it proved adequate to make the necessary compounds. The company's pioneering position in polymers and chemical engineering provided the expertise for making the polymer and spinning it into fibers. And Du Pont's rayon business had provided the company with knowledge of textiles and a working relationship with that industry. Du Pont did not make any mistakes commercializing nylon, and it was a highly successful product immediately after its introduction in May 1940. Synthetics were no longer cheap substitutes for natural materials. Because of their durability, nylon stockings commanded higher prices than silk ones.[40] Other companies had made polymers too and wanted to emulate Du Pont, but they lacked the leader's depth and breadth of capabilities. The war changed this situation completely, sweeping away barriers to innovation and replacing them with incentives.

The government's administration of the American chemical industry during World War II led to technological convergence centered around polymers made from petrochemicals. During the war the production of synthetics boomed as they were found useful as substitutes for aluminum, brass, and other strategic materials. With an assured market, chemical companies agreed to expand their production. In some cases where companies doubted that large postwar markets would exist for a particular product, the government provided funds to build the plants.[41] For example, Du Pont had worked hard to develop markets for its specialty rubber, neoprene, in the 1930s and had committed itself to a ten-thousand-ton-per-year plant in 1940. When the government requested that Du Pont increase its capacity to forty thousand tons per year, the company officials required that the government pay for the new construction and assume ownership of the plant. Du Pont was not sure how large the postwar market would be for neoprene and did not want to be stuck with large overcapacity.[42]

With other plastics the rewards of expansion clearly outweighed the risks. Both Du Pont and Rohm and Haas had developed acrylic resins, Lucite and Plexiglas, respectively, in the 1930s, but markets grew slowly. Prewar use as a denture material fulfilled a longstanding human need but did not consume large quantities of plastic. Wartime airplane construction did, however, increase production tenfold. Even more dramatic was the growth of vinyl resins, which had been invented in 1912, from 5 to 220 million pounds per year. In addition to pushing these existing materials into full commercial status, the war initiated the development of two exotic polymers for high-technology applications.[43]

Polyethylene had been discovered by ICI in the early 1930s, but attempts to tame the unruly reaction failed. ICI assigned it a low priority because no one then had any idea what to use it for. During the early stages of the war, the British discovered that polyethylene was needed to serve critical functions in radar and began to make the polymer on a small scale. Periodic explosions were accepted as part of the business. Soon two United States companies entered the field. Du Pont had acquired patents from ICI, and Union Carbide decided to ignore the patents and enter the business anyway. Carbide, which already was the leading producer of ethylene gas, captured most of the market for the new polymer. Du Pont had to spread its resources over a number of wartime projects, the biggest of which was the design, construction, and operation of the Hanford, Washington, plutonium works. A much smaller effort centered on developing Teflon, a polymer that was so unusual that standard manufacturing and processing technology could not be used. Du Pont made major strides in understanding this bizarre material and created enough polymer for critical applications in proximity fuses and the separation of U_{235} from U_{238} for the atomic bomb.[44]

The chemical equivalent of the Manhattan project was the wartime creation of the American synthetic rubber industry. I. G. Farben had developed the technology in the 1930s and commercialized it through the support of the government. In America there was no economic incentive for developing a general-purpose synthetic rubber. Until the Japanese cut off the supply of natural rubber, it had been very cheap throughout the 1930s. In addition to economic disincentives, large technological obstacles stood in the way of synthetic rubber. During the war, the combined efforts of rubber, chemical, and oil companies were necessary to accomplish the task. Remarkably, U.S. synthetic rubber production went from almost nothing to eight hundred thousand tons in two years. This required enormous quantities of butadiene, a chemical that had not been commercially manufactured in the United States before the war.[45]

The war effort put the oil companies in the petrochemicals business not only as suppliers of butadiene for synthetic rubber but also as makers of toluene for TNT and a host of other chemicals. In a related development, the use of catalytic cracking of crude oil increased to supply high-octane aviation gasoline. This new cracking technology also proved useful in the production of chemicals from oil. After the war, the oil companies would develop sophisticated techniques for making a wide range of compounds from petroleum. These would be the raw materials for the postwar boom in polymer output.[46]

World War II had set the stage for the expansion of the chemical industry. It now became possible for everyone to do what Du Pont had done with nylon before the war. Polymer know-how had diffused throughout the industry, and polymer products were now firmly established in the economy. In addition, companies could buy their intermediates from oil companies rather than manufacture them. In the postwar era, chemical companies developed many new polymers that in many applications have replaced rubber, natural fibers, paper, glass, metal, and wood. Chemical companies transformed the businesses of their customers by making them dependent upon polymer-based materials. The golden age of the chemical industry had arrived.[47]

In 1950 *Fortune* magazine, in an article entitled "The Chemical Century," asserted that chemicals had become America's premier industry, eclipsing automobiles. In the 1940s the industry had doubled its investment and tripled its sales. Dow, Monsanto, and American Cyanamid joined the earlier big three in the United States as major powers. The newcomers made big strides in the fields of plastics and fibers. Allied had lost ground by sticking to its traditional product lines while the other companies were embracing polymers made from petrochemicals.[48]

In the 1950s the chemical industry continued on the trajectory it had set in the previous decade. The old polymers, led by nylon and polyethylene, grew tremendously. The latter became the first billion-pound-per-year polymer by 1960. In addition, new important polymers were commercialized; polyester and acrylic fibers, polyurethane foams and plastics, linear polyethylene and polypropylene, and several engineering plastics that were used in structural applications. Polymers became the engine of growth for the entire industry.[49] By the late 1960s polymers probably accounted for over half the volume of the chemical industry. This growth and prosperity did not go unnoticed.

All kinds of companies invaded the chemical industry in the 1950s. They could do this because technological barriers to entry had become very low. Technology and raw materials were widely available, and markets for polymers were well established. It was relatively easy to join the game. In his book *Petrochemicals: The Rise of an Industry,* Peter Spitz shows how the construction and engineering companies, such as Stone and Webster, Kellogg, and Lummus, made it possible to buy turnkey petrochemical plants. The oil companies led by Phillips Petroleum integrated forward into plastics; Rexall Drug teamed up with El Paso Natural Gas to begin plastics production. The rubber companies, now manufacturers of polymers themselves, diversified into related plastics. And even W. R. Grace, a steamship company, radically diversified into chemicals.[50] These companies competed by adding new capacity for old polymers and introducing new ones. By 1960 many polymers had become commodities. Polypropylene was born a commodity because a number of producers installed large plants before markets had been developed. Supply consistently ran ahead of demand.

This intense competition resulted in part from the inability of research to lead to impregnable patent positions. Because everybody was doing polymer research, the chances of coming up with something completely new diminished over time. When a patented new product was introduced, competitors soon developed different products to compete for the same market. This situation led *Fortune* in 1962 to declare the entire industry mature.[51] The earlier 1950 article had hinted about the possibility that the industry might find itself in this position some day, but did not expect it to happen so fast.

In the 1960s Du Pont and the other companies sought to move into less competitive areas. One response was to integrate forward into consumer products, which had the drawback of bringing the companies into competition with their customers. Some companies such as Dow had some limited success with this strategy. Du Pont preferred to stick closer to its traditional strengths by developing a new generation of sophisticated products for high-priced specialty markets. This strategy was in line with Du Pont's belief that it had significant technological advantages over its competitors. Other companies also went the specialty route hoping that their long experience working with fabricators would give them advantages over newcomers in finding market niches. A third approach was diversification. To some extent this has been going on since the end of the war in the fields of pharmaceuticals and agricultural chemicals. Chemical companies later enviously watched the rapid growth of electronics and sought means to participate in this industry.[52]

In 1960, when fears of impending maturity were building, Du Pont launched a major preemptive program employing all three strategies discussed above. The company's Executive Committee, however, was unwilling to direct this effort leaving it largely to the research directors in the industrial departments, which were still organized along the lines of the prewar industry. Du Pont developed a flood of new products over the next several years, but most of them were in the company's traditional markets. Some such as Corfam leather substitute were disasters; others became important products that opened up a few new lines of business. In total Du Pont spent about $2 billion on new products but did not achieve the goal of reducing the company's dependence on plastics and fibers. By the late 1960s this had become a critical corporate concern.[53]

The general prosperity of the first half of the 1960s masked the increasing competitiveness of the industry. In the recession at the end of the decade, prices fell fast and far. Before any new strategies could be put in place, environmental legislation and the Arab Oil Embargo created a sense of helplessness among the industry leaders. At Du Pont, the 1970s was largely an era of reacting to external threats. Other companies probably suffered similar fates. Not until the 1980s would companies focus attention on strategies for the future. As the older businesses matured, American chemical companies attempted to shift their resources into faster growing and less crowded areas.[54]

When American firms have wanted to get out of particular chemical businesses, German and Swiss companies have frequently acquired the unwanted assets. Consequently, the American market has witnessed the growing presence of these overseas corporations. Before the war, Du Pont, ICI, and I. G. Farben generally agreed to respect each other's territory. Following 1945, the U.S. government forced the termination of prewar agreements and split up the giant German company into three smaller ones, Bayer, BASF, and Hoechst. Serious international competition for chemical markets was delayed until the 1950s when the European companies had rebuilt at home and the formation of the Common Market encouraged American firms to build plants in Europe. Because the United States accounted for one third of world chemical consumption, the American market became the major battleground.[55]

The resurgence of the German chemical firms has been remarkable. By 1980, all three firms had exceeded Du Pont in chemical sales volume. These companies have been more successful than their American counterparts because they pursued a more broad-based strategy. These foreign firms participated in the petrochemical-polymer bonanza but not at the expense of older product lines and continued diversification. After the Germans came back into the dye business in the 1950s, Du Pont's twenty-year-old technology needed updating. With more attractive investment opportunities in fibers, Du Pont did not want to put a lot of money into less profitable dyes. When the business did become unprofitable, Du Pont got out of it in 1980. In addition to holding their ground in traditional businesses, the German companies put more emphasis on diversification into agricultural chemicals and pharmaceuticals. The Germans have been able to do all these things because of long-established organizational capabilities, close cooperation between firms—increasing the scope of their enterprises—and availability of capital. High rates of depreciation and a high degree of leveraging gave the German companies the cash flow that they needed to expand across a broad front.[56]

Before World War II, the American chemical industry consisted of a diverse collection of technologies that served a broad spectrum of markets. The rapid development of petrochemicals and polymers in the war and immediate postwar years attracted the attention of the entire industry as well as outsiders. The investment of enormous sums in developing new polymers and dramatically expanding the production and marketing of older ones led the chemical companies to neglect older businesses, such as dyes and other specialty chemicals, and to limit the extent of diversification. When the polymer-based industry began to mature, chemical companies intensified efforts to move into new businesses, while shedding older commodity products. Firms such as Du Pont are taking on a new look. That company's motto is now "Better Things for Better Living"; the phrase "Through Chemistry" has been dropped. The great American chemical company might be disappearing, leaving the traditional field to the Germans, who rely on the scope of their varied chemical enterprises to give them long-enduring competitive advantages.

Notes

1. On Arthur D. Little and chemical engineering, see Terry S. Reynolds, *75 Years of Progress: A History of the American Institute of Chemical Engineers, 1908–1983* (New York: American Institute of Chemical Engineers, 1983). On the development of chemical engineering, see David A. Hounshell and John K. Smith, *Science and Corporate Strategy: Du Pont R&D, 1902–1980* (New York: Cambridge University Press, 1988), chap. 14.

2. On the importance of scale in chemical technology, see John A. Heitmann and David Rhees, *Scaling Up: Science, Engineering, and the American Chemical Industry* (Philadelphia: Beckman Center for History of Chemistry, University of Pennsylvania, 1984).

3. On the development of the lead-chamber sulfuric acid process, see John Graham Smith, *The Origins and Development of the Heavy Chemical Industry in France* (Oxford: Clarendon Press, 1979).

4. Alfred Du Pont Chandler, Jr. *Scale and Scope: The Dynamics of Industrial Enterprise* (Cambridge, Mass.: Harvard University Press, 1990).

5. On dyestuffs manufacture, see John J. Beer, *The Emergence of the German Dye Industry* (Urbana: University of Illinois Press, 1959) and Georg Meyer-Thurow, "The Industrialization of Invention: A Case Study from the German Chemical Industry," *Isis* 73 (1982): 363–381.

6. On the interdependence of chemical companies, see Jules Backman, *The Economics of the Chemical Industry* (Washington, D.C.: Manufacturing Chemists Association, 1970), pp. 91–93.

7. One of the few historical studies of catalysis is *Heterogeneous Catalysis* (Washington, D.C.: American Chemical Society, 1983).

8. For example, see "Symposium on Automatic Control," *Industrial and Engineering Chemistry* (November 1937): 1209–1213.

9. Peter H. Spitz, *Petrochemicals: The Rise of an Industry* (New York: John Wiley & Sons, 1988).

10. Backman, *Economics,* pp. 30–32.

11. Spitz, *Petrochemicals,* chaps. 6 and 7; Hounshell and Smith, *Science and Strategy,* chaps. 13, 18, and 21.

12. Hounshell and Smith, *Science and Strategy,* chaps. 13 and 21.

13. Ibid., chaps. 12 and 13; on science and synthetic dyestuffs, see Anthony S. Travis, "Perkin's Mauve: Ancestor of the Organic Chemical Industry," *Technology and Culture* 31 (January 1990): 51–82.

14. Smith, *Origins* (note 3).

15. Ibid.

16. For this interpretation, see Chandler, *Scale and Scope* (note 4).

17. Williams Haynes, *Cellulose: The Chemical That Grows* (Garden City, N.Y.: Doubleday, 1953).

18. On Du Pont and dynamite, see Hounshell and Smith, *Science and Strategy,* chap. 1.

19. Robert Friedel, *Pioneer Plastic: The Making and Selling of Celluloid* (Madison: University of Wisconsin Press, 1983).

20. On early plastics, see "What Man Has Joined Together," *Fortune* (March 1936): 69–75, 143–150.

21. On electrochemicals, see Martha Trescott, *The Rise of the American Electrochemicals Industry, 1880–1910* (Westport, Conn.: Greenwood Press, 1981) and John Servos's review, "The Development of an Industry," *Science* (26 February 1982): 1087–1088.

22. On the origins of these companies, see Williams Haynes, *American Chemical Industry: A History* (New York: D. Van Nostrand, 1945–1954), 1: 277 (Dow), and 4: 40–43.

23. On Ostwald, see Edward Farber, *Nobel Prize Winners in Chemistry, 1901–1961* (London and New York: Abelard-Schuman, 1963), pp. 36–40.

24. On the Haber-Bosch process, see Ludwig F. Haber, *The Chemical Industry, 1900–1930* (Oxford: Clarendon Press, 1971), pp. 90–95.

25. The extent of those small companies is evident from a reading of Haynes's encyclopedic history of the industry.

26. Hounshell and Smith, *Science and Strategy,* chaps. 1 and 2.

27. Sheldon Hochheiser, *Rohm and Haas: History of a Chemical Company* (Philadelphia: University of Pennsylvania Press, 1986).

28. Haynes, *American Chemical Industry,* vol. 2.

29. Hounshell and Smith, *Science and Strategy,* chap. 4.

30. Haynes, *American Chemical Industry,* 4:40–43.

31. Ibid.

32. A. D. H. Kaplan, *Big Enterprise in a Competitive System* (Washington, D.C.: Brookings Institution, 1964), pp. 144–145.

33. Hounshell and Smith, *Science and Strategy,* chaps. 4, 6, and 9.

34. Ibid.

35. On ICI, see William J. Reader, *Imperial Chemical Industries: A History* (London: Oxford University Press, 1970–1975), 2 vols. On I. G. Farben, see Haber, *Chemical Industry* (note 24).

36. Hounshell and Smith, *Science and Strategy,* chap. 12.

37. On the development of polymer science, see Yasu Furukawa, "Staudinger, Carothers, and the Emergence of Macromolecular Chemistry" (Ph.D. diss., University of Oklahoma, 1983). On Bakelite, see Friedel, *Pioneer Plastic* (note 19), pp. 103–110.

38. Hounshell and Smith, *Science and Strategy,* chap. 12.

39. On the proliferation of polymers, "A 1950 Guide to Plastics," *Fortune,* May 1950, 109–120; and "1948 . . . and the Future," *Modern Plastics* 26 (January 1949): 55–74.

40. Hounshell and Smith, *Science and Strategy,* chap. 13.

41. On plastics in World War II, see J. Harry DuBois, *Plastics History U.S.A.* (Boston: Cahners, 1972), and B. H. Weil and Victor J. Anhorn, *Plastics Horizons* (Lancaster, Pa.: Cattell Press, 1944).

42. Hounshell and Smith, *Science and Strategy,* chap. 13.

43. Ibid., chap. 21.
44. Ibid.
45. Vernon Herbert and Attilio Bisio, *Synthetic Rubber: A Project That Had to Succeed* (Westport, Conn.: Greenwood Press, 1985).
46. On the petrochemicals and World War II, see Spitz, *Petrochemicals,* chap. 3; and Harold F. Williamson et al., *The American Petroleum Industry: The Age of Energy, 1899–1959* (Evanston, Ill.: Northwestern University Press, 1963), chap. 21.
47. Hounshell and Smith, *Science and Strategy,* chap. 21.
48. "The Chemical Century," *Fortune,* March 1950, 71.
49. Spitz, *Petrochemicals,* chap. 6; and Backman, *Economics,* pp. 30–47, 115–132. See also Walter S. Fedor, "Thermoplastics: Progress Amid Problems," *Chemical and Engineering News* 39 (29 May 1961): 80–92. Reprinted in *The Chemical Industry: Viewpoints and Perspectives,* ed. Conrad Berenson (New York: Interscience, 1963).
50. Spitz, *Petrochemicals,* chap. 8; and Perrin Stryker, "Chemicals: The Ball is Over," *Fortune,* October 1961, 125–127, 207–218.
51. Stryker, "Chemicals."
52. Ibid.
53. Hounshell and Smith, *Science and Strategy,* chap. 22.
54. Ibid., chaps. 23 and 25.
55. On international competition in the chemical industry, see George W. Stocking and Myron W. Watkins, *Cartels in Action: Case Studies in International Business Diplomacy* (New York: Twentieth Century Fund, 1946). See also Lawrence Lessing, "How Du Pont Keeps Out in Front," *Fortune,* December 1962, 88–92.
56. Paul Gibson, "How the Germans Dominate the World Chemical Industry," *Forbes,* 13 October 1980, 155–164.

Chemistry and Biomedicine in an Industrial Setting: The Invention of the Sulfa Drugs

John E. Lesch

In 1939 the Nobel Prize in physiology or medicine was awarded to Gerhard Domagk, a medical researcher employed by the German fine-chemicals conglomerate I. G. Farbenindustrie. Domagk was cited for his "recognition of the antibacterial activity of Prontosil," a finding for which the first experiments were dated 1932 and the first publication 1935. Judged by its medical consequences alone, the introduction of Prontosil, the first of the sulfonamide or "sulfa" drugs, was richly deserving of recognition. In the few years since their appearance, Prontosil and related sulfa drugs had revolutionized the therapy of bacterial infections. By the early 1940s, many of the most dreaded diseases of the first half of the twentieth century, including pneumococcal pneumonia, menigococcal meningitis, most streptococcal infections (including puerperal or childbed fever and wound infections), bacillary dysentery, gonorrhea, and most urinary infections, had been brought within the range of effective chemical treatment. In 1939, however, Germany was in the grip of National Socialism and Europe was on the eve of war. Hitler had forbidden Germans to accept the Nobel Prize, and Domagk was arrested by the Gestapo for doing so.[1] By the time Domagk reached Stockholm in 1947, the sulfa drugs had been partially eclipsed—though by no means eliminated—by the still more effective antibiotics, the development of which they had helped to make possible.[2]

But for the delay of the award by Hitler and the war, the story so far seems straightforward enough: a major medical breakthrough by a research scientist is appropriately rewarded. To those most closely involved in the development of the sulfa drugs, however, it could not be so simple. There is good evidence that Domagk's Nobel award led to tension among the research staff of the company, especially between Domagk and the chemists with whom he had collaborated and who perceived Domagk's disproportionate honors as a slight to themselves. In 1945 Domagk wrote an angry letter to the company's directors, complaining that other researchers were downplaying the importance of the sulfa drugs, and that "various interest groups" were trying to break up his longstanding collaboration with the chemists Fritz Mietzsch and Joseph Klarer. Extensive negotiations Domagk carried on with the University of Heidelberg in 1946 may reflect in part tensions with the chemists over credit for the sulfa drugs. Domagk did not leave for an academic post, but tension persisted, perhaps aggravated by Domagk's collection of his Nobel Prize in 1947 and the attendant publicity. A letter of 1953 from Mietzsch—by that time director of the pharmaceutical research division—to Domagk, to take one example, complained of a press article in which Domagk was wrongly described as a chemist. In this letter and in several publications of the 1950s Mietzsch emphasized that he and other company chemists were exclusively responsible for the chemical part of the work on the sulfas.[3]

In at least one way Mietzsch's complaints were justified. In a stronger sense than is true for most achievements rewarded by the scientific community, and contrary to the premise of the Nobel Prizes in particular, the development of the first sulfa drugs cannot be credited to any single individual. It was, on the contrary, a case of industrialized invention, as that phrase has been used by historians of science and technology including J. D. Bernal and Georg Meyer-Thurow.[4] The industrialization of invention, as applied to industrial science, means several things. Technical innovations and the knowledge on which they are based are a product of cooperative research within a bureaucratic structure. Invention is routinized, that is, anticipated, planned for, and increasingly channeled through a series of steps based on past experience. The role of the individual inventor or investigator tends to be effaced. The company becomes, in a sense, a collective inventor.

The research that produced the first sulfa drugs was a collaborative effort in which biomedical researchers and chemists played distinct and complementary roles. Behind their association were the vision and experience of a company research manager, Heinrich Hörlein. Hörlein approached the problem of antibacterial chemotherapy with an expansive view of the potential uses of chemistry in medicine and systemic approach to innovation, and as director of a research organization with a record of recent successes in work on closely related problems. Behind Hörlein was the company, for which the venture into antibacterial chemotherapy was among the latest in a series of steps in a diversification, begun in the 1880s, that led from synthetic dyestuffs through pharmaceuticals to chemotherapy.

Tensions between chemists and biomedical researchers were, at least in part, surface expressions of more fundamental issues. Award of the Nobel Prize to a single individual obscured the equally important roles of the chemists, the manager, and the commercial-industrial setting in which all participants worked. The same roles tended to be effaced by the conventions of presentation of research in publications of academic science and medicine. Perception of the invention of the sulfa drugs as a case of industrialized invention was also hindered by parallel conventions that placed medicine and medical discovery in separate categories from engineering and technological innovation. Products of an industrialized system of invention, the first sulfa drugs were seen and rewarded as discoveries of academic medicine and science.

Structure of Innovation: The Pharmaceutical Research Division of Bayer/I. G. Farbenindustrie

When Gerhard Domagk joined the research staff of I. G. Farbenindustrie in 1927, his laboratory became part of an elaborate research organization that had been under development at Bayer—one of the companies that had joined to form I. G. Farben in 1925—since the 1880s. As Meyer-Thurow has shown, the establishment of research within Bayer and its contribution to the company's success were results of an evolutionary process in which legal and economic as well as scientific developments played important roles. Major components of the process included the opening of a main scientific laboratory in 1891; the building up of a research infrastructure to support the main laboratory, including a library, a literary department, a patent bureau, a control laboratory, a training laboratory, and an experimental dye house; and the increasing concentration of management on the organization of research and on the introduction of related managerial techniques, including labor contract regulations, financial incentives, conferences, regular progress reports, and creation of the role of research administrator. Research became increasingly differentiated, specialized, and decentralized with the setting up of new laboratories, including, by 1902, a pharmaceutical and a bacteriological laboratory. Largely in place by the beginning of World War I, the establishment of research within Bayer had as one major consequence the industrialization of invention, as technical innovation at Bayer became a bureaucratically managed, routinized process carried out by teams of cooperating specialists.[5]

I. G. Farben controlled not only the Bayer laboratories at Elberfeld and Leverkusen, but also others at Höchst and Mainkur. Of these the one at Elberfeld was the largest and had the most subdivisions. Its director was Heinrich Hörlein, himself a chemist and a manager of some influence in the company. As of 1927 Hörlein presided over seven laboratories, with a total staff, including assistants, of 82. There were two chemical laboratories, one of which included among its staff Domagk's collaborators Joseph Klarer and Fritz Mietzsch. There was one special laboratory, one chemical trials room, one physiological and biochemical laboratory, one pharmacological laboratory, and one chemotherapeutic laboratory. The director of the latter was Wilhelm Roehl, Paul Ehrlich's student and a Bayer employee since 1910. Domagk's laboratory, which became the eighth under Hörlein's direction, was inserted into a carefully constructed, large-scale, ongoing cooperative research effort.[6]

An internal company report of 1930, evidently prepared for the information of higher management, shows how far the pharmaceuticals division had gone by that time in routinizing invention. A flow chart (Figure 1) taken from this report gives a bird's-eye view of the development of a pharmaceutical product within I. G. Farben. The work of the scientific laboratories, such as those at Elberfeld, comprises only the early phases of the process. This is so because, as the lower third of the chart makes clear, the goal is not simply a promising compound or technique but a recognized medical technology and commercial product. Promising findings were first gathered from the company research laboratories and from outside investigators, then channeled through a set of experimental tests: pharmacological, toxicological, chemotherapeutic, physiological, and biological. Results of these tests were gathered at Leverkusen and scrutinized in conferences with the research laboratories and with scientists outside the company. The same meetings led to decisions on clinical trials, which necessarily took place outside the company. All results were continuously sorted and recorded. The whole testing process took at least one year and could take as long as three years. A key step, the examination for profitability, followed. This led into the fateful decision on whether or not to introduce the item, that is, whether or not to make it a commercial product. This was accomplished through a major conference, which also fixed the price if the decision was affirmative. The chart gives this step special emphasis. Having passed this hurdle, the product had to be prepared for the rigors of the marketplace with packaging, advertising materials, and so on. Finally it was put up for sale to wholesalers, pharmacists, and doctors. The scheme is not complete. For example, it does not mention patenting, although that was cer-

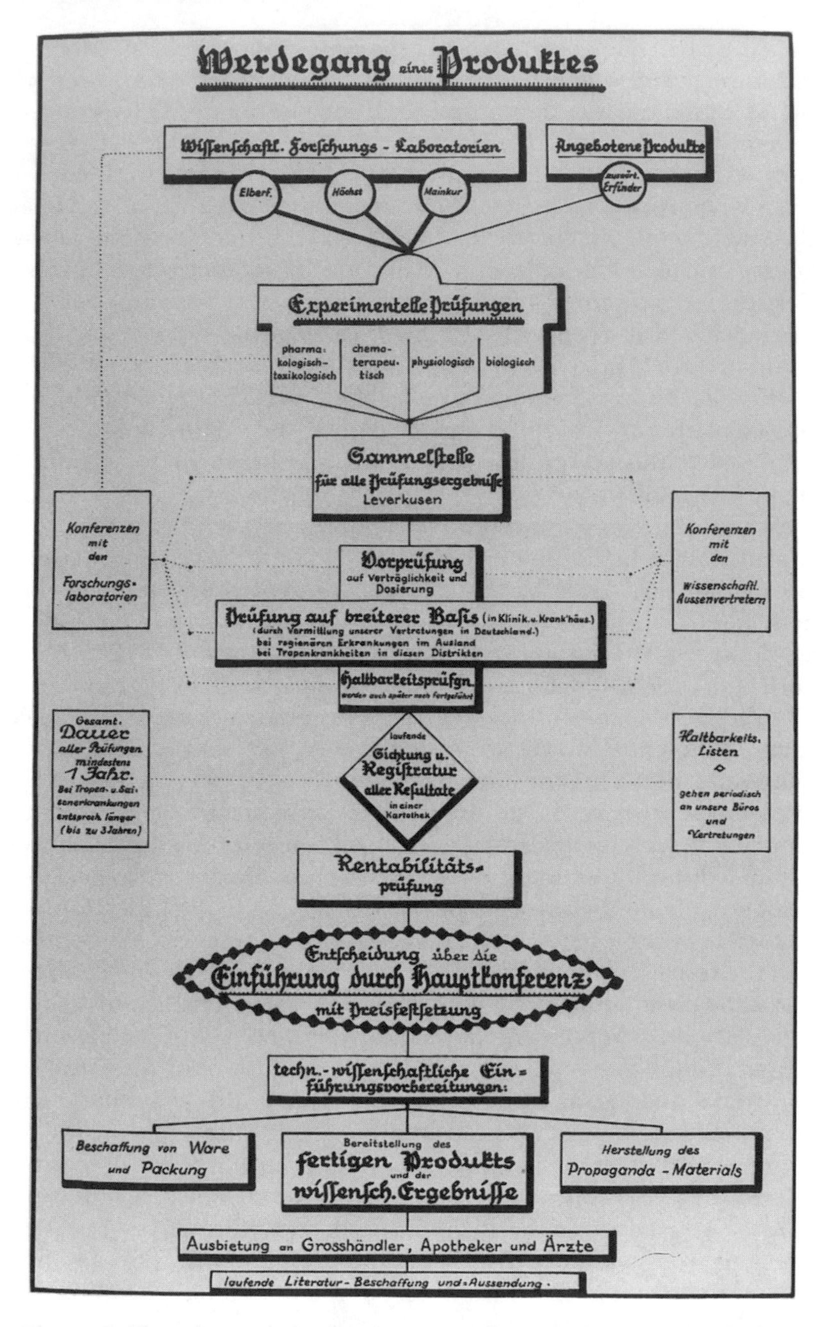

Figure 1. Flow chart on the development of a new pharmaceutical product, prepared as part of an internal company report for I.G. Farbenindustrie management, 1930. Courtesy of Bayer-Archiv.

tainly part of the determination of profitability, and the company maintained permanent offices and staff to handle patent questions. Nor does it show how pharmaceutical products, even after going on the market, could be further improved by research on chemical derivatives or mode of action. What it does reflect, despite gaps, is the company's confidence in its procedures after more than four decades of experience with pharmaceuticals.[7]

Manager of Innovation: Heinrich Hörlein

A principal architect of I. G. Farben's pharmaceutical research organization as it stood in 1930 was Heinrich Hörlein. Like the German fine-chemicals industry itself, Hörlein's early career moved between dyestuffs and pharmaceuticals. Born in 1882, he had studied chemistry first at the Technische Hochschule in Darmstadt, then at the University of Jena where he took his doctorate in 1903 under Ludwig Knorr, an organic chemist and the discoverer of antipyrine. From 1904 to 1908 Hörlein was assistant and collaborator to Knorr, and concentrated on the chemical structure of morphine. Recruited by Carl Duisberg—himself a chemist and by then director of Bayer—Hörlein joined the scientific staff of the company at Elberfeld in 1909. For several months he worked in dyestuffs chemistry. He then returned to pharmaceutical chemistry and in 1910 was instrumental in introducing Luminal (phenobarbital), a sedative that came to play an important role in the treatment of epilepsy.[8]

Hörlein rose steadily in the company. In 1911 Duisberg placed him in charge of the pharmaceutical division's chemical-scientific laboratory, and he thereby joined the governing body of the pharmaceutical research division in Elberfeld. Later he took over general direction of the existing pharmacological, chemotherapeutic, and bacteriological laboratories, in the course of time enlarging them and adding a physiological laboratory and a laboratory for experimental pathology (Domagk's, 1927). In 1914 Hörlein was named *Prokurist,* in 1919 deputy director, and in 1921 deputy member of the executive board of Bayer. In 1931 he became a regular member of the executive board (*Vorstand*) of I. G. Farben, member of the technical committee of I. G. Farben (TEA), and director of the pharmaceutical and plant-protection divisions in Elberfeld. Hörlein took over direction of the Elberfeld works in 1933, and in 1935 he became chairman of the pharmaceutical major conference of I. G. Farben, in which the directors of all pharmaceutical units, including those for research, manufacture, advertising, and marketing from Leverkusen, Höchst, and Elberfeld, took part.[9] Without exaggeration Hörlein may be called the leading figure in the management of pharmaceutical research and development at Bayer and I. G. Farbenindustrie between the world wars.

One of the roles that Hörlein took upon himself as director of the pharmaceutical research division and company manager was liaison between the company and the scientific and medical communities. By his own account he saw the care (*Betreuung*) of scientific societies within Germany as part of his life's work. At one time or another he served as treasurer of the German Chemical Society, the *Kaiser Wilhelm Gesellschaft* for the Advancement of Science, the Society of German Scientists and Physicians, the *Adolf Baeyer Gesellschaft* for the Advancement of Chemical Literature, and the *Emil Fischer Gesellschaft* (financed by the Kaiser Wilhelm Institute for Chemistry in Berlin-Dahlem). He also served as chairman of the Justus Liebig Society for the Advancement of Chemical Instruction, which distributed stipends to graduated chemists and to chemistry students.[10]

In his capacity as liaison to the scientific and medical worlds, Hörlein gave several talks to professional meetings in Germany and Britain between 1927 and 1937. In two of these, presented in 1927 and 1932—before the major work on Prontosil—he gave an expansive picture of his views on the goals, achievements, problems, and prospects of pharmaceutical research.[11] These are worth considering in detail, since they yield insights into the institutional structure of innovation at Elberfeld, and the vision on which that structure—and, not incidentally, Bayer/IGF's willingness to commit substantial resources to research over a long period—were based.

In March 1927 Hörlein spoke to a gathering of Munich physicians "on the scientific basis of the synthesis of medicines."[12] Comparing the recent sixth edition of the pharmacopoeia *Deutsches Arzneibuch* with the second edition of 1882, he drew a sweeping contrast between the overall drug treatment of the 1880s and that of 1927. Partly the result of a weeding out of traditional remedies rejected as ineffective, the contrast was most strikingly made by the addition of new, synthetic drugs. No surgeon could now imagine working without synthetically produced local and inhalation anesthesias, Hörlein pointed out, just as no internist would wish to do without aspirin, no psychiatrist would dispense with Luminal (phenobarbital, a sedative and antispasmodic), and no dermatologist could do without the salvarsans. "It is not too much to say," Hörlein concluded, "that the medicines available in the 1880s stand in relation to our modern medicinal tools as the gas lamp does to the electric light or the horse cab to the automobile." If the new medicines, like the electric light and the automobile, had come to be taken for granted, Hörlein remarked, that was only a measure of how much had been achieved in four decades.

To what could this progress be credited? First of all, according to Hörlein, to pharmacy, and especially to the success of pharmacists in the early nineteenth century in chemically isolating the pure active principles of crude plant or animal drugs and showing that they alone were responsible for medicinal effects.

In retrospect, Hörlein pointed out, the coal tar dyestuffs industry had created two essential prerequisites for the production of synthetic medicines. One of these was the insight that the properties of a dye could be changed in a determinate way by deliberately altering the dyestuff molecule. One needed only to make an analogy between dyestuff and medicinal compound to draw the conclusion that variations in physiological effect might be produced by alteration of the chemical composition or structure of a medicine. A second prerequisite was the creation of an intermediate products industry necessary to dyestuffs manufacture. These intermediates and their readily obtainable derivatives constituted a vast pool of compounds open to investigation for possible application as medicines. Formulation of the analogy and recognition of the potential of dyestuff intermediates came only in the 1880s, with the discovery of the fever-lowering properties of several compounds previously unconnected to medicine.

In Hörlein's view, one of these compounds, antifebrin (acetanilide), had provided a crucial new model for pharmaceutical chemists by showing that antipyretics could be obtained not only from the complex synthesis of quinine-like ring-form bases but also from compounds with relatively simple structures. Further impetus to the new field had come from the therapeutic success of antipyrine, a derivative of pyrazolone dyes and a product of the Höchst dyestuff works, and phenacetin (aceto-phenetidide), from the Elberfeld dyestuffs factory, in the influenza epidemic of the late 1880s and early 1890s. Two compounds of very different chemical composition had proved equally valuable. The result was that, in Hörlein's words, "a field of immeasurable extent, whose cultivation awaited, lay before researchers. . . . Were there not therefore numberless opportunities, through alteration of the known, through the formulation of homologues, derivatives, double salts, substitution products, and condensation products, and so on, to search out new remedies and medicines?"

This potential could be realized only to the extent that the effects of synthetic compounds on the human or animal organism—effects that were at best unpredictable and in many cases useless or harmful—could be systematically studied. Fortunately for the fine-chemicals industry, Hörlein pointed out, this capability was at hand in the form of experimental pharmacology, which became established as an academic discipline in the German universities from the 1870s.

As the dyestuffs industry had become convinced of the potential of synthetic medicines, the numbers of compounds available for testing had grown far beyond the capacities of university research institutes. This fact, Hörlein pointed out, together with the desire of companies "to gather as much of their own experience as possible," had led to the establishment of pharmacological and bacteriological laboratories alongside chemical ones within the larger industries. Hörlein viewed the research done in these laboratories as related to academic science in some respects but distinct from it in others. On the one hand, he argued, they could not be viewed as simply routine testing grounds for chemical preparations delivered to them. On the contrary, they were genuine "scientific research institutes, in which general questions of the day in pharmacological and physiological science are tested, new methods of investigation are invented, and extended scientific studies to set up new working theories (*Arbeitstheorien*) are undertaken." To this end the industrial laboratories followed very closely the work of academic research institutes.

On the other hand, the particular goals of the industrial laboratories might lead to emphasis on some methods over others. For example, experience had shown that whereas studies on cold-blooded animals were often sufficient for control or checking of known medicinal compounds, these methods were less satisfactory for "the discovery of unrecognized effects with new substances." Since the latter was the paramount goal of pharmaceutical research in industry, industrial laboratories were inclined to emphasize studies on warm-blooded animals. In order that these studies approach as closely as possible to the events of the organism, Hörlein continued, "in place of pharmacological investigations on isolated organs of warm-blooded animals, whenever possible the investigation is on organs in situ, in order to keep to the conditions of the physiological detoxification process through blood, liver, kidney, organs of internal secretion, and so on, as well as the contributing reflexes of the autonomic and central nervous systems."

Chemists, proceeding according to some working hypothesis or new chemical reaction, annually supplied medical researchers with hundreds of new compounds. Since each chemist wished to see each compound thoroughly tested before being rejected as therapeutically useless, Hörlein pointed out, the cost in mental energy, animals, and personnel of selection of one compound out of hundreds for clinical testing was very high. Even then, the hopes of chemists and pharmacologists might be dashed by clinical trials that showed their work to be "a useless study of a useless object." Such an effort, continued year after year, was sustained by an "unquenchable optimism."

Optimism was necessary because, Hörlein remarked, clinical trials—"the most important presupposition of the scientifically practiced synthesis of medicines"—routinely eliminated a high percentage of compounds left over from pharmacological tests. These failures at the sickbed highlighted the limits of experimental pharmacology. Laboratory investigations on animals might not correspond to the clinical picture of the illness, for example, or the tested compound might have a different decomposition in the human than in the animal organism. Whatever the reasons, Hörlein could cite ample evidence to show that a compound might have an effect in animals but not in humans, or might show no effect in animals but be active in humans. From these observations Hörlein drew two lessons. First, that in pharmaceutical research scientific theories must be viewed as working hypotheses: they are "today's mistakes" that "show the way to tomorrow's truths." Second—a point of special significance for his medical audience—the pharmaceutical industry depended on the willingness of practicing physicians to subject new medicines to clinical tests according to the most stringent criteria.

Pharmacology—the study of effects of medicines in the healthy animal—had been joined by chemotherapy, which examined the action of compounds in the infected animal, as a source and basis of synthetic medicines. Hörlein emphasized Paul Ehrlich's service in bringing about this change: his success with salvarsan, "the first conclusive evidence for the justification of this direction of research against the skeptics," his "inspiring and inspired devotion to the goal he had set" in the face of all obstacles, and his ability to marshal resources to fund the early research. Apart from salvarsan, the most striking successes of chemotherapy lay in the treatment of tropical diseases. Hörlein pointed out that in the few years since Ehrlich's breakthrough effective chemotherapeutic agents had been found for relapsing fever, sleeping sickness, yaws (frambesia), leprosy, amoebic dysentery, ancylostomiasis, malaria, kala-azar, and filaria. By developing Ehrlich's program in this direction, Hörlein asserted, the pharmaceutical industry had begun the liberation of the tropical lands from the diseases prevailing there—"one of the most important problems of civilization"—and was thereby opening the way to the economic and colonial development of these areas.[13]

Hörlein conceded that on the whole the pharmaceutical industry had been well served by the strength of German academic and medical institutions. Regrettably, the same could not be said of one key area of research, physiological chemistry. Of course German organic chemists had contributed much to the chemistry of compounds important for plant or animal physiology, such as alkaloids, sugars, proteins, enzymes, bile acids, coloring matters, and cholesterol. But the institutes in which much of this work was done had to be reserved for general chemical purposes, just as the clinics in which physiological chemistry had found a place had to preserve their emphasis on clinical medicine. Hörlein pointed out that whereas the "more physically oriented physiology" was represented in all German universities, physiological chemistry had been treated like a stepchild by the ministry of instruction, so that there were special institutes for the field only at Freiburg, Tübingen, Leipzig, Frankfurt, and Würzburg. In contrast, Hörlein claimed, every university in the Anglo-Saxon countries had a physiological-chemical institute. The result was that major findings in physiological chemistry with important medical implications, such as work on thyroxine, insulin, other hormones, and vitamins, had come almost entirely out of the Anglo-Saxon research centers. This situation could be corrected, Hörlein contended, if special institutes for physiological chemistry were created to work on such problems in a basic way. The benefit would go both to German science, which would gain a share in the research, and to the pharmaceutical industry, which would gain a new field for the synthesis of medicines.

Hormones offered an especially promising subject for such research, in Hörlein's view. To the chemist it was self-evident that the final goal was the isolation of the hormone in question in a chemically pure state, such as had already been achieved with adrenaline, from the medullary substance of the suprarenal capsule, and thyroxine, from the thyroid gland. A model for this sort of work was available in Frederick Banting and Charles Best's isolation of insulin. Since "reliable biological test objects" (animal experimental models) were also available for study of hormones of the parathyroid glands, ovaries, and posterior and anterior lobes of the hypophysis, Hörlein argued that the way was open for the physiological chemist, and at a later stage the synthetic chemist, to take on this work with some prospect of success.

Hörlein concluded his lecture by stressing the mutually beneficial interaction (*Ineinandergreifen*) of academic science and "this branch of chemical technology" involved with the synthesis of medicines. On the one hand, the introduction of synthetic medicines had, he claimed, "opened up many important biological questions and was the foundation of many purely scientific demonstrations." At the same time, the practical results of the exchange had a major impact on Germany's reputation and economic position. Abroad, "every aspirin tablet or salvarsan ampule used in the most distant countries bears witness to the high position of German science and technology," a position attested to by the fairly prompt resumption of the export of medicines after the war. At home the pharmaceutical industry employed thousands of workers and generated valuable foreign exchange. Beyond that, Hörlein argued, the pharmaceutical industry was distinguished by its ethical dimension. Hörlein told his medical audience that he gained the same satisfaction in making available an effective new medicine that they would have in successfully controlling an epidemic. "Our work proceeds in the consciousness of producing in ongoing manufacturing important remedies for the alleviation of pain, the curing of illness, and fighting infectious disease, and in the idea, in the field of research and technology to be able to harness the financial and scientific resources and possibilities of the large-scale concern for the highest cultural tasks of humanity."

Five years later, in September 1932, Hörlein spoke on the subject of medicine and chemistry to a meeting of the Society of German Scientists and Physicians in Wiesbaden. Although echoing the earlier lecture in several of its themes and examples, this talk differed especially in its sharper definition of the differences between chemotherapy and physiological chemistry and in its explication of the nature of the research process in each field and the specific problems facing it.[14]

"At no time," Hörlein began, "has chemistry influenced medical thinking and practice to so great an extent as it does today." At the same time, "never have medical problems so much interested chemists as they do today." As evidence he pointed to the exchanges taking place between chemists and clinicians at the Wiesbaden meeting on the subject of the sexual hormones, claiming that gynecology had recently "received a whole new scientific impulse from the doctrine of hormones." By the same token, Hörlein pointed out, academic chemists had begun to incorporate physiological and pharmacological problems into their advanced training and research. With the establishment of new institutions such as the Kaiser Wilhelm Institute for Medical Research in Heidelberg and the British Medical Research Council, new opportunities had been created for the cooperation of physicians and chemists. In Germany, the way for the investigation of such problems had been prepared by the long-standing interests of the best organic chemists in the study of physiologically important crude and metabolic products.

In Hörlein's view the medicine-chemistry relationship had taken three distinct forms, each associated with a separate discipline, or field of research and teaching. First historically was pharmaceutical chemistry, which had added to its early preoccupation with the chemical preparation of natural medicines the chemical production of synthetic ones. The point of contact of pharmaceutical chemistry with medicine, Hörlein noted, was pharmacology, a science devoted to "the qualitative and quantitative evaluation of these natural and artificial remedies in experimental investigation." Hörlein was quick to point out that clinical medicine remained the court of last resort and beneficiary of the application of these results to treatment of the sick person.

The bonds between chemists and medical researchers were much closer in chemotherapy, according to Hörlein. Close cooperation between the medical researcher and the chemist was, in this view, a necessary result of the need for "formation of the required test objects [*Testobjekte,* animal experimental models] that are suitable for the serial testing of chemical substances, on the one hand, and the production of these substances and their variation depending on the results of the experimental investigation and the connected clinical testing, on the other." Referring to Bayer's and I. G. Farben's experience in the discovery of new medicines for tropical diseases, Hörlein remarked that "I believe that I can say that the most successful chemotherapist in the creation of new specific remedies will be the medical man who has the best chemist at his disposal, and that the chemist will be able to tackle the synthesis of these kinds of substances with the greatest prospect of success who has the good luck to work with a medical man who is successful in the discovery and elaboration of new test objects." In Hörlein's opinion Paul Ehrlich combined in his person the qualities of an excellent medical researcher and considerable chemical knowledge, but also knew how to guarantee a frictionless cooperation between his laboratory and skillful chemists. "Recently in Anglo-Saxon countries," Hörlein continued, "the designation 'team work' has been adopted for this kind of organized cooperation on large problems the solution of which is beyond the powers of an individual researcher and which go beyond the bounds of one science, while the organiser of this type of collaborative work [*Gemeinschaftsarbeit*], in whose hands all the threads converge, is designated 'leader of the team.'"[15]

Hörlein regarded chemotherapy's struggle against infectious disease as poised between a recent past marked by a few striking successes and a future of nearly unbounded challenge. The successes had come mainly in the search for synthetic remedies for tropical, mostly protozoal, diseases. The challenges lay, on the one hand, in the chemical preparation of vaccines, serums, and antitoxins, and on the other in the identification of chemical agents effective against diseases caused by bacteria and filterable viruses. Characteristically, Hörlein saw opportunity rather than grounds for discouragement in the utter failure of antibacterial chemotherapies over the previous two decades. One source of his optimism, and of his view of the path to be followed, lay in the exemplary histories of kala-azar and malaria.[16]

Kala-azar, a protozoal infection and variety of Leishmaniasis occurring mainly in Asia, was almost 100 percent fatal to its victims before 1915, although symptomatic treatments were available. Efforts to treat the disease with antimony in the form of tartar emetic were partially successful but involved harmful side effects and a mortality rate of 15 to 25 percent. These problems were solved, Hörlein emphasized, by a cooperative effort among synthetic organic chemists, laboratory medical researchers, and clinicians. A turning point came when Wilhelm Roehl, of Bayer's chemotherapeutic laboratory, replaced the animal experimental model used in the early phase of the work, the nagana-infected mouse, with the hamster. An effective compound had been identified, and Hörlein regarded the problem of the chemotherapeutic cure of kala-azar through synthetic chemistry as solved.

Hörlein framed the problem of malaria chemotherapy as a search for "a better quinine." Efforts by Bayer chemists and medical researchers led in 1924 to Plasmoquine, a synthetic compound that made up for some of quinine's failings but lacked some of its advantages. Even as Plasmoquine was put on the market, therefore, the search for a better quinine continued. Hörlein remarked that "this task has been an essential part of the work of our research laboratories in recent years." The location of a synthetic antimalarial that incorporated the advantages of quinine while improving on it was effected by key findings in both the chemical and the animal experimental lines of research. Roehl's canary model for malaria proved incapable of defining a qualitative distinction between the effects of quinine and Plasmoquine. Following Roehl's early death in 1929, his successor at Bayer, Walter Kikuth, developed a model in which a hemoproteus infection was combined with malaria in the ricebird, allowing reliable differentiation of schizonticidal and gametocidal activity and thereby of the effects of antimalarial agents. Meanwhile the Bayer chemists Hans Mauss and Fritz Mietzsch prepared a compound of the acridine series—a dyestuff—that met the desired criteria in animal trials. Following clinical tests the compound was placed on the market in 1932 under the trade name Atabrine (*Atebrin*).

In Hörlein's view the relations between chemists and medical researchers in physiological chemistry stood in sharp contrast to those in chemotherapy. Whereas in the latter field medical researchers and chemists together sought synthetic compounds effective against pathogenic organisms infecting a host, in physiological chemistry clinicians, medical researchers, and chemists often worked sequentially in the study of substances that were normal nutrients or products of the organism's metabolism. "Decisive for the modern direction taken by this field of work," Hörlein pointed out, "is the finding that the materials [*Stoffe*] formed in the animal or plant organism, besides their static significance, also have a dynamic significance, and that in certain cases these materials also possess pharmacological effects and thereby become medicines." Best illustrated by the hormones and vitamins, the research in this field was, in Hörlein's view, typically more linear and stepwise than collaborative and interactive. The problems were usually first posed by clinicians, such as Thomas Addison or Oscar Minkowski. Discovery of the test object, or suitable animal model, was then most often the work of experimental medical researchers—for example, Banting and Best with insulin, or Allen with estrous hormone. The next step—purification of the effective substance from the animal starting material—was taken by the physiological or pure chemist, as by Furth with adrenaline, Kendall with thyroxine, or Doisy and Butenandt with the sexual hormones. The last step, synthesis, was entirely the domain of the chemist—for example, Stolz's production of adrenaline, or Harrington's of thyroxine. Hörlein pointed out that availability of the active compound in purified form often gave rise to new medical investigations, and in illustration pointed to the thousands of publications on insulin and thyroxine.

The work of medical researchers joined to that of chemists had uncovered other pharmacologically active substances produced by the organism, including liver preparation and the nucleosides. Hörlein pointed out that in the case of liver preparation chemical study was still blocked by the absence of an appropriate test object or experimental animal that would make possible extended trials to determine in which fraction the effective principle was present. He noted that the case of the nucleosides illustrated how study of normal function could take on therapeutic interest: "as physiological methods were replaced by pharmacological investigation, physiological chemistry became pharmacology, and finally therapy."

In Hörlein's view the study of vitamins, a major discovery of recent decades and a field intensively pursued as Hörlein wrote, also depended on the efforts of clinicians, experimental medical researchers, and chemists, although those efforts were not necessarily collaborative or coincident in time. Hörlein cited the recent isolation in pure form of the complex vitamin effective against beri-beri (thiamine) as an example of the common effort of a university chemical laboratory (Göttingen) and an industrial physiological laboratory (I. G. Farben/Elberfeld). He pointed out that for vitamins A and C the chemical work was now in the foreground, and that for vitamin C especially the joining of chemical research with "exactly performed animal investigations" had been shown to be crucial if premature conclusions were to be avoided. Hörlein recounted the story of the elucidation of the role of vitamin D against rickets as an illustration of the differentiation of the roles of physicians, physiologists, physiological chemists, and chemists in vitamin research. In this case too, the academic-industrial link figured in the production of the antirachitic substance in pure crystallized form through the collaboration of Adolf Windaus, a university chemist, and his former student, Otto Linsert of the biochemical laboratory at Elberfeld.

Hörlein was aware that his picture of the relations of chemists and medical researchers, though expansive, was also schematic, and that his generalizations masked many complexities and intersections among the various lines of work. Thus, for example, he predicted that the older synthetic pharmaceutical chemistry would find new opportunities in the more recent research on medicines (*Pharmaka*) formed in the body itself, or what he had termed physiological chemistry. Determination of the chemical structure of adrenaline, for example, had already led to syntheses of compounds in the adrenaline group in a search for new substances that could act on the blood vessels. Here Hörlein based his optimism squarely on an analogy between dyes and medicines. Just as synthetic chemistry had taken natural dyes and played variations on them to produce new and better colors, so too medicines produced by nature—in this case he had in mind hormones and natural nutrients such as vitamins—might be varied and surpassed. "Here also chemistry can seek to imitate the secrets of nature, so as to excel it in independent work," Hörlein remarked. "Perhaps synthetic hormones and vitamins will later excel their natural models to the same degree that today indanthrene dyes excel the dyes supplied by nature, indigo, and the purple of the ancients."

Hörlein concluded by returning once more to the centrality of physiological and chemotherapeutic experimental models (*Testobjekte*). Here his emphasis was not so much on the need to find the right experimental model for the problem at hand, as on his view that even with an imperfect or deficient model very good results could be obtained. He reminded his audience that it was studies of mouse nagana and hamster kala-azar that had made possible the finding of an effective agent against human kala-azar, and that Plasmoquine and Atabrine were developed only with the aid of experimental canary and ricebird infections. He pointed out that rat rickets was different in important points from rickets in children, but was nevertheless "the red thread that finally led to the preparation of the crystallized antirachitic vitamin." From these and other cases Hörlein drew the lesson that there were grounds for optimism even about cancer, "the disease that concerns all of us the most," in spite of the lack of an experimental model absolutely comparable with human tumors, and what appeared to be widespread skepticism among clinicians regarding investigations on animals with tumors. He told his listeners that if his discussion "should lead to greater optimism" on this point, "I would regard that as the finest result of my lecture."

It was appropriate that Hörlein should end his 1932 talk with a call for greater optimism, for that was a keynote of both lectures and indeed of his posture as manager of pharmaceutical research for I. G. Farben. As this extended paraphrase of his presentations makes clear, Hörlein's was not merely the optimism of a salesman speaking to an audience of potential buyers. It was based on a well-informed and well-articulated vision of pharmaceutical research as a domain of rational engineering or science-based technology; a synoptic view of the scientific basis, successes, and unrealized potentials of chemically oriented medicine; and a systemic approach to innovation.

Hörlein viewed the pharmaceutical industry as a branch of chemical technology, historically derived from the production of synthetic dyestuffs and distinguished by its partial incorporation of and dependence on experimental and clinical medicine. He underlined its importance as an element of the German economy and as one source of the scientific, technological, and medical reputation of Germany abroad. Like Thomas Edison, a manager of innovation in a field he compared with that of synthetic medicines, Hörlein considered his goals and problems from a technical or scientific and an economic point of view simultaneously. He paid close attention to such matters as the cost of selecting one compound out of hundreds for clinical trials and gave patenting precedence over publication.[17]

Hörlein's explicit and implicit definitions of the fields central to pharmaceutical research emphasized their character as rational engineering rather than as basic science. However much it drew upon or contributed to knowledge in chemistry, physiology, microbiology or pathology, research in pharmaceutical chemistry, physiological chemistry, and chemotherapy as Hörlein thought of them was organized around and subordinated to the practical goal of production of medically useful and commercially viable products. Pharmaceutical chemistry harnessed the methods and concepts of organic analysis and synthesis to produce new medicines. Although Hörlein was well aware of the scientific breadth of physiological chemistry as pursued in academic institutes, his own interest narrowed to the "decisive" finding that some compounds produced or taken in by the body—he thought especially of vitamins and hormones—were pharmacologically active and could therefore become medicines. As even its name indicated, chemotherapy had as primary goal the finding of chemical agents effective against systemic microbial infections. To this practical goal even its aim to understand how the chemical agents work was subordinated. Hörlein's emphasis on the ability of imperfect or deficient animal experimental models to produce valuable results was one expression of this pragmatic or engineering attitude.

Hörlein's optimism was based not simply, or even primarily, on the situation of any particular area of research considered in isolation, but on his comprehensive overview of advance in areas in which chemistry and biomedicine intersected. These areas shared a number of generic problems and solutions—for example, the need to isolate a substance (natural product, synthetic product, body substance) in chemically pure form; the need to synthesize the substance, and to do so economically if it was to go on the market; and the need for pharmacological, chemotherapeutic, toxicological, and clinical testing of the substance. Hörlein's effort to translate success in certain areas (vitamin deficiency disease, chemotherapy of protozoal infections) into optimism about possibilities in other areas (cancer, antibacterial chemotherapy) was characteristic. He regarded the chemical attack on disease as a many-fronted battle, in which there was a generally advancing line but also many points at which advance was slow or arrested. In this sense Hörlein might be said to have thought—as Thomas Hughes has shown that Edison did (see note 17)—in terms of reverse salients and critical problems. Reverse salients are areas of research and development that are lagging in some obvious way behind the general line of advance. Critical problems are the research questions, cast in terms of the concrete particulars of currently available knowledge and technique and of specific exemplars or models (such as insulin, or the chemotherapy of kala-azar and malaria) that are solvable and whose solutions would eliminate the reverse salients.

Hughes describes the components of Edison's electric lighting project as forming a system in the sense that they are interrelated and mutually dependent parts of an integrated whole. The technology that Hörlein describes does not form a system in this sense but is fragmented into distinct and not necessarily related aspects of the relation of chemistry and medicine in its practical manifestations. Hörlein has a synoptic or comprehensive view of these manifestations, but they themselves do not form a system in Hughes's sense.

What is systemic in Hörlein's way of thinking is his concept of the organizational pattern or patterns that will best facilitate the production of valuable results in the areas in which medicine and chemistry interact. A valuable result in this connection is a result of practical importance for clinical or preventive medicine and, implicitly, of commercial value for industry. Hörlein perceives a need for a set of mutually complementary institutions and trained personnel whose interaction produces the desired results. The organizational pattern that emerges more or less clearly from Hörlein's lectures is closely associated with his view of the typical phases or the cycle of development of research in chemotherapy or physiological chemistry. He sees a need for friendly and mutually supportive relations between industrial research and development organizations, academic institutions, and clinicians. He views the academic-industrial connection as crucial and mutually beneficial. Underlying this view is his definition and differentiation of the relevant disciplines, and his belief in their generally excellent condition in Germany. He sees a need for government support of appropriate institutions, especially research institutes in universities. Within industrial research organizations—and, implicitly, within academic ones—Hörlein calls for special institutional arrangements to encourage appropriate interactions between chemistry and biomedicine. Here an element of crucial—and to Hörlein, personal—importance was the role of the research manager or "team leader." When Hörlein spoke of the research done under his direction as "our work," he used the possessive advisedly to convey a strong sense of his own participation. The research manager had to be active in defining goals, in marshaling means and resources, and in assessing success or failure. He had to intervene where necessary to minimize friction between chemists and medical researchers, an especially important task for chemotherapy as a composite entity. He had to publicize the company's successes—a necessity for what was ultimately a commercial enterprise—and act as liaison between company laboratories and the academic and medical communities. Through it all he had to take a long view of the value of research, not insisting on immediate results of medical or commercial value.

As a research manager with training and experience in pharmaceutical chemistry, a lively interest in medicine, and rapport with the medical community, Hörlein was well positioned to survey the field where chemistry and medicine joined battle against disease. He could spot the points where the enemy's line was broken, and the reverse salients in his own. What he could not do—or could not do alone—was to direct the day-to-day operations of his troops, that is, to define the critical problems to be solved, to identify the terms of their solution, and to do the work that would carry the day. In the case of chemotherapy these things could be effected only by the medical researcher and the chemist, each working in his own domain, and cooperatively. For his attack on one of the most important reverse salients, the chemotherapy of bacterial infections, Hörlein called upon Gerhard Domagk, a medical researcher, and Fritz Mietzsch and Joseph Klarer, chemists. Their efforts, separately and together, produced a medical triumph: the first of the sulfa drugs. Before this *Gemeinschaftsarbeit* or collaborative effort could succeed, however, each party—the medical researcher on the one hand, the chemists on the other—had to conceive the problems to be solved, and the means to their solution, in his own terms. Let us look at each approach in turn.

The Medical Researcher's View of Innovation: Gerhard Domagk

In his 1932 lecture on medicine and chemistry Hörlein had referred to the "endless work" that still lay before researchers on infectious diseases. "The field of bacterial illnesses and those caused by filterable viruses," he had remarked, "still lies absolutely open for a collaboration of medical men and chemists."[18] In announcing Prontosil to the world three years later, and in subsequent publications of 1936 and 1937, Gerhard Domagk described a major response to Hörlein's challenge, while emphasizing the many ways in which that challenge was still to be met. At the same time, he revealed his own conception of the process of innovation in chemotherapeutic research, a conception which converged at many points with that of Hörlein but which was also stamped with the particular motives, methods, and ways of thinking of the medical researcher. For Domagk as for Hörlein, the problems of chemotherapy were ultimately problems in rational engineering, that is, the bringing of scientific knowledge to bear to achieve a practical end: a technique or substance effective in accomplishing a therapeutic task. In contrast to Hörlein's relatively evenhanded emphasis on chemistry and experimental medicine, however, Domagk conceived of the scientific knowledge most germane to solution of the problems of chemotherapy as biological and medical. For Domagk chemotherapeutic research was centered on such problems as the appropriateness of the experimental model, the uses of in vitro versus in vivo testing, the classification of microbes, and the specificity of pathogens and of host-pathogen interactions. He emphatically, if briefly, recognized the essential role of chemistry and chemists, but—whether intentionally or not—left the impression that this role was confined to the unproblematic synthesis of compounds. An independent, creative role for chemical reasoning did not appear in Domagk's account. Thus, at least implicitly or by default, Domagk assigned to chemistry an ancillary and subordinate role in chemotherapeutic research.[19]

Domagk had come to chemotherapy via medicine and experimental pathology. Born in Lagow in the province of Brandenburg in 1895, the son of a teacher, Domagk was educated at a scientifically oriented grammar school in Liegnitz (now Legnica, Poland). He made an early decision to study medicine but was interrupted in his first term of studies at the University of Kiel by the outbreak of war. Wounded in 1915 while serving in a grenadier regiment, he was transferred to the medical corps. Resuming his studies after the war, he took his M.D. at Kiel in 1921.[20]

Domagk's M.D. thesis, done under the direction of the internist Max Burger, addressed a problem in physiological chemistry, the precipitation of creatinine in humans after muscular exercise. Still at Kiel, Domagk served as assistant to the chemist Ernst Hoppe-Seyler and a pathologist, Dr. Emmerich, joining his chemical and pathological studies in an analysis of the composition of heart muscle, liver, and kidney in disease. In 1924 Domagk joined Walter Gross as privatdozent in the pathology department at the University of Greifswald. By this time Domagk had embraced Virchow's injunction that "pathological anatomy must become pathological physiology," and a dynamic or physiological conception of pathology emphasizing the chemical aspects of health and disease became the hallmark of his research. An early expression of this trend was Domagk's habilitation essay, which examined the defensive role of the reticuloendothelial system in infections. In a paper published in 1924 based on this research, he described experiments in which mice were injected with living or killed bacteria. Domagk was able to show that within a few minutes of injection amyloid precipitates could be detected around the endothelial cells involved in phagocytosis in the liver, spleen, and lungs. He concluded that the amyloid was a degradation product of the phagocytosis, and he was able to strengthen the phagocytosis by sensitization.[21]

At Greifswald, Domagk also began studies of cancer that were to continue, without dramatic result, for much of his career. He continued his work on the reticuloendothelial system, on the view that it might be possible to reinforce the body's natural defenses against infection. Domagk failed in his effort to isolate substances from the reticuloendothelial system that would harm pathogenic bacteria, but realized that synthetic compounds might be found that would damage or destroy specific cocci.[22]

This activity in the common area between medicine and chemistry drew the attention of Heinrich Hörlein, and in 1927 Domagk accepted an offer by I. G. Farbenindustrie to direct a laboratory in experimental pathology in the company's pharmaceutical research division at Elberfeld. He was thirty-two years old. At first the connection with I. G. Farben was tentative. Domagk signed a renewable two-year contract and retained his academic association with the University of Münster, where he was named extraordinary professor of general pathology and pathological anatomy in 1928. The bond with industry proved durable, however, and despite occasional longings for academia, Domagk was to remain with I. G. Farben (from 1945, Bayer again) until his death in 1964.[23]

Neither Domagk nor Hörlein shared the pessimism then current in some circles regarding the potential of antibacterial chemotherapy. They agreed on a systematic review and search for chemical agents effective against bacterial infections. Domagk, in collaboration with research division chemists, especially Fritz Mietzsch and Joseph Klarer, was to direct the effort.[24]

The research path followed by Domagk and his collaborators may be partially reconstructed from Domagk's account in his articles of 1935. A suitable experimental model (*Modellversuch*) had to be identified. The essential requirement was that host and pathogen be matched in such a way that the investigator succeeded "in infecting animals with certainty in a high percentage of cases." In addition, "the course of the infection must be as much as possible the same in all animals, so that from the observations on the treated animals and the untreated control animals we can draw reliable conclusions." Domagk found that his specifications were best met with streptococcal infections of mice and rabbits, which were almost invariably fatal.[25]

To narrow the field of compounds to be submitted to animal trials Domagk relied in part on in vitro experiments, in which the target strain of bacterium was methodically exposed to a variety of chemicals. Such "orienting" test tube trials could yield "general clues" as to which classes of bodies might contain compounds effective against certain bacteria. Domagk emphasized that these could not substitute for animal tests and could even be misleading in the sense that they could not distinguish between true chemotherapeutic agents harmless to the host and disinfectants that would kill or damage the host as well as the pathogen.[26]

Domagk's target was a deadly one. Streptococci—immobile, spherical bacteria occurring in chains—were known to be causative agents in the most dangerous and common wound infections, childbed fever (puerperal sepsis), tonsillitis, endocarditis, many kidney and joint conditions, appendicitis, and other diseases. For some of these diseases the mortality was high, as was the cost in human suffering. Domagk described the type used in his animal trials, *Streptococcus pyogenes hemolyticus,* as the most dangerous cause of wound infections in humans. The strain was taken from the blood of a patient who died of blood poisoning, or streptococcal sepsis. For his experiments Domagk made intraperitoneal injections of twenty-four-hour cultures of the bacteria in dilutions of 1:1,000 to 1:100,000. He found that ordinarily 0.3 cubic centimeter of a dilution of 1:10,000 to 1:100,000 was sufficient to kill a mouse within 24 hours. To insure that all the untreated control animals died in 24, or at most 48 hours, he gave 10 to 100 times the mortal dose to all experimental animals.[27]

With his experimental model established, Domagk proceeded in collaboration with the chemists on a search for chemotherapeutically effective compounds. Selection of classes of compounds for animal trials was not random, but was guided—according to Domagk—by "orienting" in vitro tests and by a review of the literature. One lead from the latter was the work of Adolf Feldt. In 1926 Feldt had begun publishing results of treatment of experimental streptococcal infections with organic gold compounds. He had recommended as effective two of these. Domagk confirmed the effectiveness of organic gold compounds but found that in severe streptococcal infections they could not be used for extended treatment in high enough doses without danger of gold intoxication, sometimes accompanied by skin conditions and kidney damage. He recommended their use in local treatment of streptococcus-infected wounds.[28]

By Domagk's account the disadvantages of the gold compounds had led him and his collaborators to consider classes of purely organic substances. Among these were hydroquinine compounds such as the commercial preparations Vucin and Eucupin. These were already known for their effectiveness against streptococci in vitro, and were used in deep antisepsis, the local treatment of infected tissues such as joints. Unfortunately, Domagk had to report, they failed completely in systemic infections. Much the same had to be said of a group of acridine compounds that surpassed known acridine compounds in effectiveness against streptococci in vitro but lost most of this action in infected animals.[29]

Domagk recalled that a considerable body of literature from the 1910s and 1920s pointed to the antibacterial effects in vitro of azo compounds. In work published in 1913, Philip Eisenberg had suggested the use of one such substance, crysoidine (2,4-diamino-azobenzene), against systemic infections but had failed to confirm the expected results. Several other azo compounds had been recommended for treatment of urinary tract infection, but Domagk reported that none of these were efficacious against general streptococcal or staphylococcal infections in the mouse.[30]

Whatever the limits encountered by other investigators, Domagk and his associates decided at an early date to give serious attention to the azo compounds. "In a close, several-year-old cooperative work

[*Zusammenarbeit*] with the chemists Mietzsch and Klarer," Domagk reported in 1935, "we have worked through the field of the azo compounds in their chemotherapeutic applicability." In so doing, Domagk continued, "we first chanced upon [*stiessen wir zunächst auf*] azo compounds with incomparably higher disinfection effect against streptococci than those of the above-mentioned known azo compounds." These substances too had failed, however, against systemic streptococcal infections in the mouse. "Nevertheless in the course of our investigations we came to a group of very nontoxic sulfonamide-containing azo compounds, for which indeed no disinfection value against streptococci in vitro could be discerned, but now in animal trials showed a clear effect in streptococcal sepsis of the mouse." One of this group was the hydrochloride of 4-sulfonamido-2,4-diaminoazobenzene,

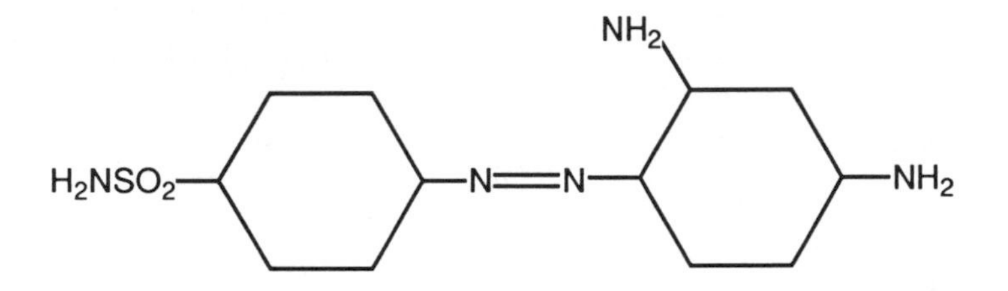

a red crystalline powder with a melting point of 247 to 251 degrees Celsius, soluble in cold water to 0.25 percent, and more easily soluble in warm water. This compound was later given the trade name Prontosil.[31]

In his 1935 papers Domagk was expansive on the toxicology and pharmacology of Prontosil, the pathology of streptococcal infections with and without treatment, and the clinical implications of use of the drug. Toxicological tests showed the harmlessness of the compound. Prontosil could be administered orally in substantial doses to mice, rabbits, and cats without producing symptoms. Mice and rabbits could also tolerate significant doses in subcutaneous injections. With oral administration of 0.1 or 0.2 grams of a 0.25 percent solution daily over fourteen days, rabbits showed no pathological changes in either blood or urine, the weight of the animals was unchanged, and their general condition remained good.[32]

Pharmacological tests, performed by research division pharmacologists Helmuth Weese and Gerhard Hecht, showed Prontosil to be "an extraordinarily indifferent compound." Even substantial intravenous injections in cats and rabbits did not affect blood pressure or heart function, Domagk reported, nor did Prontosil alter either the smooth muscle or the physiological functions of the uterus or the small or large intestines. Subcutaneous injections up to one gram per kilogram of

Nr.	Gewicht	Präparat Nr.	Präparat %	Präparat Dosis	Art der Behdlg.	21. XII.	22. XII.	23. XII.	24. XII.	25. XII.	26. XII.	27. XII.	28. XII.
201	14 g					m	kr.	kr.	†				
202	14 g					m	†						
203	14 g	Anfangskontrollen				m	†						
204	17 g					m	†						
205	19 g					m	†						
206	14 g					m	m	†					
303	18 g	Prontosil	0,01%	0,2	per os	m	m	m	m	m	m	m	m
304	19 g			0,2		m	m	m	m	m	m	m	m
305	18 g			1,0		m	m	m	m	m	m	m	m
306	14 g			1,0		m	m	m	m	m	m	m	m
307	16 g		0,1 %	0,2		m	m	m	m	m	m	m	m
308	15 g			0,2		m	m	m	m	m	m	m	m
309	17 g			1,0		m	m	m	m	m	m	m	m
310	17 g			1,0		m	m	m	m	m	m	m	m
311	14 g		1,0 %	0,2		m	m	m	m	m	m	m	m
312	17 g			0,2		m	m	m	m	m	m	m	m
313	18 g			1,0		m	m	m	m	m	m	m	m
314	14 g			1,0		m	m	m	m	m	m	m	m
315	18 g					m	kr.	†					
316	16 g					m	†						
317	15 g					m	†						
318	14 g	Endkontrollen				m	†						
319	15 g					m	†						
320	14 g					m	†						
321	15 g					m	†						
322	17 g					m	†						

m = munter, kr. = krank, † = tot

Figure 2. Table showing results of administration of Prontosil to mice infected with hemolytic streptococci, dated December 20, 1932. Source: Gerhard Domagk, "Ein Beitrag zur Chemotherapie der bakteriellen Infektionen," *Deutsche Medizinische Wochenschrift* 61 (1935): 250–253 (on 252).

body weight were not injurious to animals, and intravenous injections produced no tendency to thrombus formation as a result of damage to blood vessel walls.[33]

In the first published article Domagk included a table showing the dramatic results of one experiment in which the dye was administered to infected mice. (Figure 2). Most of the untreated control animals were dead within forty-eight hours of infection, whereas all treated animals survived.

The pathological findings were equally dramatic. Domagk included in his first article three pictures of smear preparations made from the peritoneum of infected mice (Figure 3). That from an untreated animal twenty-four hours after infection (a) was dominated by the presence of masses of cocci and decomposing body cells. The second preparation (b), made from a similarly infected animal treated with Prontosil, twenty-four hours after infection, showed intact body cells and no cocci. In the third preparation (c), made from a treated infected animal forty-eight hours after infection, cocci were absent and body cells showed evidence of previous phagocytosis, a defensive measure by which the cells engulf and destroy foreign particles. Pathological examination showed the devastating results of untreated infections. "If we examine the organs of untreated animals," Domagk remarked, "we find in the liver, spleen, kidneys, heart and many other organs typical signs of a severe general infection: colonies of bacteria in the swollen endothelium of the liver and spleen, in the renal glomeruli and in the heart muscle, leukocytic infiltration in the most various organs as well as numerous disintegrated leukocytes in the splenic pulp." By contrast, in all animals successfully treated with Prontosil these findings were absent when treatment was begun in time and in sufficient dosage.[34]

Domagk reported the results of chemotherapeutic trials in which the strain and kind of bacteria were varied. He had obtained similar favorable results in treating mice infected with streptococci isolated from erysipelas. He had found that in rabbits suffering from chronic streptococcal infections associated with joint swellings and sometimes with endocarditis, large oral doses of Prontosil usually produced a significant improvement. Domagk emphasized his conclusion that "chemotherapeutically, Prontosil and numerous other specific substituted azo compounds exhibit a nearly elective action on streptococcal sepsis in the mouse." He had also tried Prontosil against staphylococcal infections in rabbits, however, with good effect in acute progressive infections with abscess formations in kidney and heart muscle, and in chronic staphylococcal arthritis with joint swelling and abscess formation. He found, however, that the favorable effects of Prontosil were less certain and regular in staphylococcal than in streptococcal infections. He was not able to observe any effect of Prontosil on pneumococcal infections of animals.[35]

Laboratory findings had been confirmed in the clinic. Domagk noted that the specificity of action of Prontosil made it more important than ever that physicians recognize the specific nature of the infection in

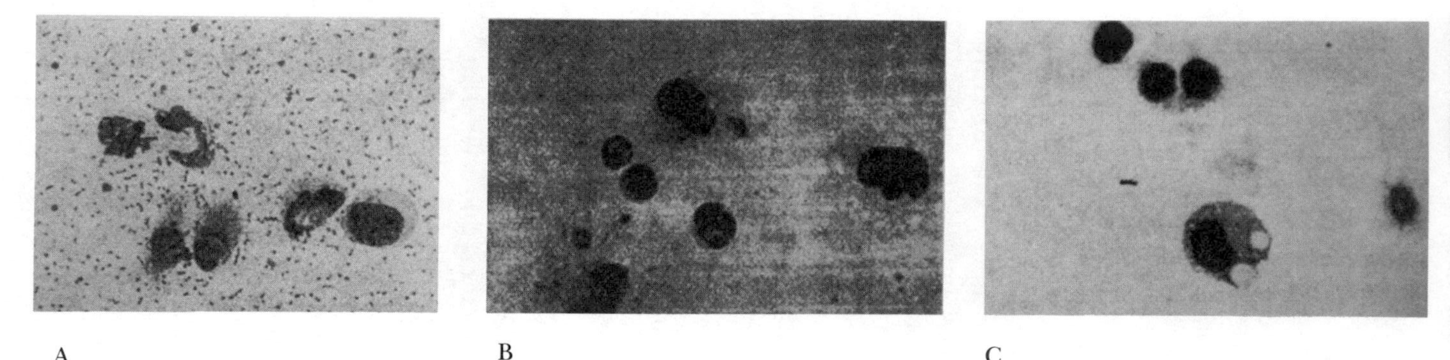

Figure 3. Smear preparations from three mice infected with hemolytic streptococci. (a) "Control mouse infected intraperitoneally with streptococci. Peritoneal smear at 24 hours. Massive numbers of cocci and abundant exudate of decomposed cells together with small numbers of intact monocytes." (b) "Mouse similarly infected intraperitoneally with streptococci and treated with subcutaneous *Prontosil*, 24 hours after infection. Enlarged monocytes, well preserved leukocytes and lymphocytes. Neither free nor intracellular cocci." (c) "Mouse infected with streptococci intraperitoneally and treated with subcutaneous *Prontosil*, 48 hours after infection. Lymphocytes and scattered monocytes with protoplasmic vacuoules, which indicated previous phagocytic activity. No cocci!"

Source: Gerhard Domagk "Ein Beitrag zur Chemotherapie der bakteriellen Infektionen," *Deutsche Medizinische Wochenschrift* 61 (1935): 250–253 (on 251). English translation from Gilbert B. Forbes and Grace M. Forbes, "An Historical Note on Chemotherapy of Bacterial Infections," *American Journal of Diseases of Children* 119 (1970): 10.

each patient, and that they do so as early as possible. Because some streptococcal infections were difficult to diagnose in early stages with available methods, Domagk called for closer cooperation between clinicians and bacteriologists to improve diagnostic technique as a prerequisite to effective chemotherapy.[36]

Little definite could be said about the mode of action of Prontosil. Domagk emphasized that Prontosil did not have a specific effect on either streptococci or staphylococci in vitro, and therefore that "it acts as a true chemotherapeutic agent only in the living organism." On the other hand its specificity of action argued against "a nonspecific, general, and indirect action of prontosil on the host organism, perhaps a nonspecific activation of the reticuloendothelial system."[37]

Domagk's presentation of the path to Prontosil is a selective one. It cannot be taken at face value as a reconstruction of the research, but does—together with other remarks in his articles of 1935, 1936, and 1937—place in relief several of the qualities of Domagk's concept and practice of chemotherapy.[38] Among these is the character of chemotherapy as a form of science-based technology or rational engineering. Domagk's primary goal was not creation of new knowledge of nature but identification of compounds effective in the treatment of bacterial infections, a practical aim. Since the medicinal drug was also to be a commercial product, patenting took precedence over publication. On Christmas day 1932, Domagk's collaborators, Fritz Mietzsch and Joseph Klarer, applied for a German patent on Prontosil. Not until more than two years later, in February 1935, a month after the patent was granted, did Domagk's first article appear in a medical journal.[39] When, between 1935 and 1937, Domagk, Mietzsch, and Klarer succeeded in locating three compounds with apparent advantages over Prontosil and its constituent compound sulfanilamide (trade name: Prontosil Album), the substances were first described in German patents.[40]

The technological character of Domagk's project is also expressed in his systematic but strongly empirical research procedure. While he strove to build a scientific chemotherapy grounded in exact knowledge of the biology of host and pathogen, Domagk had to concede that "here we stand only at the very beginning of knowledge. Until now [1935] we have been much too dependent on empiricism." And "with most chemotherapeutic agents, to be sure, we still do not know approximately how they work, not even with salvarsan."[41] Domagk's language is suggestive. They "worked through" the field of the azo compounds until they "chanced upon" ones with exceptional disinfectant powers. Then in subsequent investigations they "came to" the group of compounds from which the one that became Prontosil was singled out.[42]

The practical goals of chemotherapy have a narrowing and shaping effect on the field of investigation. This effect may be seen, for example, in Domagk's dual classification of bacteria. In one division the criteria are morphological and biological. Bacteria appear in one of three primary forms, as cocci (globular or spherical); as bacilli (in rod form); or as spirilla (in spiral or screw form). In an alternative classification, in contrast, morphological criteria are subordinated to pathological ones. Not all bacteria are pathogenic. The chemotherapist therefore first separates those that are not pathogenic from those that are. The latter are then subdivided into two groups, those that exert their harmful effects through secretion of toxin, such as diphtheria, tetanus, and botulism bacilli; and those that, by exciting inflammation or destroying tissues, create damage everywhere they are spread by the blood and lymph routes, such as streptococci, staphylococci, pneumococci, the tubercle bacillus, and many others. Domagk regarded the former group as less dangerous because their effects could often be prevented by use of antitoxin sera. When it was a question of subdividing the streptococci, Domagk cited Max Gundel's classification "on the basis of biological tests" but departed from Gundel in viewing pneumococci as a separate group because "it appears to be expedient on biological and chemotherapeutic grounds."[43]

In the differing criteria employed in classifications may be seen one indication of the divergence that opens up in Domagk's research between the bacteriologist as biologist and the chemotherapeutic researcher. The latter regroups the phenomena of the field according to criteria defined by the instrumental goals of medicine, or what may be termed technological rather than scientific goals. One result is that a large portion of the bacteriological realm, the nonpathogenic, is made irrelevant to investigation. For what remains, a hierarchy of importance is established defined by the degree of danger posed to the host organism, given the state of existing techniques. A further narrowing factor is that the host organism is almost exclusively conceived as human, although veterinary applications are presumably to be sought as well.

The shaping and directing effect of distinctively chemotherapeutic goals may be seen also in Domagk's emphasis on the phenomenon of specificity of bacteria as a target for future research. It was well known, Domagk pointed out, that bacteria manifested distinct differences in biological capabilities, for example in their metabolic needs, in their abilities to decompose animal tissues, or in their preference for definite places in the host organism for their colonization, such as the heart valves, joints, or brain. "But just these specific characteristics," Domagk remarked, "appear significant to me also for the possibility of a chemotherapy of bacterial diseases. . . . The knowledge of the specific characteristics of every type of bacterium will be the exact foundation for a scientifically well-founded chemotherapy of bacterial infections." Domagk does not call for intense study of all aspects of bacterial physiology, but for concentration on what differentiates one kind of bacterium from another, that is those aspects of bacterial physiology that promise to offer points of attack to chemotherapy.[44]

Another basis for differentiating bacteria of potential use to chemotherapy lay in their physical characteristics. Domagk observed for example that among the theories of the Gram stain then current, that based on colloid chemistry enjoyed a certain popularity. Other promising physical traits included the binding of certain parisitotropically active dyes, and the property of bacteria in an electric current to move from cathode to anode. Domagk pointed out that the latter characteristic was shared by metal colloids and by numerous dyes, especially acidic dyes of a certain molecular size. Metal colloids, acidic dyes, and bacteria likewise were all subject to phagocytosis or accumulation in certain cells of the organism, especially the reticuloendothelial cells.[45]

If in Domagk's presentation chemotherapeutic research can be seen as a form of rational engineering or science-based technology, it is equally clear that for him this was first and foremost a biological and medical endeavor. The specifically medical and biological qualities of Domagk's approach may be seen in his motives, his methods and concepts, and his view of the relationships of laboratory medicine, clinical medicine, and chemistry.

Humane and scientific considerations gave the problem of bacterial infections enormous significance for medicine, but according to Domagk it was his conviction of the known and potential specificities of bacteria that opened the field to him. The early successes of chemotherapy had been won against infections caused by protozoans and spirochetes, not bacteria. In Domagk's view this was because the former kinds of organisms were relatively well differentiated, "and the further developed such a pathogen is, the more points of attack it appears to offer to chemotherapy." Because bacteria could also be differentiated on morphological, physiological, chemical, and physical grounds, however, a bacterial chemotherapy could be imagined. "All of these considerations gave me the certainty," he remarked, "that a chemotherapy of bacterial infections was also in the realm of possibility and that it must be worthwhile to intensively take this field of work in hand, although all prior investigations made this field appear rather hopeless." To these motives must be added Hörlein's leadership—mentioned elsewhere by Domagk—and the examples of successful protozoal and spirochetal chemotherapy. Behind the visible humane and scientific motives, of course, lay others no less real but usually absent from published discussions: ambitions for corporate and personal gain and glory for those who would solve a major problem of health and disease.[46]

Given the known and probable specificities of bacteria, Domagk doubted that a pantherapeutically effective medicine against bacterial infections would ever be found. This expectation was supported by prior experience with both protozoal and bacterial infections, for which specific chemical agents were required against specific types of microbes. Domagk emphasized that the specificity of the host-pathogen–chemical agent relationship presented a challenge to the "investigative physician," who needed to quickly and accurately identify the type of bacterium involved in a given infection. Domagk also pointed out that the kind of germ used was but one of several variables to be contended with in chemotherapeutic investigations. Other variables included the numbers of germs used (dosage), the virulence of the germs, and the resistance, size, breed, and sex of the experimental animals. These variables were so many determinants of the biological specificity that came into play in a chemotherapeutic investigation and that necessarily limited the generality of its conclusions. In the course of such an investigation these biological variables would be confronted by chemical variables in the form of compounds supplied by the organic chemist. Domagk was aware of this, but his nearly exclusive emphasis on biological factors argued implicitly for the relative importance of the medical side of the medical-chemical interaction in chemotherapeutic research.[47]

One methodological implication of biological specificity was the need for controls—animals experimentally infected but not treated chemically—at the beginning and end of each investigation. Since the control animals were chosen to resemble the other experimental animals as closely as possible, they could be viewed as differing in only one variable, namely, that they were untreated. The use of controls placed that one variable in clear relief. Thus one reason for the importance attached by Domagk to the choice of experimental model, and the qualities of the model he emphasized: certainty of infection in a high percentage of cases, and a high degree of uniformity of the course of infection in all animals.[48]

Animal experimentation was one of two paths taken by Domagk in the search for effective chemotherapeutic agents. The other was in vitro or test tube trials in which bacteria were directly exposed to selected compounds. Domagk was careful to specify the rationale and uses of in vitro tests. Whereas their utility in the search for disinfectants—substances that would kill bacteria outside the body—was direct and clear, the assistance they could offer in chemotherapeutic investigations was at best indirect and suggestive. Domagk observed that "for disinfection we seek remedies that damage and destroy as many different germs as possible at the same time, which is usually only possible when we hinder the most important life processes of the bacteria, for example by denaturing of the protoplasm, which is why most of the usual disinfection agents like alcohol, mercury compounds, phenol, and so on are strong protein-precipitating agents." But for just this reason, Domagk pointed out, most disinfection agents also damaged the cells of warm-blooded animals, with the exception of skin-surface epithelium. Thus many compounds that were powerful disinfectants proved damaging or useless as chemotherapeutic agents. Nevertheless Domagk was convinced that substances shown to be effective against bacteria in vitro could be valuable indicators of chemotherapeutically-promising classes of compounds. "In my opinion," he wrote, "it pays to test these preparations and related substances in a chemotherapeutic investigation, *even when the related substances do not show noteworthy results in vitro*" [my emphasis]. Just such systematic in vitro trials of the azo compounds had led Domagk and his collaborators first to new and powerful disinfectants without effect in the living organism, then to related sulfonamide-containing compounds with no in vitro effect but with powerful action against systemic streptococcal infections. Among the latter was Prontosil.[49]

Behind Domagk's conviction of the need for animal trials lay a fundamentally biological conception of the infection process and a view of chemotherapy as a complement to the organism's own defenses. Domagk saw infection as a "struggle between the macro- and the microorganism" in which one or the other would eventually prevail. The aim of chemotherapy was not to effect a cure alone but to come to the aid of the host in its struggle with the pathogen. In considering the mode of action of Prontosil and related compounds, for example, he had concluded that it would "best be explained by the hypothesis that the substances or transformation products attack the cocci in the organism, but that in many cases they do not themselves kill the bacteria but only damage them enough that they are susceptible to attack by the organism, and indeed especially by phagocytosis by white blood corpuscles and other inflammation cells." Among the latter, Domagk had a special interest in the reticuloendothelial system, the subject of his habilitation essay and the object of an early attempt to extract an antibacterial substance.[50]

Chemotherapy could complement but not substitute for the body's own defenses. "We can also expect no more effect from these new compounds," Domagk remarked, "when the situation of the patient is from the first entirely hopeless, and all natural defensive forces in the organism are already completely exhausted." The course of an infection therefore depended as much on the reaction-state of the host organism as on the number and virulence of the pathogens. Domagk emphasized, for example, that it made a great deal of difference whether or not the host had been previously infected with a specific causative agent. Prior infection in some cases sensitized the host in such a way that reinfection with the same pathogen provoked a greatly strengthened defensive reaction from the host, a reaction—characterized as allergic—that could itself be harmful.[51]

In line with this reasoning Domagk felt justified in asserting that chemotherapy could act not only by attacking the causative agents of disease but also by strengthening the host organism. Chemotherapy should be combined with serum therapy or with active immunization: "chemotherapy and serum therapy do not exclude one another; they should supplement one another." Domagk cited recent experience with vitamins A and C as evidence suggesting a promising role for vitamins in increasing the organism's powers against infections.[52]

Domagk's biological concept of infection informed and was informed by clinical experience. He pointed out, for example, that streptococci show an organotropy, or selective affinity, for specific tissues of the host, settling by preference in the joints or on the heart valves. Once settled in these secondary foci, they were much less accessible to the physician's grasp. The aim of therapy therefore had to be early destruction of bacteria at the primary focus (point of entry, bloodstream) by surgical means (such as tonsilectomy, appendectomy) or by chemotherapy. Animal experimentation had revealed a pitfall of chemotherapeutic investigation that might also become a hazard of clinical practice. In severely infected animals, Domagk pointed out, it often happened that so much toxin was released into the bloodstream by disintegration of bacteria in the course of chemotherapy "that even with the complete destruction of all of the germs by the chemotherapeutic agent almost all the animals perish, and this can simulate ineffectiveness of the preparation used." Animal experiments had shown that the effectiveness of a medicine could also be obscured by its manner of administration. Domagk observed that if too little of the medicine was given at widely spaced intervals, all the germs might not be destroyed and the infection might return. Therefore "on the basis of our experimental experience I would think that a short treatment with a sufficient dosage is more suitable than a long treatment with a small dosage." These examples begin to suggest the ways in which Domagk believed that a biologically informed conception of the complexities of the infection process and of the host-pathogen-drug interaction—and therefore the role of the medical researcher with experience in animal experiments—was crucial to the development of an effective chemotherapy.[53]

Domagk viewed chemotherapy from the vantage point of the medical laboratory, but also saw the research as a seamless union of that approach, based on in vitro and in vivo animal tests, with those of chemistry and clinical medicine. He emphasized the independent role of the clinic. In Domagk's view, for example, only closer cooperation between clinicians and bacteriologists could assure the necessary early determination of the specific nature of infections. Just as in vitro tests alone could not substitute for trials in living animals, the latter could not replace clinical experience. "Experiment can and should give only the direction for therapeutic leadership in practice," Domagk remarked. Domagk reported that efforts to use Prontosil and related compounds in other than streptococcal infections, and attempts to find compounds more effective than Prontosil, had revealed a certain looseness in fit between laboratory and clinical results. Positive results in animal trials were not necessarily confirmed in the clinic. On the other hand, the weak performance of a compound in animal tests might be followed by outstanding clinical results. Such divergences underlined the need for clinical trials.[54]

The position of chemistry in Domagk's picture of chemotherapy was equally positive but was qualified by the phrasing and exclusions of his presentation. On the one hand he referred frequently and warmly to the essential roles played by chemists in chemotherapeutic research generally and in the development of Prontosil and related compounds in particular. Whatever importance Domagk attached to his own conviction of the possibility of a bacterial chemotherapy, he also recognized that "an intensive cultivation of this field became possible only through the full engagement of chemists who were not easily denied and who were enthusiastic for this work," namely, "Dr. Mietzsch and Dr. Klarer, who did not give up the systematic cultivation of streptococcal and staphylococcal infections with me in spite of years of shared disappointments." Domagk's terms are revealing. He referred repeatedly to his "cooperative work" (*Zusammenarbeit*), "chemical-medical cooperative work" (*chemisch-medizinische Zusammenarbeit*), or collaborative work (*Gemeinschaftsarbeit*) with Mietzsch and Klarer, in such a way as to indicate unequivocal recognition of the interdependence of chemistry and laboratory medicine. He explicitly credited Mietzsch and Klarer with producing Prontosil and related compounds. In 1935 he lectured on the chemotherapy of bacterial infections at a meeting of the Society of German Chemists and published the lecture in a chemical journal, *Angewandte Chemie*. In terms that echoed the great nineteenth-century physiologists, he endorsed "exact experiment" and association with the physical sciences as the keys to medicine's future: "the therapy of infectious diseases will be able to be placed on firm foundations only when medicine is lifted out of its present condition of inexactness through the still closer cooperative work of chemists and medical men." As was the case with Hörlein, Domagk's confidence in chemistry and optimism about its future role in medical research extended beyond chemotherapy into other areas, such as vitamins or immunology.[55]

At the same time, the language and format of Domagk's presentation tended to obscure the importance of chemical reasoning and to assign to the chemists a subordinate role in chemotherapeutic research. In describing the breakthrough achieved in the fall of 1932, for example, Domagk wrote that "the situation was first fundamentally altered when I succeeded in demonstrating an unequivocal chemotherapeutic action of sulfonamide-containing azo compounds in the streptococcus-infected mouse. Now a systematic chemical cultivation of this field could begin." Later in the same paper Domagk remarked that subsequent experimental work "always confirmed the observation established by me in common work with Mietzsch and Klarer, that for the chemotherapeutic effect the sulfonamide group is typical of all active compounds, and indeed in definite chemical positions." Even when Domagk used the first-person plural the effect was sometimes ambiguous, with the phrasing possibly but not certainly meant to include the chemists, as in "we demonstrated the chemotherapeutic effect of the Prontosil compounds found by us on the mouse infected with hemolytic streptococci or on the rabbit infected with streptococci."[56]

Often Domagk did not give a clear picture of the chemical reasoning behind the testing of particular compounds or classes of compounds. For example, the reader is left with the impression that azo compounds were chosen for trial simply because some of them previously had been shown to be effective disinfectants. No reason is given by Domagk for examining sulfonamide-containing azo compounds. Thus "in the course of our investigations we came to a group of very nontoxic sulfonamide-containing compounds," and so forth.[57]

The public format of Domagk's presentation—academic scientific papers in medical or scientific journals—itself may have militated against perception of the collaborative nature of the research as carried out in a highly structured industrial setting. At the head of these papers Domagk was called "Prof." or "Prof. Dr." and associated with the Section, or Institute for Experimental Pathology and Bacteriology at I. G. Farbenindustrie, Wuppertal-Elberfeld. In accordance with academic usage he assigned credits, or acknowledged debts, primarily in the form of citations to other literature. His own work appeared as among the most recent links in a chain of academic medical investigations. The largely unpublished and simultaneous work of his chemist-collaborators, and the more comprehensive research organization of which Domagk and the chemists were a part, lay outside this form of presentation, and therefore, despite Domagk's references to Klarer and Mietzsch, largely outside the perception of his readers.

At the end of one of his papers Domagk appealed to the younger generation of doctors to take in hand the research problems of chemotherapy as evidence of their enthusiasm for the medical calling.[58] The implicit if unintended message of this appeal, and of Domagk's publications of the 1930s, was that chemotherapy was a field of medical research in which chemists played an essential but subordinate role. Hörlein, who once described pharmaceuticals as "a branch of chemical technology," held a different view. So too, more emphatically, did the chemists.

The Chemist's View of Innovation: Fritz Mietzsch

Domagk's chemist-collaborators in the research that led to the sulfa drugs were Joseph Klarer and Fritz Mietzsch. Although both men left accounts of the work, it was Mietzsch who was most prolific and whose publications offer the fullest chemistry-centered complement to Domagk's biomedically oriented story. In several articles of the 1950s, Mietzsch outlined the historical development of chemotherapeutic research at Bayer, placing the sulfas in a larger chronological and chemical context. Mietzsch's account pointedly acknowledged the essential roles played by medical researchers in chemotherapeutic advance, and Domagk's role in recognition of the first sulfas in particular. Mietzsch's emphasis was on chemistry, however, and the selections, exclusions, and form of his reconstructions make it clear that for him it was primarily chemical ideas, chemical constraints, and chemical opportunities that drove and directed the research process. In Mietzsch's story the role of biomedical considerations was simplified and pushed into the background. At least implicitly or by default, Mietzsch's historical reflections assigned to biomedicine a subordinate and ancillary position in chemotherapeutic research.

In contrast to Domagk, who came to chemotherapeutic research via medicine and experimental pathology, Mietzsch's point of departure was dye chemistry. Born in 1896 in Dresden, Mietzsch proceeded from a Dresden gymnasium to studies in that city's Technische Hochschule, which he entered in 1915. Mietzsch's training, like Domagk's, was interrupted by war service. He returned to finish in 1922 with work completed under Walter König "on the parallelism of cationic and anionic halochromism in polymethine dyes from polyvalent phenols and indols." König, who was director of the Dresden Laboratory for Dye Chemistry and Dyeing Technology, left a lasting impression on Mietzsch: in later life he kept at hand in his desk, and consulted, his notebook of König's lectures.[59]

Following a short period as assistant in König's laboratory, Mietzsch, on his mentor's recommendation, joined Bayer at Leverkusen in early 1923. As a newly-hired dye chemist Mietzsch took the mandatory professional development course (*Ausbildungskurs*) directed by Richard Kothe and Oscar Dressel, with whom Hörlein had collaborated in his early work at Bayer on dye chemistry. Among the results of the course for Mietzsch would have been a lively awareness of the importance of the azo dyes and their initial and intermediate products. In 1924, a year after Mietzsch joined Bayer, the company's search for a "better quinine" reached a first culmination with the synthesis by the Elberfeld team—Werner Schulemann, Fritz Schönhöfer, and August Wingler—of the compound later termed Plasmoquine, the first purely synthetic antimalarial. By that time Mietzsch was working in the company's triphenylmethane dye division in Leverkusen. It occurred to him that substances resembling quinine and therefore possibly having antimalarial action might be synthesized by hydrogenation of polymethine dyes out of 6-methoxylepidine. The suggestion caught the attention of Hörlein, and in the fall of 1924 Mietzsch made the decisive move from Leverkusen to Elberfeld, and from dye chemistry to pharmaceutical chemistry.

At Elberfeld Mietzsch concentrated at first on extension of the principle of action of basic alkylation—used by Schulemann, Schönhöfer, and Wingler in the synthesis of Plasmoquine—to other classes of compounds. Application of basic alkylation in the acridine series led first to a compound with promising action in malaria. This finding prompted systematic research with chemist Hans Mauss and physician Walter Kikuth on basic alkylated acridine compounds, resulting in synthesis in 1930 of the powerful antimalarial subsequently known under the trade name Atabrine. In 1927, the year that Domagk joined Bayer, Mietzsch and Joseph Klarer began systematic study of the basic alkylated azo dyes. The compounds they prepared were submitted to a variety of in vitro and animal trials in the laboratories of Domagk and other medical researchers in Elberfeld. It was this line of work that led in 1932 to the group of compounds recognized by Domagk to have a marked action on systemic streptococcal infections. Among these was Prontosil. In the 1940s Mietzsch, together with chemists Hans Schmidt and Robert Behnisch, developed the thiosemicarbazones, partly as an extension of sulfonamides research. The effectiveness of these compounds against tuberculosis was established by Domagk, and they entered commerce and clinical practice under the trade names Conteben and Solvoteben.[60]

Honors and increased stature within the company followed Mietzsch's successes in chemotherapeutic research. Among the former were the Emil Fischer medal of the Society of German Chemists awarded in 1934 for the work on synthetic antimalarials, and an honorary doctorate of medicine from the University of Münster in 1945 for synthesis of the sulfonamides. In 1937 Mietzsch was named to the governing body of the pharmaceutical research division, and in 1939 he was given the title *Prokurist.* When Heinrich Hörlein retired in 1949, Mietzsch took his place as director of pharmaceutical research in Elberfeld and held this position until his death in 1958.

In his articles of the 1950s Mietzsch presented the history of chemotherapy as predominantly a study of classes of chemical compounds: their synthesis, variability, therapeutic properties—defined in broad terms—and relationships to one another. By this account the origins of chemotherapy lay in dye chemistry. Mietzsch observed that in the late nineteenth century, recognition of the selective and specific affinity of synthetic dyes for certain cells and cell constituents, and the finding that certain dyes had strong bacteriocidal activity, had made synthetic dyes the first working theme (*Arbeitsthema*) of the new science of chemotherapy.[61]

Mietzsch pointed out that chemotherapeutic research at Elberfeld could be traced back to Paul Ehrlich. In 1911 Hörlein had hired Ehrlich's student and collaborator Wilhelm Roehl to implement Ehrlich's program at Bayer. The small research team soon began to develop its own plans. By Mietzsch's account, the first of these took as point of departure the known effectiveness of the azo dyes trypan red, trypan blue, and afridol violet against trypanosomal infections in mice. The aim was to find compounds similar in effectiveness but colorless. Working with Roehl, the chemists Dressel and Kothe identified the urea-sulfonic acids as satisfying these criteria. Among them was the compound first synthesized in 1916, and tested and placed on the market by 1923 as Bayer 205 or Germanin, the first effective chemical therapy for sleeping sickness.[62]

As the Bayer 205 story illustrates, a major theme of Mietzsch's history was definition of the chemical relationships that linked one group of chemotherapeutic or pharmacologically active compounds to other groups, sometimes in unpredictable and unexpected ways. Further illustration of the theme could be found in the connections between several classes of chemotherapeutic agents, including the basic al-

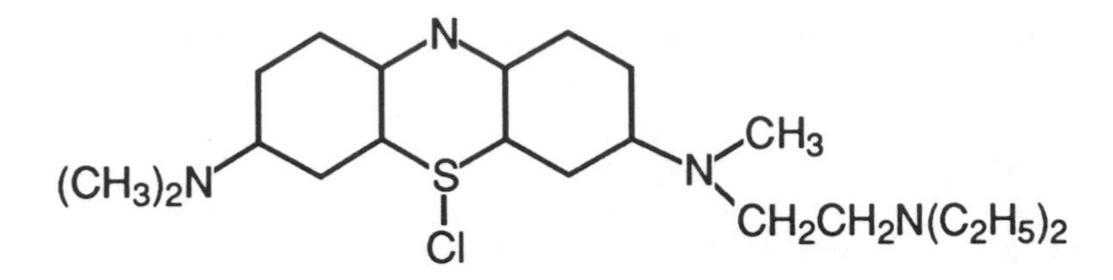

Figure 4. Basic alkylated methylene blue. After formula given in Fritz Mietzsch, "Neuere Entwicklung der synthetischen Chemotherapeutika," *Österreichische Chemiker-Zeitung* 53 (1952): 177–187 (on 178). Labels and bonds are as indicated in the original.

kylated compounds, the sulfonamides, and the thiosemicarbazones. Mietzsch proposed "to show how they have arisen from one another [*wie sie auseinander hervorgegangen sind*] and how they have developed in close contact with new pharmacologically active synthetic medicines."[63]

The importance of the principle of basic alkylation had first been perceived in pharmacological research, Mietzsch pointed out. It had then been transferred to chemotherapy with great success, especially in development of compounds against tropical protozoal infections. Finally, more recently, it had been reintroduced into pharmacology with very useful results. Mietzsch illustrated the early application of basic alkylation by reference to Novocain (procaine hydrochloride) and other local anesthetics and to various compounds with action on the heart and uterus. But the full significance of the principle had become evident, Mietzsch argued, only when aminoalkyl residues were introduced into an amino group joined to a ring. A crucial step was taken by Schulemann, Schönhöfer, and Wingler when they had the idea of converting methylene blue—known to be effective in quartan malaria—into a stronger basic derivative. The resulting diethylaminoethyl derivative of methylene blue proved more effective than the parent compound (Figure 4). The chemists had "thereby opened the era of synthetic antimalarial drugs in Elberfeld." With the search for "a better quinine" as a point of reference for Bayer's antimalarial research, the next step was transfer of basic alkylation to compounds structurally related to quinine, the aminoquinolines. It was by this path, Mietzsch pointed out, that Schulemann, Schönhöfer, and Wingler, working with Roehl, arrived at Plasmoquine.[64]

Mietzsch entered his own story with his account of yet another transfer of the principle of basic alkylation, this time to the acridine series. Roehl was aware that neither Plasmoquine nor quinoplasmine—the other synthetic antimalarial available by 1924—provided a complete solution to the malaria problem. This was because these agents acted on the sexual but not the asexual forms of the parasite. The task therefore was to find a chemical effective against the asexual form. In his work with Ehrlich, Roehl had become familiar with trypaflavine, an acridine compound used as an antiseptic. He conceived the dual ambition of finding another acridine compound effective as an antimalarial—"Malaflavine" is the name he coined in confident anticipation—and a third active against streptococcal infections. Roehl's early death in 1929 prevented his personal achievement of either goal.[65]

Mietzsch collaborated with Roehl on the acridine series from 1924, and in 1926 they were joined by the chemist Hans Mauss. Mietzsch reported that the initially modest results were improved "when, in an approach to the chemical constitution of quinine, 9-aminoacridine was made and the diethylaminoethyl butyl residue (an alkyl radical), already recognized as optimal in Plasmoquine, was used." Even this was not enough, however, and the chemists decided to undertake a systematic study of the influence of the position and nature of nuclear substituents in the acridine ring. The result was recognition of the special importance of introduction of a chlorine atom in the 6-position of the ring. The compound so obtained in 1930 proved to have an action qualitatively similar to quinine but quantitatively much superior. This was Atabrine, a major new synthetic antimalarial, introduced into commerce and medicine in 1932[66] (see Figure 6, top).

By the time Atabrine was synthesized Mietzsch and Klarer had already embarked on the line of research that would lead to Prontosil. By Mietzsch's account it was a program with important chemical lines of continuity with the work already described. "The discovery of the therapeutic effect of the sulfonamides," Mietzsch wrote, "also goes back in the last analysis to the study of basic alkylated compounds." Plasmoquine and Atabrine had shown that powerful therapeutic effects could be obtained by use of an amino group substituted with an aminoalkyl residue. In 1927 Mietzsch and Klarer began to apply the principle of aminoalkylation to the aminoazobenzol series. As we have seen, azo dyes and related compounds with bacteriocidal action in vitro (for example, crysoidine, pyridium, neotropine) had been known for the preceding two decades (Figure 5), making the group a promising target for investigation. According to Mietzsch, however, the azo compounds were even more appealing on strictly chemical grounds. "To us in Elberfeld," he wrote, "precisely the field of the azo compounds seemed especially suitable for the systematic study of the influence of substituents on chemotherapeutic action, because by the choice of different coupling and diazo components in which a basic alkyl residue was contained, numerous combinations were made possible." Creation

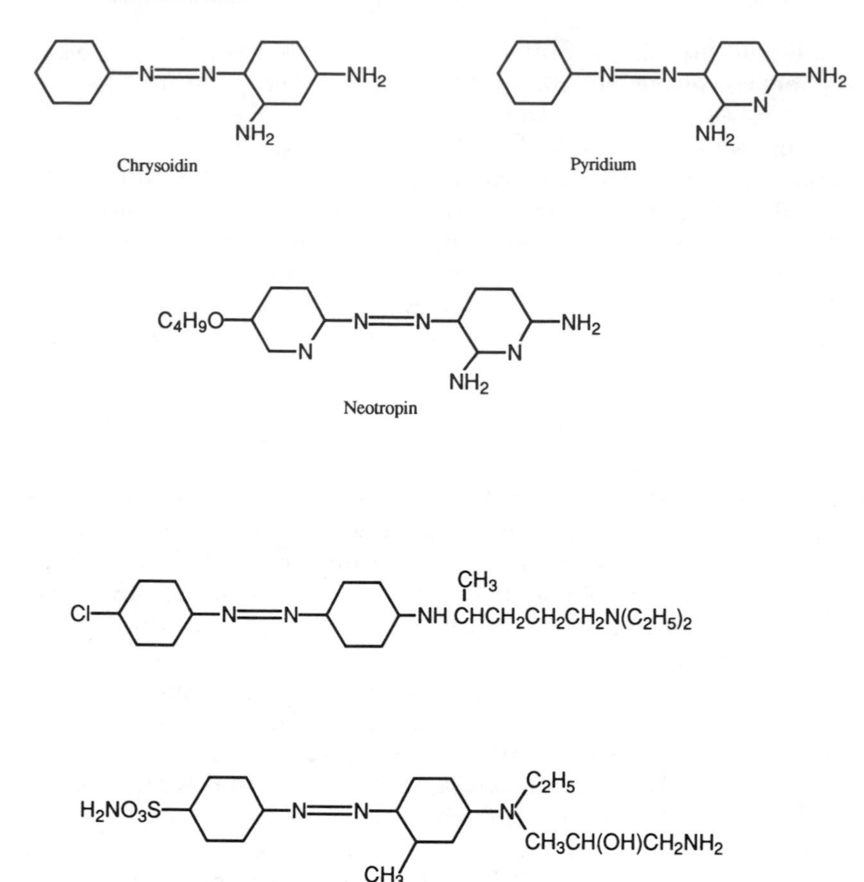

Figure 5. Crysoidine, pyridium, neotropine, and two unlabeled azo compounds. After formulas given in Fritz Mietzsch, "Entwicklungslinien der Chemotherapie (Vom chemische Standpunkt gesehen)," *Klinische Wochenschrift* 29 (1951): 125–134 (on 129). Labels and bonds are as indicated in the original.

of such derivatives was also easier, and more easily controlled, than was the case in the quinoline and acridine series. Mietzsch recalled that when he began this work with Klarer, they believed that compounds of this type would find chemotherapeutic applications, "without of course at first being able to predict the specific indications," that is, what compound would act on what pathogen.[67]

The search for new chemotherapeutic agents now became—in Mietzsch's presentation—an organic chemist's puzzle. "Here also the law of the right man at the right place holds," he wrote, "that is, translated into chemical language, the right substituents in the right position on the azo group, when it is a question of bringing out the slumbering chemotherapeutic characteristics in these systems." Klarer and Mietzsch prepared numerous basic alkylated amino azo compounds. Once again the example of a known effective compound of a different type proved useful. In analogy with Atabrine, the chemists introduced a chlorine atom in one p-position of the azobenzene, and a diethylaminoethyl residue in the other p-position. The resulting compound showed a certain action on trypanosomes, and the p-position proved to be especially favorable in other cases. Thus the p-position methoxy group, combined with a diethylaminoethyl residue, had an action on bird malaria; the p-position chlorine atom, combined with a primary aminooxypropyl residue, showed good bacteriostatic action on coli bacteria; and the p-position iodine atom had an influence on rat leprosy (Figure 6).[68]

It was in the course of this chemically well-defined research program that the first sulfonamide compounds were prepared. "As we tested in this way all the available substituents," Mietzsch recalled, "we remembered the earlier sulfonamide-containing azo dyes, which had also come out of work at Elberfeld. Transferred to our chemotherapeutic problem that meant the use of 4-aminobenzenesulfonamide [sulfanilamide] as diazotizing component." The compound then produced with sulfanilamide and aminooxypropylethyl-m-toluidine was turned over to Domagk (see Figure 6, bottom). It was the first compound to show a striking therapeutic action in mice infected with streptococci, "and thereby," Mietzsch noted, "the way to sulfonamide therapy was opened up, a way along which moved all later work at home and abroad."[69]

Emphasis was now shifted to study of sulfonamide-containing azo compounds. Variations on the new theme began to distinguish what was essential to the therapeutic action from what was not and, in Mietzsch's words, "the path that previously had led from the simply constructed crysoidine to the complicated aminoazobenzenes would be retraced backwards." The chemists soon found that, despite the importance attached to basic alkylation, the aminoalkyl residue in the amino group was not necessary to therapeutic action, and that the methyl group could be replaced on the benzene nucleus by an aliphatic amino group. The simplified compound that resulted was Prontosil. The chemists were further able to show that the basic m-phenylenediamine could be replaced by acid coupling components. One compound produced in this way was Prontosil solubile, which had the therapeutic advantage over Prontosil of greater solubility. Finally, through the work of the Institut Pasteur team of Jacques and Thérèse Trefouel, Frederic Nitti, and Daniel Bovet in 1935, it was learned that the second benzene nucleus

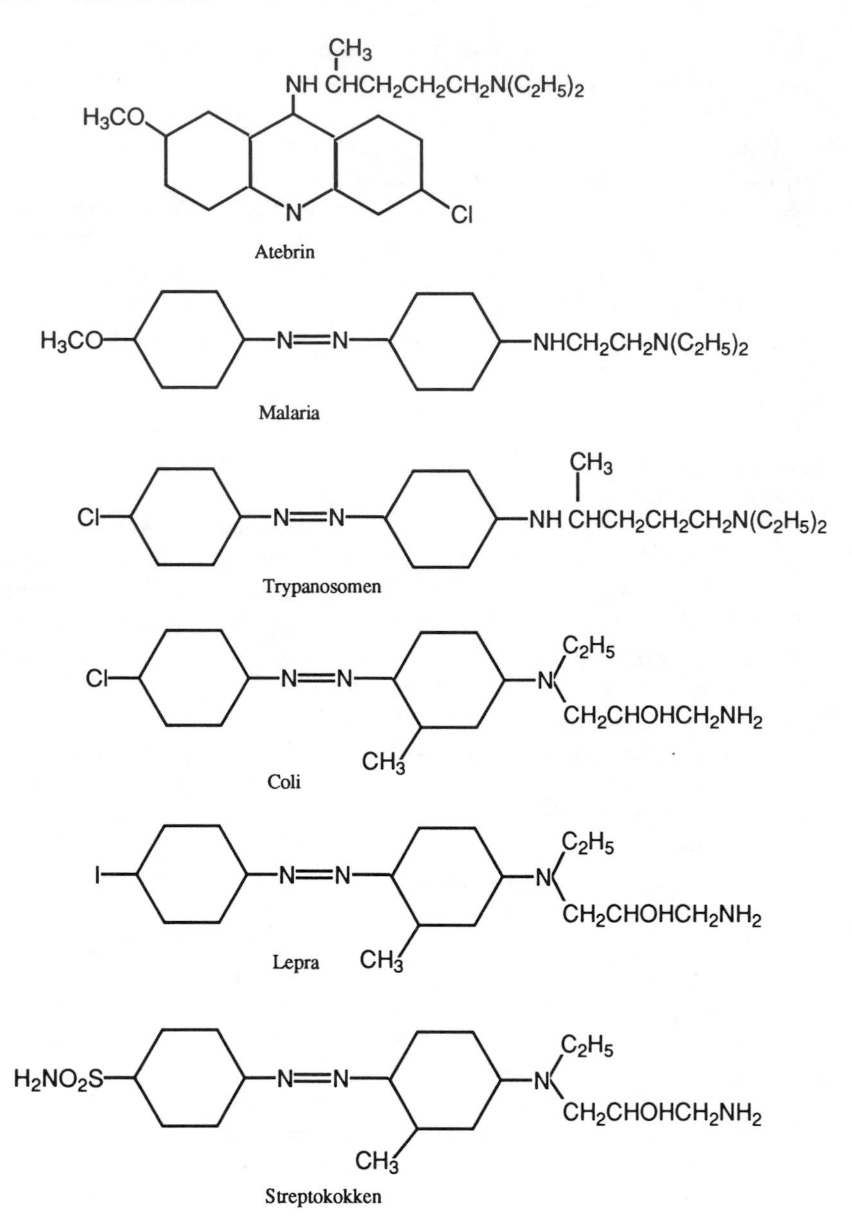

Figure 6. Formulas showing relationships of Atabrine and several azo compounds developed at Elberfeld, 1927–1932. After Fritz Mietzsch, "Chemie und wirtschaftliche Bedeutung der Sulfonamide," *Arbeitsgemeinschaft für Forshung des Landes Nordrhein-Westfalen* 31 (1954): 7–32 (on 11). Labels and bonds are as indicated in the original.

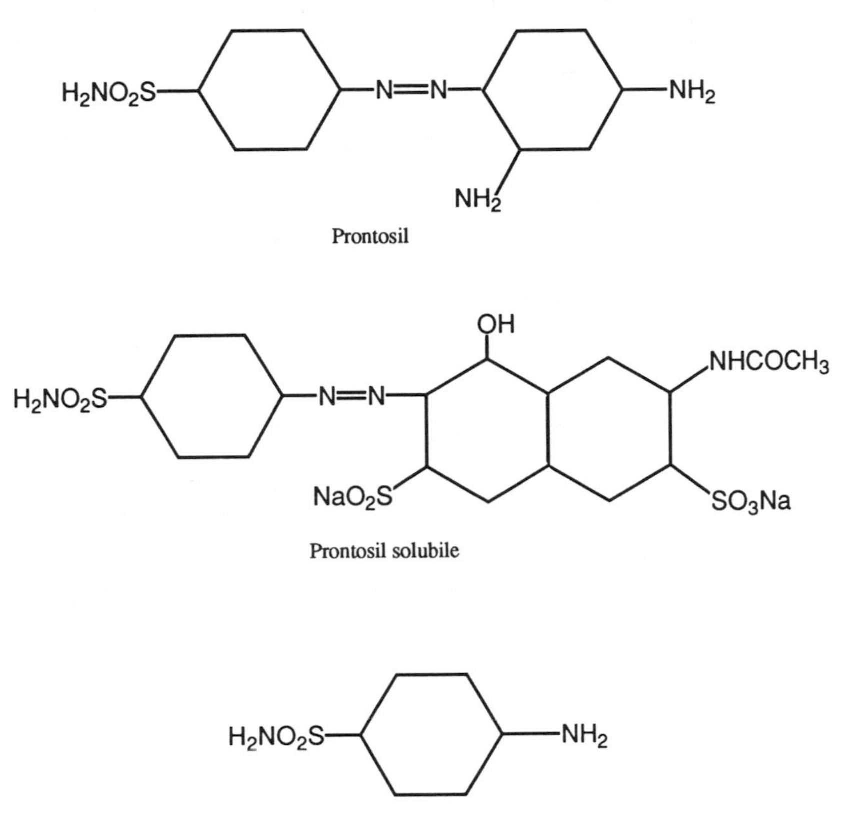

Figure 7. Prontosil, Prontosil solubile, and Prontalbin (sulfanilamide). After Fritz Mietzsch, "Entwicklungslinien der Chemotherapie (Vom chemischen Standpunkt gesehen)" *Klinische Wochenschrift* 29 (1951): 125–134 (on 129). Labels and bonds are as indicated in the original.

could be left out and that the resulting "fragment" compound, sulfanilamide (I. G. Farben trade name: Prontosil album or Prontalbin) possessed therapeutic action comparable to that of Prontosil (Figure 7).[70]

Mietzsch carried his outline of the development of chemotherapy through the later history of the sulfonamides, continuing to emphasize the chemical determinants of the research and the chemistry that bound the sulfonamides to other—in this case subsequently developed—classes of synthetic medicines. Among the latter were the sulfones, research on which was revived in the late 1930s by the success of the sulfonamides, and which provided new weapons against leprosy;

and the thiosemicarbazones, the important tuberculostatic effects of which were first recognized in the 1940s through their use as intermediate products in the preparation of certain sulfonamides.[71]

Mietzsch's presentation of the history of chemotherapeutic research at Elberfeld was laced with references to the medical researchers. From his vantage point as witness to a long series of successes in that research, and as director of the pharmaceutical research division in the 1950s, he had a clear perception of the need for a "spatially and intellectually close collaboration between the chemist who has turned to medical problems and the chemically minded medical man." Mietzsch specified the several roles of the medical researcher. Animal experiments were necessary because in vitro trials alone were unreliable, especially in bacterial infections, and because "the behavior of the preparation in the presence of serum, its reabsorption, its decomposition, its excretion and accumulation in the living organism is decisive for its chemotherapeutic value." The causative agents of infectious diseases had to be identified and their life cycles studied. Pathogens had to be transferred from human to animal hosts, and animal experimental models so constructed that chemical preparations could be tested under conditions that were as close as possible to those of human infections. Special techniques had to be developed, such as Roehl's esophageal sound method for oral administration of medications in canaries with infections similar to malaria. In at least one passage, Mietzsch emphasized the role of the medical researchers Roehl, Domagk, and Kikuth in solving the problems of identifying effective antimalarial and antistreptococcal agents. He recognized that Domagk had come to Elberfeld with the conviction that "something fundamentally new" would be required against bacterial infections, and referred to Domagk's "direction-setting" work on sulfonamides. Referring to the work on new synthetic antimalarials before and during World War II, he acknowledged the debt this research owed to new knowledge of the biology of the causative agent of malaria and to new animal tests.[72]

For all that, Mietzsch's narrative left little doubt that the primary engine of chemotherapeutic research was chemistry. All of the major turning points he described were characterized in chemical terms. He gave a leading role to the principle of basic alkylation. It was Schulemann, Schönhöfer, and Wingler's introduction of aminoalkyl residues into an amino group joined to a ring, in converting methylene blue into a stronger basic derivative, that "opened the era of synthetic antimalarial drugs at Elberfeld." In Mietzsch's narrative, Plasmoquine was the result of transfer of the principle of basic alkylation to the aminoquinolines, and the latter were chosen for the effort because of their structural similarity to quinine. Atabrine resulted from transfer of the principle of basic alkylation to the acridine series. The sulfonamides emerged from study of basic alkylated compounds and from use of azo dyes in chemotherapy. Azo dyes were appealing to Mietzsch and Klarer above all for a chemical reason, their ease of variability. The "decisive turning point" came when a sulfonamide group was introduced in the p-position. Thus the decisive turning point in the invention of the sulfonamides was characterized chemically and located in the chemical side of chemotherapeutic research at Elberfeld.

Mietzsch's overview of the past of chemotherapy and anticipation of its future were couched in chemical terms. "A survey of the structural formulas used," he wrote, "shows that a series of ring systems that are obviously well reabsorbed, transformed, and excreted by the human body, appear again and again. On the basis of a few fortunate ideas," he continued, "these ring systems are then further substituted, so that in this way synthetic elements [*Aufbauelemente*] which have already been proven to be profitable somewhere in chemotherapy, are partially used." The future of chemotherapy, Mietzsch thought, would depend on gaining further knowledge of the chemistry of life processes and on perfection of the methods of organic chemical synthesis. The yield would be ever more specifically acting and nontoxic compounds.[73]

Mietzsch was at pains to display the mental effort behind the chemical side of chemotherapeutic research. The research process is not merely trial and error, the blind grinding out and testing of one compound after another. Patterns emerge. A new chemical procedure or phenomenon is found that leads to a useful result in the sense of an effective chemotherapeutic agent. The finding is then applied to other classes of compounds in the hope of getting such a useful result with them. In one series in which this is tried, a promising compound is identified. That leads to more systematic work on that series, which in turn may lead to finding a successful chemotherapeutic agent. Thus the research process on the chemical side is not random, although it does have a trial-and-error quality. The trial and error takes place within certain boundaries that are determined by some combination of what it is possible to do chemically and previous experience with certain chemical processes or certain classes of compounds in regard to identification of successful chemotherapeutic agents. The trial-and-error process appears to have two phases: an initial, relatively loose phase in which promising leads are sought within a fairly broad domain of possibilities; and a second, more intense and systematic phase in which a lead that has been identified is followed up, for example, within a well-defined class of compounds.

For Mietzsch the search for a chemotherapeutically effective compound therefore involved a narrowing down to the right kind of chemical specificity: the right series, the right position on the compound, the right substituent. For the chemists the problem was one of chemical variation and chemical specificity, with the goal of achieving a precise match with the target organism.

Central to Mietzsch's message was the conviction that he and other Bayer chemists were pursuing not merely a chemical, but a specifically chemotherapeutic research program. Their role was not simply instrumental but involved deliberate judgments and choices based on rational and goal-directed considerations, chemical reasoning in the service of chemotherapeutic goals. Mietzsch saw the chemotherapeutic research process from the standpoint of a chemist's—specifically a dye chemist's—training, a chemist's way of thinking, and a chemist's vivid intuitive sense of the creative process in the chemical, as distinct from the biomedical, side of the research. Given the shortcomings of Domagk's presentation of the chemists' role, and the disproportionate glare of publicity that fell on the biomedical side of the research through award of the Nobel Prize to Domagk alone, a compensatory effort also lay behind Mietzsch's emphasis. As he put it in the gentlest possible terms at the end of one article, Domagk's Nobel Prize also meant "for those of the Bayer company active in this research a recognition of the scientific standards of this workplace before the whole world." What he might have said was that the achievement for which Domagk was honored was a collaborative one in which the chemists played a coequal, and in some respects a leading, role.[74]

Conclusion: Chemotherapy as Industrialized Invention

Comparison of the three images of chemotherapy presented by Hörlein, Domagk, and Mietzsch, respectively, yields a sharper picture of the development of the first sulfa drugs as a case of industrialized invention. The most important elements of this picture are a systemic approach to innovation; the technological momentum embodied in continuities between dyestuff chemistry and pharmaceutical chemistry and between the different chemical lines of pharmaceutical research; the optimism of the Bayer/I. G. Farben industrial team amidst a general atmosphere of pessimism; and the tensions created by the persistence of academic values and practices in the midst of a commercial and industrial institutional setting.

We have seen that Hörlein viewed bacterial chemotherapy as a reverse salient in a broadly advancing front along which chemistry and medicine joined battle with disease. Working within and expanding a well-developed industrial research organization, he embraced a systemic approach to innovation. He expected technically and scientifically novel results to emerge not from individuals working alone but from groups of specialists working cooperatively under a team leader, and supported by a strong research infrastructure. He saw and expected patterns in the work, with innovation occurring in typical steps or phases depending on the fields or kinds of problems involved. He acted the role of research manager to the fullest: setting aims, concentrating resources and personnel, assessing results, and representing his organization's efforts to the medical and scientific communities. It was Hörlein who brought Domagk, Mietzsch, and Klarer into the pharmaceutical research division, and it was in close association with him that their research program in bacterial chemotherapy was formulated and conducted.

For Hörlein's program to succeed, however, critical problems had to be defined and solved by the medical researcher and the chemist, each working in his own domain and with his own concepts, methods, and habits of thought. As we have seen, both medical researchers and chemists confronted the problem of specificity. The difference was that in one case the specificity was biological; in the other case, chemical. Biological specificity meant identification of the difference between host and pathogen, such that a purchase was offered to attack the pathogen without harming the host. It meant, in addition, definition of differences between the kinds of bacteria, differences that could inhere in subtle metabolic, nutritional, or other aspects of the organism. Chemical specificity meant conceiving and synthesizing all possible variants on a given kind of compound—that is, by starting with a nucleus of some sort and building around it; or in a more general sense, the division and subdivision of chemical compounds—in this case mainly or exclusively organic ones—into classes and sub-classes.

Both the medical researcher and the chemist ostensibly had the same goal, namely, to identify a specific compound that would attack a specific pathogen that was harming a specific host. But to reach the goal two complementary but distinct efforts had to be mounted in which each investigator wrestled with his own problems of specificity. The medical researcher had to find an experimental model that was adequate to the pathogen in the sense that experimental infections could be consistently produced with that type of pathogen, and that was adequate to the host in the sense that it was the same pathogen that attacked the human host and against which a chemotherapy was sought. He also had to take into account the reaction of the host organism, both to the pathogen and the changes it induced, and to any chemical agent introduced as a therapeutic measure. The medical researcher also had to consider specificity in the sense of the particular locale (tissue, organ, cells) attacked by the pathogen, and the pattern of distribution of the chemical agent in the host organism.

The chemist had to decide—with or without the aid of the medical researcher—which class or classes of compounds to try against the kinds of infections in question. His criteria for doing so might be chemical—for example, ease of variability by changes in coupling components or substituents, or prompted in some way by prior experience with living organisms, such as in vitro activity against bacteria. Once a promising compound was identified by the medical researcher, the chemist's task was to produce variations around this starting point with the aim of increasing the specificity of the match between the chemical agent, the pathogen, and the host. Here indeed was a signal difference between the two components of chemotherapeutic research. Whereas the medical researcher took the variability given in nature, tried to analyze it in detail, and identified points where chemical therapy might find a purchase, the chemists produced variability by synthetic manipulation.

Chemotherapy so conceived and practiced cannot be regarded as a unitary scientific discipline in the academic sense. The nature of its problems necessitates cooperation between chemists and medical researchers, the latter variously identifying themselves as pathologists, bacteriologists, physiologists, or in other ways. Neither the chemist nor the medical researcher by himself has the necessary experience to conduct research in the field. Chemotherapy in Hörlein's conception therefore is defined not so much by a move for autonomy vis-à-vis other disciplines—as, for example, physiological chemistry or biochemistry vis-à-vis physiology—as by the necessary cooperative linkage of two or more disciplines that continue to retain distinct identities while interacting at specific sites on their boundaries. That is, chemotherapy can be characterized as a permanent cooperative association of two or more disciplines around certain problems.[75]

The problems were first defined by Ehrlich. To some degree, as Hörlein pointed out, Ehrlich embodied in himself the two necessary components of the field, the medical and the chemical, although even in his case he relied on chemists for part of the work. But the fine-chemicals industry, because of its prior commitment to pharmaceuticals, already had in place one of the institutional arrangements required by Ehrlich's program, that is, an organization in which chemists and medical researchers were in cooperative association. What was needed was the addition of a new component that would emphasize the particular aspect of the medical-chemical interaction defined by Ehrlich. At Bayer, for example, this took the form of the new chemotherapy laboratory added in 1910, or of Domagk's laboratory set up in

1927 and expanded in 1930. In the setting of the fine-chemicals industry the components of Ehrlich's program were more clearly differentiated in both methodology and organization. Thus the primary aim of chemotherapy—essentially an engineering goal—could be most effectively realized in a setting that was interdisciplinary in the sense of juxtaposing more than one field in a single institutional setting; and industrial, because of the massive resources that could be brought to bear to set up and maintain a differentiated research infrastructure and—in the case of the fine-chemicals industry—to make available thousands of synthetically produced chemicals and the expertise to vary these compounds in determinate ways to achieve the ends in view.

We have already seen that the vision and practice of the Bayer/I. G. Farben pharmaceutical research division under Hörlein's direction can be illuminated by use of Hughes's concepts of reverse salients and critical problems. It might also be argued that pharmaceutical research at Bayer/I. G. Farben, and chemotherapeutic research in particular, were products of, and were incorporated into, the technological momentum of the fine-chemicals industry. In Hughes's formulation a technological system is said to have momentum when it has mass and velocity. The research and development enterprise within Bayer/I. G. Farben had mass in the sense of an aggregate of devices, processes, artifacts (for example, compounds), organizational structures, trained personnel, and so on, in which substantial capital had been invested. It had velocity, analyzed into a rate of innovation—not yet studied quantitatively but, to all appearances, at least steady—and direction. The direction of the research enterprise was a product both of goals imagined and set by researchers and managers and of responses to opportunities opened up by the results of prior research and development efforts. The endeavor was driven in part by economic motives that favored diversification to maximize return on capital investment and to cushion fluctuations in demand for any one kind of product. As Meyer-Thurow has shown, the perceived need for diversification was one factor behind the institutionalization of research within Bayer from the 1880s. Finally, a powerful component of the momentum of the research enterprise within Bayer/I. G. Farben was the culture that had developed around it by the early twentieth century. Many individuals and institutions in the medical and academic scientific communities were committed in one way or another to its goals, methods, and products.

In several ways the research that produced the first sulfa drugs may be viewed as an instance of this technological momentum. The movement from synthetic dyestuffs to pharmaceuticals to chemotherapy to bacterial chemotherapy may be regarded as a series of steps in the diversification of an industrial enterprise and its research establishment. Continuity from dyestuffs chemistry to pharmaceutical chemistry was expressed in personnel, in concepts and methods, and in the use of specific products. Hörlein, Mietzsch, and Klarer were trained in dyestuffs chemistry and practiced it before moving to pharmaceuticals. On its chemical side, chemotherapeutic research involved methods of synthesis, and concepts of specificity and variability taken over from dyestuffs chemistry. Hörlein drew a direct analogy between the natural/synthetic dichotomy in dyestuffs and the natural/synthetic dichotomy in medicinal substances. The azo dyes were tried as medicines, and one of them became Prontosil. According to Mietzsch there was a thread of continuity in the research at Elberfeld running from the basic alkylated compounds through the sulfonamides to the sulfones and thiosemicarbazones, with each development drawing on previous ones. Hörlein was acutely aware of the need for a culture of pharmaceutical research, both in the sense of a network of supporting institutions and in the sense of a strong constituency in the medical and scientific communities, a constituency he worked assiduously to build.[76]

Analysis of pharmaceutical research at Bayer/I. G. Farben in terms of technological momentum goes far to explain the optimism that prevailed within the Elberfeld group in the 1920s and early 1930s. That mood is in need of explanation, since by most accounts the prevailing attitude in the medical and scientific communities was one of pessimism regarding bacterial chemotherapy.[77] Hörlein could be optimistic because he approached bacterial chemotherapy not as an individual working on a discrete problem within a single discipline, but as a manager of a well-developed research organization with a systemic approach to innovation, a wide-ranging vision of the possibilities of chemistry in medicine, and a record of success with closely related problems.

In 1937 Hörlein wrote a letter politely rejecting a proposal that Gerhard Domagk be made director of the pathological section of the Kaiser Wilhelm Institute for Medical Research in Heidelberg. His reason was that further breakthroughs in the field of the sulfa drugs were imminent, and that the best way to achieve them would be to "let the apparatus created by us work as quietly as possible." His emphasis was not on Domagk but on the "apparatus" of which Domagk was but one part. When, two years later, the Nobel committee awarded to Domagk its prize in physiology or medicine for his recognition of the antibacterial activity of Prontosil, it placed its emphasis on the individual, not the apparatus. This divergence exposes the lack of congruence between the industrialized system of invention that we have examined, on the one hand, and the individualized system of credits carried over from earlier patterns of invention and from academic science, and embodied in the Nobel Prizes, on the other.

In its allotment of credits to individuals, the Nobel Prize tends to efface the roles of colleagues and supporting institutions. This distortion is especially egregious in a case like Domagk's, in which these factors were significant and unmistakable. Domagk's role was decisive, but so too was that of the chemists Klarer and Mietzsch. And all three worked within an "apparatus" not of their own making. The biochemist Paul György drew a logical, if unexpected, conclusion, when he stated in 1947 that Hörlein rather than Domagk should have received the Nobel for Prontosil. For the first sulfa drugs were the product of an organized, collaborative effort, of which Hörlein had been the architect.[78]

Related to this incongruity between credits and the system of invention is the near invisibility of important aspects of the commercial and industrial setting of the research in published accounts by Hörlein, Domagk, and Mietzsch. All three tend to present their work in general, and that on the sulfas in particular, as a scientific, medical, and humanitarian enterprise. Only occasionally and incidentally do they refer to such issues as profits, manufacturing and marketing costs, and patenting, although Hörlein especially as company manager must have been closely involved in these questions. Domagk's published papers, in particular, tend to present his work as continuous with an academic medical tradition. As a result, not only is the industrial setting of the work effaced and the full role of the chemists obscured, but the engineering character of chemotherapeutic research is never fully apparent.[79]

We have seen that this research may best be characterized as a form of rational engineering or science-based technology. What emerges at Elberfeld is a form of rational engineering with both chemical and biomedical components, and in which the biology is as important as the chemistry. This was difficult to perceive at the time and afterward, for several reasons. We are now accustomed to apply "biotechnology" only to recent developments in molecular genetics and recombinant DNA. Before the advent of genetic engineering in the 1970s, the categories of medical research and engineering were largely disjunct. This was so not only because of the biological aspect of medical as distinct from other kinds of engineering, and not only because of the historically distinct professional identities and institutions of physicians and engineers, but also because of the strongly ethical quality of the normative self-image of physicians. The special ethical status ascribed to medicine by academic scientists and physicians alike was expressed in the resistance of both groups to the patenting of medically related discoveries or inventions and in their distrust—especially marked in English-speaking countries until the interwar period—of pharmaceutical companies and of the profit motive.[80]

The placement until recently of medicine and biology in a separate conceptual box from engineering helps to account for the awkwardness that may be felt in calling medical findings inventions, even when they have been deliberately sought as practical results and are of great practical consequence. The term discovery is far more common in the medical and scientific literature and seems to situate the finding, device, or product to which it refers more comfortably in the domain of academic science.

There is a common denominator in the incongruity between credits and the system of invention, the effacement of the commercial and industrial setting in publications, and the residual awkwardness of viewing and describing medical research as a form of engineering. All are expressions of the persistence of an academic scientific and medical culture in the face of an institutional reality in which commercial and industrial motives and organizational structures materially support and give form to research problems and results. The sulfa drugs were products of an industrialized system of invention, but their appearance was described and rewarded in the language and practices of academic science.

Research for this paper was supported in part by the National Science Foundation (SES-8111985 and SES-8409848) and the National Institutes of Health (NIH-R01 LM4231). Special thanks are due to Peter Göb, Michael Pohlenz, and the staff of the Bayer-Archiv, Leverkusen, for assistance with documents, and to Glenn Bugos, Matthias Dörries, Peter Morris, Wolfgang Sadee, and Jeffrey Sturchio for comments on earlier versions.

Notes

1. Erich Posner, "Domagk, Gerhard," *Dictionary of Scientific Biography* (New York: Scribner's, 1971), 4:153–156.

2. Posner, "Domagk," pp. 154–155. On the medical impact of the sulfa drugs, see Harry F. Dowling, *Fighting Infection: Conquests of the Twentieth Century* (Cambridge, Mass. and London: Harvard University Press, 1977), pp. 108–124.

3. Bayer archives, file Prof. Dr. Domagk, Gerhard. See esp. letter of Hörlein to Prof. Richard Kuhn, 21 Jul 1937; letter of Domagk to directors of I. G. Farbenindustrie, 21 Dec 1945; memo signed Haberland, 14 Oct 1946; and letter of Mietzsch to Domagk, 13 Jul 1953. For Mietzsch's articles of the 1950s, see below.

4. Georg Meyer-Thurow, "The Industrialization of Invention: A Case Study from the German Chemical Industry," *Isis* 73 (1982): 363–381.

5. Meyer-Thurow, "Industrialization of Invention," esp. pp. 367–379. The need for careful organization of industrial research laboratories in industries dependent on technological innovation is emphasized by Herman Daems, "The Rise of the Modern Industrial Enterprise: A New Perspective," in *Managerial Hierarchies: Comparative Perspectives on the Modern Industrial Enterprise*, ed. Alfred D. Chandler, Jr., and Herman Daems (Cambridge, Mass.: Harvard University Press, 1980), pp. 203–223. On I. G. Farben, see Gottfried Plumpe, *Die I. G. Farbenindustrie AG: Wirtschaft, Technik und Politik 1904–1945* (Berlin: Duncker & Humblot GmbH, 1990); and Peter Hayes, *Industry and Ideology: I. G. Farben in the Nazi Era* (Cambridge: Cambridge University Press, 1987). On the early development of pharmaceutical research at Bayer, see the printed but unpublished work in the Bayer archives, *Geshichte und Entwicklung der Farbenfabriken vorm. Friedr. Bayer & Co. Elberfeld in den erster 50 Jahren* (Munich: Klischees und Druck von Meisenbach Riffarth & Co., 1918), pp. 407–424.

6. Bayer archives, 166/2, "Gliederung der pharmazeutischen Abteilung 'Bayer-Meister Lucius'" (1927), pp. 1, 25–29.

7. Bayer archives, 166/3, *Aufgaben und Organisation der Pharmazeutischen Abteilung Bayer Meister Lucius,* LeverKusen, 1930.

8. Heinrich Hörlein, "Lebenslauf" (undated, but grouped with materials related to the Nuremberg trials). Bayer archives, Personalia, Hörlein, H., 271/2.

9. Hörlein, "Lebenslauf"; career summary titled Hörlein, Philipp Heinrich (1959) in Bayer archives, Personalia, Hörlein, H., 271/2.

10. Hörlein, "Lebenslauf," p. 2.

11. In a paper presented at the chemistry section of the British Association meeting at Nottingham in September 1937, Hörlein said that by the time of his September 1932 lecture on "Medizin und Chemie" (see below, note 14), the "first positive results" had already been obtained on the chemotherapy of bacterial infections. Heinrich Hörlein, "The Development of Chemotherapy for Bacterial Diseases," *The Practitioner* 139 (1937): 635–649. My copy is a reprint paginated 1–16. The statement referred to is found on page 2. Joseph Klarer, who prepared the compounds tested by Domagk, dated the first positive results from the summer of 1932. Joseph Klarer, "Lebenslauf," 1945. Bayer archives, Personalia, Klarer, Dr. Joseph, 271/2. Daniel Bovet's reconstruction, based in part on documents in the Bayer archives, indicates that "promising" results had been obtained by the late summer of 1932. Daniel Bovet, *Une chimie qui guérit: histoire de la découverte des sulfamides* (Paris: Editions Payot, 1988), p. 110. Bovet's book is the most comprehensive history of the sulfa drugs.

12. Heinrich Hörlein, "Ueber die wissenschaftlichen Grundlagen der Arzneimittelsynthese," *Münchener medizinischen Wochenschrift* 19 (1927): 801. My copy is a reprint paginated 1–15.

13. On the origins of Ehrlich's program for chemotherapy, see Timothy Lenoir, "A Magic Bullet: Research for Profit and the Growth of Knowledge in Germany around 1900," *Minerva* 26 (1988): 66–88; and John Parascandola and Ronald Jasensky, "Origins of the Receptor Theory of Drug Action," *Bulletin of the History of Medicine* 48 (1974): 199–220.

14. Heinrich Hörlein, "Medizin und Chemie," lecture given at the meeting of the Society of German Scientists and Physicians, Wiesbaden, 28 September 1932. Printed in *Medizin und Chemie* (1933. Abhandlungen aus den

Medizinisch-chemischen Forschungsstatten der I. G. Farbenindustrie Aktiengesellschaft). My copy is a reprint paginated 1–19.

15. Hörlein, "Medizin und Chemie," pp. 3–4. Hörlein gives the phrases "team work" and "leader of the team" in English. He notes their origins in the language of sports and their recent use in England and America especially to designate cooperation in neighboring biological fields.

16. Hörlein, "Medizin und Chemie," pp. 15–19. On the chemical work, see below. On the biological background, see P. C. C. Garnham, "History of Discoveries of Malaria Parasites and of Their Life Cycles," *History and Philosophy of the Life Sciences* 10 (1988): 93–108.

17. On Edison, see Thomas P. Hughes, *Networks of Power: Electrification in Western Society 1880–1930* (Baltimore: Johns Hopkins University Press, 1983), pp. 18–46.

18. Hörlein, "Medizin und Chemie," pp. 10–11.

19. Gerhard Domagk, "Ein Beitrag zur Chemotherapie der bacteriellen Infektionen," *Deutsche Medizinische Wocheschrift* 61 (1935): 250–253; Domagk, "Chemotherapie der bakteriellen Infectionen," *Angewandte Chemie* 46 (1935): 657–667; Domagk, "Chemotherapie der Streptokokkeninfektionen," *Klinische Wochenschrift* 15 (1936): 1585–1590; Domagk, "Weitere Untersuchungen über die chemotherapeutische Wirkung sulfonamidhaltiger Verbindungen bei bakteriellen Infektionen," *Klinische Wochenschrift* 16 (1937): 1412–1418.

20. For Domagk's biography see Leonard Colebrook, "Gerhard Domagk 1895–1964," *Biographical Memoirs of Fellows of the Royal Society* 10 (1964): 39–50; Ralph E. Oesper, "Gerhard Domagk and Chemotherapy," *Journal of Chemical Education* 31 (1954): 188–191; and Posner, "Domagk." For additional references on Domagk, see Bovet, *Une chimie qui quérit,* pp. 151–153.

21. Gerhard Domagk, "Untersuchungen über die Bedeutung des retikuloendothelialen Systems für die Vernichtung von Infektionserregern und für die Entstehung des Amyloids, "*Virchows Archiv für pathologische Anatomie und Physiologie und für klinische Medizin* 235 (1924): 594–638; Oesper, "Gerhard Domagk," pp. 188–189.

22. Oesper, "Gerhard Domagk," pp. 188–189.

23. Posner, "Domagk," p. 154; Bayer archives, file Prof. Dr. Domagk, Gerhard.

24. Oesper, "Gerhard Domagk," pp. 188–189; Colebrook, "Gerhard Domagk," p. 39; Posner, "Domagk," p. 154.

25. Domagk, "Chemotherapie der bakteriellen Infektionen," p. 659; Domagk, "Ein Beitrag," p. 250.

26. Domagk, "Chemotherapie der bakteriellen Infektionen," p. 659.

27. Domagk, "Ein Beitrag," p. 251; Domagk, "Chemotherapie der bakteriellen Infektionen," pp. 658–659.

28. Domagk, "Ein Beitrag," p. 250; Domagk, "Chemotherapie der bakteriellen Infektionen," p. 659. The timing of Feldt's initial publication on gold compounds (1926) suggests that his work may have been one factor prompting Hörlein to raise I. G. Farben's commitment to antibacterial chemotherapy, for example by hiring Domagk the following year. Direct evidence for this is not yet available.

29. Domagk, "Ein Beitrag," p. 250; Domagk, "Chemotherapie der bakteriellen Infektionen," p. 660.

30. Ibid.

31. Ibid.

32. Ibid.
33. Domagk, "Ein Beitrag," p. 251; "Chemotherapie der bakteriellen Infektionen," p. 660.
34. Domagk, "Ein Beitrag," p. 251.
35. Domagk, "Ein Beitrag," pp. 252–253; "Chemotherapie der bakteriellen Infektionen," p. 660 (quote).
36. Domagk, "Ein Beitrag," p. 252.
37. Domagk, "Ein Beitrag," p. 252.
38. For a reconstruction of the chronological sequence of the work that led to Prontosil, see Bovet, *Une chimie qui quérit* (note 11), pp. 103–116.
39. Domagk, "Ein Beitrag"; Bovet, *Une chimie qui quérit,* p. 116.
40. Domagk, "Weitere Untersuchungen," p. 1413.
41. Domagk, "Chemotherapie der bakteriellen Infektionen," pp. 657–658, 659.
42. Ibid., p. 660. See also above.
43. Ibid., pp. 657–658.
44. Ibid.
45. Domagk, "Chemotherapie der Streptokokkeninfektionen," p. 1585.
46. Domagk, "Chemotherapie der bakteriellen Infektionen," p. 657 (quote); Domagk, "Chemotherapie der Steptokokkeninfektionen," p. 1585.
47. Domagk, "Chemotherapie der bakteriellen Infektionen," pp. 657–658; Domagk, "Weitere Untersuchungen," pp. 1414–1417; Domagk, "Chemotherapie der Steptokokkeninfektionen," pp. 1585–1587.
48. Domagk, "Weitere Untersuchungen," pp. 1414–1417.
49. Domagk, "Chemotherapie der Steptokokkeninfektionen," p. 1586 (quotes, my emphasis); Domagk, "Chemotherapie der bakteriellen Infektionen," pp. 657, 660.
50. Domagk, "Chemotherapie der bakteriellen Infektionen," p. 659 (quote); Domagk, "Weitere Untersuchungen," p. 1417 (quote). Another reference to the importance of the reticuloendothelial system is Domagk, "Chemotherapie der Steptokokkeninfektionen," p. 1585.
51. Domagk, "Weitere Untersuchungen," p. 1417 (quote); Domagk, "Chemotherapie der bakteriellen Infektionen," pp. 658–659.
52. Domagk, "Chemotherapie der bakteriellen Infektionen," pp. 659, 664 (quote), 666.
53. Ibid., p. 658; Domagk, "Weitere Untersuchungen," pp. 1415 (quote), 1416 (quote).
54. Domagk, "Chemotherapie der bakteriellen Infektionen," p. 658; Domagk, "Weitere Untersuchungen," pp. 1412–1413, 1415 (quote).
55. Domagk, "Chemotherapie der Streptokokkeninfektionen," pp. 1586 (quote), 1587; Domagk, "Chemotherapie der bakteriellen Infektionen," pp. 657, 660, 661, 664, 666 (quote); Domagk, "Weitere Untersuchungen," pp. 1412, 1413.
56. Domagk, "Chemotherapie der Streptokokkeninfectionen," pp. 1586 (quote), 1587 (quotes); Domagk, "Weitere Untersuchungen," p. 1413.
57. Domagk, "Chemotherapie der bakteriellen Infektionen," p. 660 (quote).
58. Domagk, "Chemotherapie der Streptokokkeninfectionen," p. 1590.
59. "60 Geburtstag, Direktor Professor Dr. ing. Dr. med. e. h. Fritz Mietzsch." Stamped 9 July 1956. 4 pages typescript. Bayer archives, Personalia, 271/2 (hereafter "Fritz Mietzsch"); "Wer ist's," *Der Chemiemarkt,* 12 April 1956. The following summary of Mietzsch's biography is based on these sources.

60. "Fritz Mietzsch," pp. 2–3; "Wer ist's," p. 1. At Hörlein's invitation Klarer joined the I. G. Farben research laboratories at Elberfeld in 1927, following training in dye chemistry under Hans Fischer at the Technische Hochschule, Munich. Joseph Klarer, "Lebenslauf."

61. Fritz Mietzsch, "Entwicklungslinien der Chemotherapie (Vom chemischen Standpunkt gesehen)" *Klinische Wochenschrift* 29 (1951): 125–134, on p. 125.

62. Fritz Mietzsch, "Beiträge zur Entwicklung der Chemotherapie aus den Laboratorien der Farbenfabriken Bayer," *Arzneimittel-Forschung* (September 1956): 503–508.

63. Mietzsch, "Entwicklungslinien," p. 125.

64. Ibid., pp. 125–126 (quote p. 127).

65. Mietzsch, "Beiträge," pp. 505–506; Mietzsch, "Entwicklungslinien," p. 127.

66. Mietzsch, "Beiträge," pp. 505–506.

67. Mietzsch, "Entwicklungslinien," p. 129; Mietzsch, "Beiträge," p. 507. Klarer also emphasized the crucial role of basic alkylation. Klarer, "Lebenslauf."

68. Mietzsch, "Chemie und wirtschaftliche Bedeutung der Sulfonamide," *Arbeitsgemeinschaft für Forshung des Landes Nordrhein-Westfalen* 31 (1954): 7–32, quote on pp. 10–11. See also Mietzsch, "Entwicklungslinien," p. 129.

69. Mietzsch, "Chemie und wirtschaftliche Bedeutung der Sulfonamide," pp. 11–12; Mietzsch, "Entwicklungslinien," p. 129; Klarer, "Lebenslauf." Klarer dated the preparation of the first sulfonamide-containing azo compound with marked action on streptococcus-infected mice, labeled Kl. 695, to the summer of 1932. He prepared the simpler compound that eventually took the trade name Prontosil, but was at first labeled Kl. 730, in the fall of 1932.

70. Mietzsch, "Entwicklungslinien," p. 129; Mietzsch, "Chemie und wirtschaftliche Bedeutung der Sulfonamide," pp. 12–13; Klarer, "Lebenslauf."

71. Mietzsch, "Entwicklungslinien," pp. 130–131; Mietzsch, "Chemie und wirtschaftliche Bedeutung der Sulfonamide," pp. 13–25.

72. Mietzsch, "Neuere Entwicklung der synthetischen Chemotherapeutika," *Österreichische Chemiker-Zeitung,* 53 (1952): 178–187, quotes on p. 178; Mietzsch, "Beiträge," pp. 503, 505–506, 508; Mietzsch, "Entwicklungslinien," p. 127.

73. Mietzsch, "Neuere Entwicklung," p. 187.

74. Mietzsch, "Beiträge," p. 508. Bovet, who tends to analyze the research from the chemists' point of view, credits Mietzsch with the initiative that led to Prontosil. See *Une chimie qui quérit,* pp. 137–143.

75. Further study of the boundary character of research in chemotherapy in various institutional settings is needed. Suggestive discussions of very different examples are Glenn Bugos, "Managing Cooperative Research and Borderland Science in the National Research Council, 1922–1942," *Historical Studies in the Physical and Biological Sciences* 20, no.1 (1989): 1–32; and Susan Leigh Star and James R. Griesemer, "Institutional Ecology, 'Translations,' and Boundary Objects; Amateurs and Professionals in Berkeley's Museum of Vertebrate Zoology, 1907–1939," *Social Studies of Science* 19 (1989): 387–420.

76. Consideration of the relationship between synthetic drugs and dyes in terms of technological momentum deserves more ample discussion than can be given here. For a comparable case, which involves a different technology and a different branch of I. G. Farben in the same period but which highlights

the roles played by the company's need to obtain maximum return on its investment in techniques, personnel, and equipment; analogies drawn between earlier successful projects and later ones; and continuity of leadership, see Thomas Parke Hughes, "Technological Momentum in History: Hydrogenation in Germany 1898–1933," *Past and Present,* 44 (1969): 106–132. For other definitions of technological momentum see Hughes, *Networks of Power,* esp. pp. 15, 140; and Hughes, "The Evolution of Large Technological Systems," in *The Social Construction of Technological Systems,* ed. Wiebe E. Bijker, Thomas P. Hughes, and Trevor Pinch (Cambridge, Mass.: MIT Press, 1987), pp. 51–82 (esp. 76–80). Jürgen Kocka has pointed to "the availability of both a pool of highly qualified and expensive technical know-how that could be put to many uses and relevant methods and machinery" as an important stimulus to diversification in the chemical and electrical engineering industries. See his "The Rise of the Modern Industrial Enterprise in Germany" (note 9), in Chandler and Daems, *Managerial Hierarchies,* pp. 77–116, on 87–88.

77. Dowling, *Fighting Infection* (note 2), pp. 106–107.

78. Paul György, affidavit dated 9 July 1947, submitted as testimonial evidence for the Nuremberg trials. Bayer archives, Personalia, Hörlein, H. 271/2. Both Hörlein, an upper-level manager, and Mietzsch, who succeeded him as director of the pharmaceutical research division, were rewarded within the company. The comparison of internal and external rewards merits further study.

79. As noted above, Hörlein occasionally refers to the economic impact of pharmaceuticals in general terms. Mietzsch, too, briefly discusses the economic dimension in "Chemie und wirtschaftliche Bedeutung der Sulfonamide." Neither directly addresses the bearing of commercial considerations on research policy.

80. On these attitudes and the beginning of change between the world wars, see John P. Swann, *Academic Scientists and the Pharmaceutical Industry: Cooperative Research in Twentieth-Century America* (Baltimore: Johns Hopkins University Press, 1988), esp. pp. 24–56; and John Parascandola, "The 'Preposterous Provision': The American Society for Pharmacology and Experimental Therapeutics' ban on industrial pharmacologists, 1908–1941," in *Pill Peddlers: Essays on the History of the Pharmaceutical Industry,* ed. Jonathan Liebenau, Gregory H. Higby, and Elaine C. Stroud (Madison, Wis.: American Institute of the History of Pharmacy, 1990), pp. 29–47.

Defining Chemistry: Origins of the Heroic Chemist

Robert Friedel

> After having made a few preparatory experiments, he concluded with a panegyric upon modern chemistry, the terms of which I shall never forget:—
>
> "The ancient teachers of this science," said he, "promised impossibilities, and performed nothing. The modern masters promise very little; they know that metals cannot be transmuted, and that the elixir of life is a chimera. But these philosophers, whose hands seem only made to dabble in dirt, and their eyes to pore over the microscope or crucible, have indeed performed miracles. They penetrate into the recesses of nature, and show how she works in her hiding places. They ascend into the heavens: they have discovered how the blood circulates, and the nature of the air we breathe. They have acquired new and almost unlimited powers; they can command the thunder of heaven, mimic the earthquake, and even mock the invisible world with its own shadows."[1]

Victor Frankenstein thus records "the words of fate, enounced to destroy" him early in Mary Shelley's 1819 classic. In these same words Shelley herself provided a remarkably astute glimpse at the changing public image of chemistry at the beginning of the nineteenth century. Building on the work of the eighteenth-century chemists, culminating with the new nomenclature of Lavoisier, theorists like Proust and Dalton and experimentalists such as Gay-Lussac, Berzelius, and Davy gave credence to the claim of Frankenstein's professor Waldman that "Chemistry is that branch of natural philosophy in which the greatest improvements have been and may be made. . . ." The progress of the science, however, was clearly based on a recognition not only of its power but also of its limits—"the modern masters promise very little." While Victor Frankenstein was to go on to claim a great deal indeed for his own chemical endeavors, Shelley touched on a key element of the science's prestige in remarking on its retreat from the promises of the alchemists.

A century later, however, new promises characterized the chemist's public image. Gone was the reticence of Shelley's Waldman, replaced by pictures like the following, from John K. Mumford's 1924 *Story of Bakelite*:

> He no longer sat in his office, waiting for somebody to bring him a job of analysis. His business was to make a new world. It was a job of synthesis. He was the Builder. The architect might design skyscrapers; the Chemist would create new substances, and create them out of anything. He had made an inventory and turned the world's waste heaps into colossal assets. Once more, mankind was rich, and getting richer.[2]

A bit more purple than most, perhaps, this description is still representative of an important and common presentation of the twentieth-century chemist. While this image could be dismissed as largely the creation of enthusiastic publicists, promoters of products and businesses who find it worthwhile to make the chemical industry out as the creator and purveyor of miracles, to so dismiss it would be to miss an important clue to understanding the changing place of the chemist and of chemistry in the modern scheme of things.

How is it, we might ask, that such a grand view of the field and its practitioners should have enough plausibility to be at the center of a portrayal of the chemist and of chemical products in the early twentieth century? Certainly, we cannot say that the view is odd or that it is representative of only a passing enthusiasm of post-World War I America. Most of us, after all, have grown up listening to America's largest chemical company promise "Better Things for Better Living Through Chemistry." We are hardly startled, even in our more cynical day, to hear promises of chemical miracles to come, and indeed we have no problem with the young student of chemistry setting out on his or her education and career with the hope, even expectation, that the creation of some novel substance would be the eventual outcome of a life's work. Such deeply ingrained attitudes are an important part of chemistry's and the chemical industry's claims for support and even indulgence from society. As such, we need to understand their origins and their changing status if we are to successfully grasp the history of the relationship between the discipline and profession of chemistry, on the one hand, and the larger society that maintains and uses them, on the other.

These perspectives on chemistry and its place in society are not simply the creation of advertisers and promoters; they are rather the product of a century of developing attitudes, rooted in responses to the "Chemical Revolution" of the late eighteenth and early nineteenth centuries and cultivated through decades of attempts to define and explain both the field and the usefulness of chemistry. To be sure, from the beginning of this process, there *were* promoters—individuals who saw their purpose in presenting chemistry to the public as going beyond simple education and including the cultivation of a wider appreciation of not only why chemistry was to be studied but also why it should be supported. Most of the people who presented and interpreted chemistry for the wider public were promoters in this sense. Such promotion did not represent a conscious or coordinated effort but instead constituted the vanguard of the ideas about what chemists do and what chemistry is that came to be so common in our century.

In tracing the emergence of the modern attitude, one does not find a steady or even progression of ideas. As chemists, somewhat self-consciously, sought both intellectual definition and a public image in the nineteenth century, various spokesmen reached for different ways of portraying their field. Their efforts revealed tensions and apparent contradictions that did not always diminish quickly or easily. The most important of these tensions was that between "science" and "art." The close relationship between chemical knowledge and its application for practical purposes was often remarked on, but different attitudes prevailed about the implications of this relationship for the advancement and understanding of chemistry as a discipline. For many, the close connection between theory and practice was simply one of the field's great recommendations, but for some it posed barriers to a more profound appreciation of chemical knowledge as a contribution to what was still often termed "philosophy." Another, less obvious, tension existed in efforts to characterize chemistry's fundamental approaches—what methods could be called most essentially *chemical?* Here the most common references were to *analysis* and *synthesis*. Of course, there was nothing necessarily exclusive about one over the other, but a careful reading of descriptions of the field suggests indeed that now one, and now the other, were seen as best embodying chemistry's primary means for understanding the universe.

Yet one other tension existed, sometimes in the background, at times more clearly at the front, in characterizations of chemistry's central aims. This dichotomy, perhaps the most difficult for us to appreciate, was in some ways at the heart of the process of forging the final shape and meaning of chemistry. Here the tension was between a focus on the old and on the new, between a concern for understanding how and why the world was put together in a particular way and for remaking the world in new ways. This division mirrors the others, for "analysis" and "science" could be arrayed with the traditional and philosophical against "synthesis," "art," and preoccupation with novelty and practical application. These dualities do not represent different choices or paths, but rather alternate sides of the same body of thought. The question was not usually "in what direction is chemistry to go," but rather "what should be our science's *primary* undertaking—what are the most important and distinctive features of chemistry?"

By looking at a variety of published statements, from textbooks, encyclopedias, and magazine articles, it is possible to understand how these sources of tension worked themselves out in the course of the nineteenth century. The gleaning here can only provide an impressionistic picture, but that should be at least a beginning in tracing the origins of chemistry's modern image. The sources used here have some important limitations that should be specified at the outset. They are all in English and in fact are generally either from American sources or from material readily available to American readers. They are not results of a systematic, much less comprehensive, survey of possible sources, but rather materials readily at hand (with the resources of the Smithsonian Institution and the Library of Congress, to be sure), of sufficient quantity and quality at least to suggest the range of ideas and statements that characterized chemical thought from the early nineteenth century to the years just after World War I.

The implications of chemistry's combination of "science" and "art" were the subject of discussion from the very earliest efforts to bring the fruits of the Chemical Revolution to a wider audience. The English chemist William Henry published his *Elements of Experimental Chemistry* in 1799, and by 1814 it was already in its third American (and sixth English) edition. In the introduction, he attempted to define his field as clearly and simply as possible, and then he launched directly into a discussion of chemistry as science and as art:

> It has not been unusual to consider chemistry, under the two-fold view of a science and of an art. This arrangement, however, appears to have had its origin in an imperfect discrimination between two objects, that are essentially distinct. Science consists of assemblages of facts, associated together in classes, according to circumstances of resemblance or analogy. The business of its cultivators is, first to investigate and establish individual truths, either by the careful observation of natural appearances, or of new and artificial combinations of phenomena produced by the instruments of experiment. The next step is the induction, from well ascertained facts, of general principles or laws, more or less comprehensive in their extent, and serving like the classes and orders of natural history, the purposes of an artificial arrangement. Of such a body of facts and doctrines, the science of chemistry is composed. But the

employment of the artist consists merely in producing a given effect, for the most part by the sole guidance of practice or experience. In the repetition of processes, he has only to follow an established rule; and in the improvement of his art, he is benefitted generally by fortuitous combinations, to which he has not been directed by any general axiom. An artist, indeed, of enlarged and enlightened mind, may avail himself of general principles, and may employ them as an useful instrument in perfecting established operations: but the art and the science are still marked by a distinct boundary. In such hands, they are auxiliaries to each other; the one contributing a valuable accession of facts; and the other, in return, imparting fixed and comprehensive principles, which simplify the process of art, and direct to new and important practices.[3]

In the Baconian tradition to which any respectable British scientist would attach himself, Henry has no problem acknowledging chemistry's utility for all sorts of purposes (and in fact he discusses this at length later in his introduction), but he is at pains to separate these applications from the science itself.

Why should Henry feel the need to do this—to distinguish his science from its applications? Part of the answer can be gleaned from the 1812 "Introductory Lecture" (in chemistry, at Dickinson [then, Carlisle] College, Pennsylvania) of his contemporary, Thomas Cooper, who had emigrated to America in 1793 and was a close friend of Joseph Priestley:

Until these last forty years, chemistry could hardly be called any thing but a collection of detached facts: much indeed was known in the workshop of the manufacturer, and much by the experimentalist. But until the popular compilations of Macquer and Beaume [sic] in France, the rapid and important discoveries of Scheele, Priestley, Cavendish, and Black, and the beautiful system of Lavoisier, chemistry could hardly be called a science: for it had ascertained no laws, by which traditionary processes could be explained, or material improvements suggested. Let any one examine the state of the arts and manufactures, fifty years ago, and compare it with the situation of the present day, and it will be found, that during these fifty years, more improvements have been made, originally suggested by chemical theories, and pursued under the guidance of chemical knowledge, than in two thousand years preceding. At present, there is not a manufacturer of note in England, who is not more or less acquainted with chemistry as a regular branch of education and study.[4]

The implicit conflation of chemical science and chemical practice is clear here, and it was unquestionably prevalent. And why not, for the close perceived linkage between chemical knowledge and its practical uses was surely in the interests of all promoters of the science. Cooper's promotion of the science of chemistry, as far as it went, was enough to raise some suspicions in the minds of Americans concerned to defend what they saw as a democratic tradition of practical science. Thomas Jefferson, for example, wrote this warning note to Cooper:

You know the just esteem which attached itself to Dr. Franklin's science, because he always endeavored to direct it to something useful in private life. The chemists have not been attentive enough to this. I wished to see their science applied to domestic objects, to smelting, for instance, brewing, making cider, to fermentation and distillation generally, to the making of bread, butter, cheese, soap, to the incubation of eggs, etc. And I am happy to see some of these titles in the syllabus of your lecture. I hope you will make the chemistry of these subjects intelligible to our good housewives.[5]

Perhaps the best testimony to this intermingling of theory and application in comprehending chemistry is from the most general presentations of the subject, as in the article on chemistry in the *Encyclopedia Americana*'s 1830 edition.[6] After extensive discussion of how the subject is defined, especially as distinguished from natural philosophy, there are descriptions of the various *kinds* of chemistry, beginning with "philosophical," but going on to include meteorological, geological, physiological, pathological, therapeutic or pharmaceutic, hygietic (hygiene), and agricultural chemistry. Then the most clearly practical side of the field is presented with no suggestion that it is any less "scientific" than any other:

Chemistry, finally, exerts an influence on the routine of domestic life, and on the arts. It simplifies and regulates the daily offices of the housekeeper; renders our dwellings healthy, warm, light; assists us in preparing clothing, food, drink, &c.: it teaches the best way of making bread; preparing and purifying oils; of constructing bakehouses, ovens and hearths; of bleaching and washing all kinds of stuff; of producing artificial cold, &c. The application of chemistry to the arts and manufactures is, however, still more important and extensive. Here its aim is to discover, improve, extend, perfect, and simplify the processes by which the objects to be prepared may be adapted to our wants.

This approach to defining chemistry and its place in the scheme of things mustn't be dismissed as only for the simplistic. Some of the most eminent promoters of chemistry found it useful to obscure, rather than emphasize, the distinction between the "philosophical" and the "practical." None other than Benjamin Silliman, the greatest champion of science, and especially chemistry, in the early Republic, took this approach. In his *Introductory Lecture* of 1828, he acknowledged that "Chemistry is distinguished as *an art* or a collection of arts, from *chemistry* as a science," but he went on to emphasize the extent to which the two went hand-in-hand: ". . . science without art is inefficient, art without science is blind."[7] In the Introduction of his 1831 *Elements of Chemistry* he remarked on the extent to which it was in practice impossible to separate the two aspects of chemistry:

The arts are all either mechanical or chemical, and not unfrequently both are involved in the same processes. The practices of the arts may be regarded as experiments in natural philosophy and chemistry. The object of the arts is usually gain; but he, or any other person, who views the facts correctly, may reason upon them advantageously, and thus obtain important instruction.[8]

Even more pointed was the remark of one of Silliman's reviewers, that "you must convince men that Chemistry will enable them to increase their wealth, before they will consider the study of it worthy of their attention."[9]

Americans were certainly not oblivious to the tension between the pursuit of theoretical chemistry and the application of chemical methods and knowledge to practical ends. The more "philosophical" side of chemistry was recognized by some as lagging behind the areas of the science that related more closely to experiment and application. Benjamin Silliman's own Yale colleague Denison Olmsted wrote, for example, in 1826 in Silliman's *American Journal of Science*, "Although the field of experimental chemistry, has been crowded with ardent votaries, and although every corner of it has been hunted by competitors eager for discovery, the philosophy of chemistry appears not to have been cultivated with equal zeal or ability."[10] One result of this neglect, according to Olmsted, was the difficulty in simply defining chemistry. Even those definitions that succeeded in distinguishing chemistry from natural philosophy generally gave, in Olmsted's words, "very little information respecting the appropriate business of the chemist." This problem of defining chemistry, addressed by almost every textbook and encyclopedia article on the subject, was symptomatic of the widespread feeling that, while new discoveries were appearing at an impressive rate, the assimilation of these discoveries into a coherent body of knowledge was badly lagging.[11]

It was possible to define chemistry in ways that more clearly spelled out the "business of the chemist." In the choice of definitions, the promoters of the science could suggest the most appropriate ends for chemistry, and a careful reading of these provides clues to how the duties and capabilities of the chemist were perceived in the course of the nineteenth century. Appended to this paper is a list of twenty simple statements defining chemistry, whose sources range from the 1814 edition of William Henry's popular *Elements of Experimental Chemistry* to the *Textbook of Chemistry* of William A. Noyes, published in 1919. Represented in addition to textbooks and popular introductions to the subject are a few encyclopedia entries. Just a few words can be singled out from this sampling to demonstrate the key elements of chemistry's definition in this period. The most important word, from Henry to Noyes, is *composition*. The chemist wants to know, before anything else, what things are made of. Sometimes this concern is more specifically focused on *changes* in composition, and such changes are pointed to as being fundamental to the distinction between chemical and other (especially physical) phenomena. The second most common word is *properties,* and the traditional descriptive role of chemistry is never abandoned. Changing properties are almost as important as changing composition and are sometimes the criteria for identifying chemical action. *Change* itself is the third key word that makes its way into many definitions, and it is sometimes implied in others by reference to "conversion" or "new combinations." The persistence of these words testifies first to the unchanging core of chemistry's identification; it is in shifts of emphasis and aim that can be seen how in fact the image of the field changed in the course of the century.

The key shift can perhaps be characterized as movement from a passive vision to an active one. The concern of the early nineteenth-century chemist for composition, properties, and even change is expressed by *viewing* nature. Phrases like "the changes of composition that occur" (Henry, 1814), "changes . . . which they produce" (Olmsted, 1826), or "the natural changes which take place" (Comstock, 1839) all posit the chemist as an observer of natural happenings, who does his job by standing by to recognize and record nature in action. Later phrases, such as "the experimental examination of the properties of elements" (Roscoe, 1869), "to determine the limits within which such changes are possible" (*Encyclopedia Americana,* 1903), and "Chemistry is concerned with those [changes] that result in new substances" (Hessler and Smith, 1914) all begin, at least, to suggest a figure who puts his hand into the action. This transition is, admittedly, subtle, but a closer look at what is being said about chemistry and chemists will make it more evident. (See appendix to this chapter for more definitions of chemistry during the nineteenth and early twentieth century.)

From the earliest discussions of what chemists did, reference was made to the two key, complementary, techniques of the field: *analysis* and *synthesis*. The description of these techniques and their purposes, however, changed in ways reflective of shifting attitudes. The 1830 *Encyclopedia Americana*'s article on chemistry, for example, included this explanation:

> Chemistry has two ways of becoming acquainted with the internal structure of bodies, *analysis* and *synthesis* (decomposition and combination). By the former, it separates the component parts of a compound body; by the latter, it combines the separated elements, so as to form anew the decomposed body, and to prove the correctness of the former process.[12]

Note that the use of synthesis is simply to confirm the results of analysis, not to create new substances. John Comstock's much reprinted *Ele-*

ments of Chemistry spoke of techniques in a similar vein: "All chemical knowledge is founded on *analysis* and *synthesis,* that is, the decomposition of bodies, or the separation of compounds into their simple elements, or the recomposition of simple bodies into compounds."[13] Another popular English work with numerous American editions, George Fownes's *Elementary Chemistry,* treated the subject this way: "There are two distinct methods of research in chemistry: the *analytical,* or that in which the compound is resolved into its elements, and the *synthetical,* in which the elements are made to unite and produce the compound."[14] The use of the definite article—"the compound"—to describe the product of synthesis (as well as the context of Fownes's remarks, in which he is discussing Cavendish's experiments with water) make it clear that the function of synthesis is fundamentally no more than as a confirmation of the results of analysis.

In the second half of the nineteenth century, however, a new meaning was given to the idea of chemical synthesis, and this suggested a much broader scope of action for the chemist. The rise of the coal tar dye industry and related applications of organic synthesis were, it may be assumed, the primary sources of this departure. A particularly explicit discussion of the implications of the new way of looking at the field appeared in a *Popular Science Monthly* article on "Synthetic Chemistry" in 1874:

> Chemistry has been called *the* analytical science, and undoubtedly with justice in the past, since the most exact processes with which it deals are still those which go technically by the name of analysis; but, of recent years, the arts have been enriched by many perfumes, colors, and drugs, which are the results of careful and laborious construction on the part of the manufacturer, operating under chemical laws. Gradually there has arisen a new branch of the science, whose aim it is to produce artificially new compounds out of old material. . . .[15]

A decade later, even bolder declarations of chemistry's new scope could be found, such as in Brown University professor John Appleton's *Beginner's Hand-Book of Chemistry* (1885): "The chemical laws now known suggest the possibility of producing *artificially* a great multitude of substances not yet recognized, and even more than have yet been produced in the great laboratory of nature."[16] Although some of the earlier celebrators of the triumphs of organic synthesis saw laboratory creations as necessarily limited to either nature's own products or close imitations thereof, toward the end of the century such limitations quietly faded away.

This new view was much more than simply a response to the remarkable feats of such chemists as Hoffmann, Baeyer, Fischer, and others, who kept a stream of artificial dyes, perfumes, flavorings, and drugs pouring out of (largely German) laboratories in the last third of the nineteenth century. The idea of "heroic synthesis" also marked what might be called the "secularization" of chemistry. As such, it was part of the much broader secularization of all science, and indeed the triumph of materialism in general, that characterized the century of Laplace, Lyell, Darwin, and Tyndall.[17] It is perhaps not well appreciated the extent to which chemistry, like natural history, was seen earlier in the century as a means for appreciating the world's divine harmony. At first, references to this role could be found coming from some of the most eminent men of the field. Humphry Davy, for instance, referred to the uses of chemistry for "perceiving in all the phenomena of the universe the designs of a perfect intelligence."[18] In his *Elements* William Henry spoke in the same vein of the moral usefulness of chemical study:

> The possession of the general principles of chemistry enables us to comprehend the mutual relation of a great variety of events, the form a part of the established course of nature. It unfolds the most sublime views of the beauty and harmony of the universe; and developes a plan of vast extent, and of uninterrupted order, which could have been conceived only by perfect wisdom, and executed by unbounded power.[19]

Some popular works in this period were explicitly aimed at promoting this view, such as Samuel Parkes's *Chemical Catechism,* wherein it was remarked that "every thing around us bears evident marks of the skill and beneficence of its Omnipotent Author."[20] Even as late as 1861, the founder of Harvard's chemistry department, Josiah P. Cook, Jr., devoted several public lectures to explication of the "Proofs of God's Plan in the Atmosphere and its Elements." At the outset of these he declared, "The illustrations of the attributes of God, which may be drawn from the constitution of matter, are conveniently divided into two classes,—first, those which appear in the adaptation of various means to a particular end, and, second, those which are to be found in the unity of plan according to which the whole frame of nature has been constructed."[21]

After this time, as excursions into Natural Theology became less respectable for men of science, sentiments such as these were relegated to books aimed exclusively at popular audiences. So a very simple 1873 text entitled *Fourteen Weeks in Chemistry* ended by declaring, "Thus does Nature attest the sublime truth of Revelation, that in all, and through all, and over all, the Lord God omnipotent reigneth."[22] With the demise of its theological justification (however limited its use might actually have been in academic circles), chemistry, in the eyes of some, faced a problem in both image and practice. If part of the aim of the science was not the exposition of the divine plan of matter, then was chemistry reduced completely to the service of practical ends? The rejection of theological utility in the life sciences had been accompanied (and in part accomplished) by the grand vision of Darwinism. Physicists had long been able to speak of the broader philosophical implications of their science. But chemists, despite the wonderful progress in theory at midcentury, were still self-conscious about the shortcomings of "philosophical chemistry." In such a context, then, was the field to be viewed by the world as simply the servant of material progress? The ever-growing prestige of *synthesis* over *analysis* could only emphasize such a vision.

These anxieties were given remarkable explicitness by one of the great champions of the field in late-nineteenth-century America, Ira Remsen. Shortly after taking up his post as professor of chemistry at the new Johns Hopkins University, Remsen published an article in *Popular Science Monthly* on "The Science *vs.* the Art of Chemistry." Here he returned to the problem that had bothered William Henry three-quarters of a century before:

> The attitude of the world in general toward chemistry is peculiar, and, as this paper is intended to show, it is not what it ought to be. This is due in turn to a peculiarity of the science itself, which distinguishes it from most other sciences. We refer to its close connection with matters of every-day experience, and of practical importance. . . . owing to this close connection, the unscientific world has grown into the habit of considering the practical problems as the problems *par excellence* of chemistry; and, having once recognized *some* object of the science, they inquire no further, and hence they fail to recognize its most important and only legitimate object.[23]

That object, Remsen eventually concluded, is the pursuit of natural laws, in particular, "the laws of combination and decomposition of bodies." But before getting to that point, he bemoaned the extent to which the applications of the new synthetic chemistry seemed to make the comprehension of chemistry's true goal more elusive than ever:

> The public knows when a new dye is discovered; it knows when the poison has been found in some strange stomach; it knows when a new milk for babies has been concocted; it knows when precious metals have been detected in the depths of the earth; it knows all these things because it is promptly informed in regard to them; and it is right and good that the information should be given, and that these things should be known. It is plain, however, that a thousand dyes might be discovered; that a thousand murderers might be brought to justice through the aid of the chemist; the varieties innumberable of milk for babes might be concocted; or that mines upon mines of gold might be unearthed without the slightest ennobling or elevating influence being exerted upon the mass of mankind.[24]

One of the consequences of this state of affairs, Remsen went on to say, was that chemistry was held in low esteem in intellectual circles. He complained that even those who offered to enlighten the public in open lectures were guilty of pandering to sensationalism, filling their talks with flashy demonstrations for entertainment rather than "elevating mankind." Finally, the pernicious effect of those misplaced priorities boded ill for the future of the science:

> . . . there are very few positions in the country which enable their encumbents to devote themselves to the pure science of chemistry without obliging them at the same time to look for additional means of support to that furnished by the positions themselves. This additional means of support can usually be found most readily in the practice of the art of chemistry. Too often, time that could and would be devoted to grappling with the problems of the science is given up to the art in order to keep the purse supplied. . . . the students who are placed under the influences mentioned are not stimulated, as they should be, to consider the higher questions of the science, but go out into the world only to keep alive the popular and erroneous idea concerning the nature of chemistry.[25]

The dangers of "the popular and erroneous idea" were to concern Remsen for much of the remainder of his long career. He repeated the themes of his *Popular Science Monthly* article in 1888 in an address at Lehigh University on "The Relations between Chemical Science and Chemical Industry." As Owen Hannaway has pointed out, "Remsen's naive view displayed no glimmer of a comprehension of the delicate interplay of research and industrial development on which modern commercial technology even then was coming to be based."[26]

The technological and economic significance of chemistry was of course to become ever more apparent in the coming decades. Well before the "Chemists' War" of 1914–18 brought the national and strategic importance of the field into sharp public focus, the ever-broadening application of advanced chemical science to the problems of industry was evident to many. An impressionistic but persuasive indicator of this comes from the progressively more extensive offerings in chemistry in institutions dedicated to the best training of "practical" men and women, such as the Massachusetts Institute of Technology. From the earliest date, MIT expected its students to take chemical instruction, but the steady expansion of both the required instruction and the facilities for it testified unambiguously to the field's growing stature as a key part of the modern technological order. The 1867–68 *Annual Catalogue,* for example, described a one-year course for all students, devoted exclusively to "Qualitative Analysis." Ten years later, a full four-year course in chemistry was available, and by the late 1880s a considerable variety of chemical courses was offered, including "Sanitary Chemistry," "Industrial Chemistry," and "Organic Chemistry," in addition to the inorganic and analytical courses, and the facilities had expanded to include fourteen laboratories and accommodations for hundreds of students.[27] The growth of MIT and the greater elaboration of its curriculum in general accounted for much of this expansion, but the ever longer descriptions of the chemical offerings made it clear that the field itself was seen as one of increasing importance and promise. None of this can come as news to anyone familiar with chemistry's late-nineteenth-century history, but it provides further emphasis to the observation that much of the world was accommodating itself readily to the conditions that made Ira Remsen so uncomfortable.

By the first decades of the twentieth century, the expanded role of the chemist in the economy and society was widely recognized. Books with titles like *Chemistry in Daily Life* and *Chemistry in Modern Life* were common, written by the most eminent chemists as well as lesser ones.[28] To be sure, the Great War highlighted particular features of chemistry's importance, but the capacity of the chemist for changing the world around him was already widely noticed. The Nobel Prizes awarded to Emil Fischer, Adolph von Baeyer, and Victor Grignard were early indications that remaking the world was indeed one of the acknowledged tasks of the chemist; and not just of the "art" of chemistry, but of the "science" as well. Not many chemists would have praises sung quite as flamboyantly as was done for the inventor of Bakelite, but the heroic chemist was not an ephemeral figure, the momentary product of war's anxieties or of industry's boasts. The heroic side of chemistry's image was firmly rooted in the science's century of self-examination and self-definition.

Notes

1. Mary Wollstonecraft Shelley, *Frankenstein or The Modern Prometheus* (Oxford: Oxford University Press, 1980, from 1831 edition), pp. 47–48.

2. John K. Mumford, *The Story of Bakelite* (New York: Robert L. Stillson, 1924), p. 24.

3. William Henry, *The Elements of Experimental Chemistry* (Boston: Thomas & Andrews, 1814; 3d American, fr. the 6th English ed.), 1:10. The very popular *Elements* went through 11 editions in 30 years.

4. Thomas Cooper, *Introductory Lecture* (Carlisle, Pa.: Archibald Loudon, 1812; reprint, New York: Arno Press, 1980), pp. 94–95. On Cooper, see Edgar F. Smith, *Chemistry in America* (New York: D. Appleton & Co., 1914; reprint, New York: Arno Press, 1972), chap. 6; see also John C. Greene, *American Science in the Age of Jefferson* (Ames: Iowa State University Press, 1984), pp. 170–171; Greene discusses here the generally utilitarian appeal of the first American promoters of chemistry.

5. Thomas Jefferson, in letter to Thomas Cooper, quoted by Hugo Meier in

"Thomas Jefferson and a Democratic Technology," in Carroll Pursell, *Technology in America* (Cambridge, Mass.: MIT Press, 1990), p. 23.

6. *Encyclopedia Americana,* s.v. "Chemistry" (Philadelphia: Carey & Lea, 1830) 3:124–128. This same article appears, unchanged, in the 1848 edition.

7. Benjamin Silliman, *An Introductory Lecture, Delivered in the Laboratory of Yale College, October, 1828* (New Haven, Conn.: H. Howe, 1828), p. 37, as quoted in John C. Greene, "Protestantism, Science and American Enterprise: Benjamin Silliman's Moral Universe," in *Benjamin Silliman and His Circle,* ed. Leonard G. Wilson (New York: Science History Publications, 1979), p. 14.

8. Silliman, *Elements of Chemistry in the order of the Lectures given in Yale College* (New Haven, Conn.: H. Howe, 1831), as quoted in W. C. Fowler, "Silliman's Chemistry," *North American Review* 34(1832): 82.

9. Fowler, "Silliman's Chemistry," p. 90.

10. Denison Olmsted, "On the Present State of Chemical Science," *American Journal of Science* 11 (1826): 349.

11. See, e.g., the article "Chemistry," in the *London Encyclopædia* (London: Thomas Tegg, 1829), 5:353–560. This lag is discussed by George Daniels as a general impression in early nineteenth-century America that "Baconian" science had produced a "deluge of facts" without sufficient synthesis; George H. Daniels, *American Science in the Age of Jackson* (New York: Columbia University Press, 1968), chaps. 5 and 6, esp. pp. 102–107.

12. *Encyclopedia Americana,* s.v. "Chemistry," 1830 ed., 3:125.

13. John L. Comstock, *Elements of Chemistry* (New York: Robinson, Pratt & Co., 1839), p. 9.

14. George Fownes, *Elementary Chemistry, Theoretical and Practical* (Philadelphia: Blanchard & Lea, 1857), p. 115.

15. John W. Langley, "Synthetic Chemistry," *Popular Science Monthly* 5 (May 1874): 39–40.

16. John H. Appleton, *Chemistry [Beginner's Hand-Book of Chemistry]* (Providence, R.I.: Providence Lithograph Co., 1885), p. 8.

17. A particularly forceful expression of the implications of the materialist viewpoint for science was offered by John Tyndall in his presidential address before the British Association for the Advancement of Science in 1874; this is reprinted in *Victorian Science,* ed. George Basalla et al. (Garden City, N.Y.: Doubleday & Co., 1970), pp. 441–478.

18. Humphry Davy, "A Discourse Introductory to a Course of Lectures on Chemistry . . . 1802," in *Collected Works of Sir Humphry Davy* (London: Smith, Elder & Co., 1839; reprint, New York: Johnson Reprint Co., 1972), 2:326.

19. Henry, *Elements* (note 3), p. 12.

20. Samuel Parkes, *The Chemical Catechism* (New York: Collins & Co., 1821), p. 218, note b.

21. Josiah P. Cook, Jr., *Religion and Chemistry: or, Proofs of God's Plan in the Atmosphere and its Elements* (New York: Charles Scribner & Co., 1865): "The Graham Lectures: On the Power, Wisdom, and Goodness of God, as Manifested in His Works" (first delivered at the Brooklyn Institute, 1861), 3:5.

22. John Dorman Steele, *Fourteen Weeks in Chemistry* (New York: A. S. Barnes & Co., 1873), p. 240.

23. Ira Remsen, "The Science *vs.* the Art of Chemistry," *Popular Science Monthly* 10 (April 1877): 691.

24. Ibid., p. 692.

25. Ibid., pp. 695–696.

26. Owen Hannaway, "The German Model of Chemical Education in America: Ira Remsen at Johns Hopkins (1876–1913)," *Ambix* 23 (1976): 157.

27. Massachusetts Institute of Technology, *Annual Catalogue:* Third (1867–68), Thirteenth (1877–78), Twenty-third (1887–88), and Thirty-third (1897–98). Thanks to Helen Samuels and the staff of the MIT Archives for making this material available.

28. Ernst Lassar-Cohn, *Chemistry in Daily Life,* trans. M. M. Pattison Muir (Philadelphia: J. B. Lippincott Co., 1905); Svante Arrhenius, *Chemistry in Modern Life,* trans. and rev. Clifford S. Leonard (New York: D. Van Nostrand Co., 1925). It should be noted that books of this type were not new in the twentieth century; James F. W. Johnston's *The Chemistry of Common Life,* for example, was published in 1855.

Appendix: Some Definitions of Chemistry: 1814–1919

Chemistry, therefore, may be defined, as that science, the object of which is to discover and explain the changes of composition that occur among the integrant and constituent parts of different bodies. (William Henry, *The Elements of Experimental Chemistry* [Boston: Thomas & Andrews, 1814; 3d American, fr. the 6th English ed.])

Chemistry investigates the composition of bodies, and the changes of constitution, which they produce by their action on each other. (Denison Olmsted, "On the Present State of Chemical Science," *American Journal of Science* 11 [1826]: 350)

"Chemistry," says Jacquin, "is that branch of natural philosophy which unfolds the nature of all material bodies, determines the number and properties of their component parts, and teaches us how those parts are united, and by what means they may be separated and recombined." (*London Encyclopædia.* London: Thomas Tegg, 1829, article: "Chemistry," vol. 5, pp. 353–560)

. . . the science which teaches the nature of bodies, or rather the mutual agencies of the elements of which they are composed, with a view to determine the nature, proportions and mode of combination of these elements in all bodies. (*Encyclopedia Americana.* Philadelphia: Carey and Lea, 1830, article: "Chemistry," vol. 3, pp. 124–128)

Chemistry is that science which investigates the composition of all bodies, and the laws by which it is governed. (Benjamin Silliman, *Elements of Chemistry in the Order of Lectures Given in Yale College.* New Haven, Conn.: H. Howe, 1831. 2 vols.)

Chemistry is that science which investigates the composition and properties of bodies, and by which we are enabled to explain the causes of the natural changes which take place in material substances. (John L. Comstock, *Elements of Chemistry* [New York: Robinson, Pratt & Co., 1839])

There are *four leading questions* which the chemist puts to the different natural bodies . . .

a) Of what are they composed?

b) What changes do bodies undergo, when placed in contact with other bodies?

c) What useful applications can be made of chemical theory and practice?

d) What are the causes of chemical changes, and according to what laws do they take place?

(Julius A. Stöckhardt, *The Principles of Chemistry, Illustrated by Simple Experiments,* trans. C. H. Peirce [Cambridge, Mass.: John Bartlett, 1852], pp. 9–11)

Chemistry, or that department of physical science which recognises the nature and composition of bodies. . . . (*Chambers's Information for the People,* 5th American ed. [Philadelphia: J. W. Moore, 1853], 1:230)

The Science of Chemistry has for its object the study of the nature and properties of all the materials which enter into the composition or structure of the earth, the sea, and the air, and of the various organized or living beings which inhabit these latter. (George Fownes, *Elementary Chemistry, Theoretical and Practical* [Philadelphia: Blanchard & Lea, 1857])

This globe, and every thing appertaining to it, is composed of substances, which exist either in a compound or simple state. It is the object of the scientific chemist to investigate the properties of these substances, and to show their action upon each other. By this science, therefore, compound bodies are reduced to the simple elements of which they are composed, or new combinations formed. (Edward Hazen, *The Panorama of Professions and Trades* [Philadelphia: Uriah Hunt & Son, 1863], p. 131)

Chemistry teaches us the intimate and invisible constitution of bodies, and makes known the compounds which may be formed by the union of simple substances, the laws of their combination, and the properties of the new compounds. (Benjamin Silliman, Jr. *First Principles of Chemistry,* 15th ed. [Philadelphia: Thomas Bliss & Co., 1864], p. 15)

The science of chemistry has for its aim the experimental examination of the properties of the elements and their compounds, and the investigation of the laws which regulate their combination one with another. (Henry E. Roscoe, *Lessons in Elementary Chemistry* [London: Macmillan, 1869], p. 6.)

Chemistry treats of the composition of bodies and the specific properties of matter. (John Dorman Steele, *Fourteen Weeks in Chemistry* [New York: A. S. Barnes & Co., 1873])

The science of chemistry is that particular science which treats of the action of bodies upon each other, in so far as this action causes a change in the composition of the bodies. (Ira Remsen, "The Science vs the Art of Chemistry," *Popular Science Monthly* 10 [April 1877]: 694)

Sixty-seven elements have been proved to exist. . . . The art by which these and all other compound substances are resolved into their elements is termed Chemistry, a name derived from the Arabic word *kamai,* to conceal. The *art* of chemistry also includes the construction of compounds from elements, and the conversion of substances of one character into those of another. The general principles or leading truths relating to the elements, to the manner in which they severally combine, and to the properties of the compound substances formed by their union, constitute the *science* of chemistry. (John Attfield, *Chemistry: General, Medical, and Pharmaceutical* [Philadelphia: Henry C. Lea, 1879], p. 13)

Chemistry, then, may be most simply defined as that branch of natural science which considers (1.) The combination of two or more substances to form a third body, with properties unlike either of its components; and (2.) The separating from a compound substance of the more simple bodies present in it. (*The International Cyclopedia,* s.v. "Chemistry" [New York: Dodd, Mead & Co., 1885], 3:740)

Chemistry deals with certain portions of one class of material phenomena. The mark of this class of phenomena is, *change of properties accompanying change of composition.* (M. M. Pattison Muir and Charles Slater, *Elementary Chemistry* [Cambridge: Cambridge University Press, 1887], p. 1)

The object of chemistry is to discover the laws which govern the union or the decomposition of substances, and to determine the limits within which such changes are possible. (*Encyclopedia Americana,* s.v. "Chemistry" [New York: The Americana Company, 1903], vol. 4, not paginated)

Both Physics and Chemistry deal with matter and its changes. But Physics considers especially the changes in which the composition of the substance is not altered, which Chemistry is concerned with those that result in *new substances.* (John C. Hessler and Albert L. Smith, *Essentials of Chemistry* [Boston: B. H. Sanborn & Co., 1914], p. 9)

Roughly speaking, physics treats of energy and chemistry of matter. Thus chemistry tells us of the properties and composition of substances, as of water, of iron or of sulfur, of the action of substances on each other and of the changes in composition which they undergo in a great variety of circumstances. (William A. Noyes, *A Textbook of Chemistry* [New York: Henry Holt & Co., 1919], p. 5)

III
Public Interface

Challenge to Preserve
Challenge to Public Outreach
Challenge to Public Policy

CHALLENGE TO PRESERVE

Documenting Modern Chemistry: The Historical Task of the Archivist

Helen W. Samuels

Theodore William Richards (1868–1928), the first American chemist to receive the Nobel Prize, spent almost all his career at Harvard University teaching and working with graduate students in his own laboratory. Robert Burns Woodward (1917–1979), the 1965 recipient of the Nobel Prize in chemistry, was also based at Harvard University, but at the same time he collaborated with research groups at Charles Pfizer & Company and the Eli Lilly Company, and directed the Woodward Research Institute at the University of Basel.[1] Documenting these careers presents archivists with two very different problems. Assembling the documentation of scientists who work in large teams at several industrial and academic settings is more complex than gathering the records of a scientist who spent a career at one site. The challenges created by the size and complexity of modern documentation have forced archivists to rethink the methods they use to identify records of enduring research value. This paper addresses some of the problems and proposed solutions. Because the solutions entail historical research, the paper also explores archivists' changing use of history.

The archival profession in the United States is primarily the offspring of the American historical community and to this day remains heavily influenced by its antecedents.[2] Our training and clientele bind us to our roots. A knowledge of history and historical methods is considered an essential tool for the archivist. Traditionally, those skills are used as archivists prepare material for historical researchers. Collections are examined, arranged, and described in finding aids that include biographies for papers of individuals, administrative histories for the records of institutional offices, and explanations of the scope and content of the collections. Finally, subject and narrative entries are prepared for published guides and card catalogues.

For these tasks archivists use their historical skills for descriptive purposes. The intent is to create products that facilitate the use of the collections. Now, however, historical skills are needed to solve another aspect of archival work, namely, the selection of materials. In this case, analytic rather than descriptive skills are required.

The contrast between the respective careers of Richards and Woodward reflects the changes that have taken place in the conduct of science and implies the resultant changes that archivists have made in their efforts to document scientific activities. In the past, archivists, especially academic archivists, assumed that by gathering the teaching and research material that a faculty member left at the university they would have adequately documented a career. Now the records are dispersed and duplicated among the numerous institutions and the individuals with whom the faculty member worked. Locating this dispersed record and assuring its preservation in appropriate repositories presents a new challenge for the archivist.

The efforts of the Center for History of Physics at the American Institute of Physics (AIP) demonstrate the evolution of archival thinking. In the early 1960s, when the Center first set out to document physics, the "major emphasis was on individuals because most work in physics before World War II was done by individuals working alone or perhaps with a few others."[3] The AIP's activities were therefore devoted to locating and placing the papers of prominent physicists in academic archives or other appropriate repositories. The extensive historical studies conducted by the Center staff of the subfields of physics, and especially their shift in attention to the post-World War II years, necessitated a redefinition of their approaches. Confronting "big physics" required dealing with industrial and government laboratories and large teams of scientists. The AIP therefore undertook a series of historical and archival research projects, first of government laboratories, now of multi-institutional research, to help them formulate methods to adequately document these complex phenomena. Their findings have stimulated new approaches to documenting physics and have influenced the archival profession as a whole.

A further examination of the problems that archivists face as they try to document modern physics, chemistry, or any other aspect of modern society is in order. The problems are inherent in the nature of modern institutions and the records they produce.

MIT receives research funds from the National Institutes of Health (NIH), the National Science Foundation (NSF), Ciba-Geigy, and Exxon. MIT faculty members oversee research activities and consult for governments and industry. Several institutions and individual scientists work together on an experiment and jointly publish the results. Where, then, is the documentation of these activities? Appropriately, it is dispersed. The records documenting the decision to fund an MIT research project are located at the granting agency, but the records of the actual research are in the laboratories where the work was done. Records mirror the society that creates them. Archivists recognize that records of individual careers and cooperative ventures are dispersed among the relevant institutions. The challenge is to transform this perceived problem into a possible solution.

The documentation of modern chemistry resides in many forms of evidence: published monographs, technical reports, and journal articles; data (numerical, visual, and so on); administrative correspondence and research notes (on paper and in machine-readable form); equipment and chemical samples; and visual documentation of the research process. The documentary record is integrated, with each form of evidence contributing information to our total understanding. While this may appear obvious to the historical researcher who uses multiple forms of evidence, this integration of information is not reflected in curatorial practice.

Librarians, archivists, museum curators, sound and film specialists, and data archivists share the responsibility to select, describe, and make information available. For the most part, however, each curator adopts a format-specific approach. Archivists, for instance, generally do not consider the availability of information in published, visual, or other forms when they select manuscripts or archival records. The development of semisynthetic penicillin is documented not only in John Sheehan's research notes but also in the technical reports issued by his laboratory. Only recently has advice on an integrated approach to the selection of the records of modern science been made available to the archival community.[4]

Legislators and lawyers also contribute to our documentary woes. Environmental and public health concerns generate an ever increasing number of laws and regulations that burden public and private institutions who must comply with reporting requirements. Academic institutions and the chemical industries are faced with regulations specifically designed to control the production and disposal of chemicals as well as more general regulations that control such things as the minimum wage, access for the physically handicapped, and affirmative action. Government regulations often require the creation and retention of specific documentation, but the records created and managed through this process are often not the records desired for long-term historical research. In addition, the short retention periods mandated by governments do not encourage archival programs. Lawyers, nervous about the availability of information that can damage their clients, exacerbate this problem by encouraging the destruction of records as soon as they have fulfilled their minimum legal requirements.

Compounding these problems are the overwhelming volume of modern records and the preservation problems associated with the poor quality of paper and the impermanence of magnetic media. A recent study of the selection of modern records dramatized the problems posed by sheer bulk. The author calculated the amount of paper and machine-readable records currently produced in the United States and the number of archivists available to examine and select the small portion that should be retained.[5] It is clearly impossible for archivists to see, much less evaluate, all of this information. We need new techniques that enable us to analyze and plan the selection of the documentary record[6] (see Figure 1).

A concept called *documentation strategies* was proposed in the mid-1980s to respond to these problems. A documentary strategy is an analytic, planned approach to solving problems posed by the complexity and volume of modern documentation. The key elements in this approach are an analysis of the universe to be documented, an understanding of the inherent documentary problems, and the formulation of a plan to assure the adequate documentation of an ongoing issue or activity, or a geographic area. The strategy is designed, promoted, and implemented by records creators, administrators (including archivists), and users. It is an ongoing cooperative effort by many institutions and individuals to ensure the archival retention of appropriate documentation through the application of redefined archival collecting policies, and the development of sufficient resources. The strategy is altered in response to changing conditions and viewpoints.[7]

When first proposed, the concept of documentation strategies generated both skepticism and concerns. Who will select the topics to be examined, and who will be involved? How can we afford to carry out these activities without outside funding? Why should the archivist of an institution care about such cooperative projects? The published literature on the topic,[8] combined with a series of seminars and pilot projects, has clarified the potential usefulness of this concept and begun to suggest how these ideas might enhance archival practice.

Documentation strategies attempt to alter the sequence of activities and the assumptions that archivists make about selection. Traditionally, archivists launch collecting projects by conducting surveys of records. Surveys, however, provide information only about what records have survived. They do not suggest what documentation is required for an adequate record, nor do they help establish documentary goals.

Figure 1. Completed new drug application for Clinoril, submitted in June 1976. It consisted of 307 volumes, one of the largest NDAs submitted by Merck. Photo reprinted by permission of Merck & Co., Inc.

The documentation strategy approach proposes an altered sequence: first, historical investigation is required to understand the phenomenon being documented and the associated documentary problems. Only then, with the shared expertise of creators, users, and curators, can documentary goals be established and a reasonable plan constructed.

A strategy is launched by an individual or institution to remedy the poor documentation for a specific sector of society. The institution that launches the effort need not be an archival repository, and the prime mover does not have to be an archivist. The Billy Graham Center at Wheaton College in Wheaton, Illinois, launched its strategy to improve the documentation of the evangelical movement in the United States.[9] The National Air and Space Museum initiated its effort to improve the documentation of the aerospace industry. The New York State Archives' strategy was launched to improve the documentation of the western counties of the state.[10] And one of the primary reasons the Beckman Center launched its "Polymers and People" project was to improve the documentation of polymer science and technology.[11]

Once the topic is identified, advisors and participants are assembled to guide the effort. Creators (scientists, administrators), users (historical researchers), and curators of the records (archivists, librarians, museum staff) are needed to provide historical knowledge about the topic and its documentation, and to influence those who create, house, and fund archives. The Billy Graham Center called upon church administrators, historians and sociologists of the evangelical movement, as well as archivists and librarians, to participate in its effort. An institution—again, not necessarily an archival repository—is required to provide a home and staff for this ongoing activity, to coordinate the project, and to carry out the historical and archival investigations.

Then research is initiated to achieve an understanding of the phenomenon to be documented. What is the history of the evangelical movement, the aerospace industry, polymer science, or the counties in western New York? Guided by those knowledgeable about the topic, research starts in the published histories and other readily available materials. The published evidence provides a summary history of the topic as well as an understanding of the questions that historical researchers have asked. Recognizing, however, that the existing published works have limitations based on the sources available to the authors, the project participants also seek other ways to study the topic.

The use of a variety of analytic techniques provides a fuller knowledge of the purposes, functions, and special characteristics of the topic of the documentation strategy. Functional analysis, for example, is a particularly useful tool. Archivists normally examine an institution by studying its administrative structure. Functional analysis provides a means to shift the focus from a given structure to the purposes and activities of the institution.

What are the functions of an aerospace company, a county government in western New York, or a university? Functional analysis is intended to answer these questions by providing a broad understanding of all of the primary purposes or functions of the institution. In recent years three functional guides have been prepared to help archivists understand the activities and documentation that comprise modern science and technology,[12] the high-technology companies,[13] and colleges and universities.[14] The advantage of this approach is that all activities are explored. For instance, whereas college and university archives contain considerable documentation of administrative activities, the functional guide discusses the other functions of an academic institution: to teach, conduct research, socialize, provide public service, confer credentials, and maintain culture.[15] Functional analysis reinforces the need to document all facets of an institution. College archivists must document the public service functions of their institution as well as its research, teaching, and administrative activities.

Functional analyses are extremely important to documentation strategies in that they provide examinations and evaluations of the available evidence. As the component functions or activities of a given endeavor are explored, it becomes clear that each presents a different documentary problem. The administration of a church produces many records, including the minutes of the governing board as well as financial and personnel transactional records, but little evidence exists of parishioners' personal beliefs or role of the individual in worship services. Colleges and universities also produce voluminous administrative records, but little evidence exists to document the learning process in the classroom and laboratory, and the personal development of the students throughout the educational and socialization process. A documentation strategy requires an understanding of these problems, so that voluminous records can be weeded appropriately while an effort is made to create a record of the intangible processes.

An example from science and technology clarifies this point. In *Appraising the Records of Modern Science and Technology,*[16] the structure and content of the technical-report and scientific journal literature are discussed, as these published sources are the most pervasive form of documentation of these fields. At the same time, the study discusses those activities that are poorly documented in the published literature, such as the actual sequence of events of an experiment, the assignment of responsibilities to the research staff, and the problems of building and running equipment.[17] The recommendations in the MIT appraisal study address the relative values of the types of records available to document these activities (research notes, personnel files, and photographic records) and suggest when evidence must be created because no documentation was created naturally during an activity. A knowledge of the functions and activities that comprise modern science and technology, and the evidence of those activities, proves useful, I believe, to those who undertake documentation strategies in any area of modern science and technology, including chemistry.

With a knowledge of the phenomenon in question and an understanding of the associated documentary problems, goals can be formulated to assure the documentation of the topic. What institutions, activities, events, and individuals are to be documented? What material is required to document them adequately? As topics are chosen to be documented, it must also be recognized that other areas will not be documented. Accordingly, the deliberations must be well documented so that future researchers will understand the decisions and actions of the documentation strategy team.

Another facet of the documentary plan is the coordination of the dispersed record. The historical and archival studies provide information about who creates records and for what purpose. Ideally, government and other funding agencies could help coordinate the retention of dispersed records by providing guidance on records retention when they award a grant or a contract. In the absence of such a plan, the documentation strategy team should provide this guidance. Responsibility for specific parts of the integrated record must be assigned to and accepted by specific institutions. Automated networks of information about archival holdings can provide information about these dispersed collections to the institutions involved and to historical researchers.

The home base for the documentation strategy monitors the successes and failures of the plan and makes adjustments as needed. When emergencies arise, repositories must be found for endangered collections. The background historical and archival understanding, however, suggests how much of the collection should be rescued and where it could be placed. As institutions, laws, and regulations change, the documentation strategy must be modified.

The pharmaceutical industry serves as a useful hypothetical example of how documentation strategies work.[18] Stimulated by concerns voiced by the Beckman Center, the pharmaceutical industry itself, or one lone archivist, an advisory committee is formed to define the project, rally funds and political support, and guide the intellectual efforts. The advisory committee is composed of corporate executives, government officials, and scientists to represent the creators of the records; archivists and librarians to represent the curators; and historians of science and technology to speak for the potential users.

The scope of the documentary effort is defined by determining what is encompassed in the phenomenon known as the pharmaceutical industry. What companies and other institutions are involved? What types of activities are included in pharmaceutical work? What time periods and countries are to be examined? The documentation strategy team uses a variety of analytic methods to answer these questions. To start, a typology of the pharmaceutical industry facilitates this analysis and clarifies the actors and activities to be examined.

There are four main types of pharmaceutical companies:

1. *Ethical companies*—Research-based drug companies that market their products to doctors, hospitals, pharmacists, and health maintenance organizations (HMOs). Examples of these firms are Merck & Co., Eli Lilly and Company, and The Upjohn Company.

2. *Over-the-counter companies (OTC)*—Marketing-based companies that convert brand-name drugs to over-the-counter products that are sold directly to consumers. Examples of these companies include McNeil Consumer Products Co., Bristol-Myers Squibb, and Warner Lambert.

3. *Generic companies*—Marketing-based companies that convert proprietary pharmaceutical products to generic drugs after the patent protection has expired. Generic drugs share the same markets as the products of the ethical drug companies. Examples of these companies include Mylan Laboratories, Quad, and Bolar.

4. *Start-up biotechnology and experimental companies*—research-based companies that use new techniques such as genetic engineering or structure-based design to develop drugs. Examples of these firms are Genentech, Amgen, Cetus, Biogen, and Vertex.

Another way to use this typology is to examine the assets or skills required by the pharmaceutical companies. Each of the four categories in the typology has a different combination of the following assets: research and development capabilities, marketing skills, successful drugs on the market, new drugs in the pipeline, a history of getting drugs approved through the regulatory process, and the ability to use new techniques to develop drugs.

Using the typology and an understanding of these assets can guide an examination of the historical events that have altered and will continue to alter this typology. The hundred-year history of the pharmaceutical industry can be divided into several periods. The first forty years were characterized by growth and the rapid development and marketing of new products. The imposition of government regulations during the Depression transformed the process of developing, testing, and marketing new products. Government regulations were expanded as a result of the Kefauver hearings of 1962 that required drugs to be not only safe but also efficacious. Toxicology tests, clinical studies, and data gathering and reporting activities are now regularized and enlarged. The regulations have significantly affected the length of time and expense required to bring new drugs to the market.

The last twenty years have been characterized by mergers as companies seek the diverse skills they need to survive, and the revenue required to develop and market new products. Merrell Dow, a company with a research program, marketing sales force, and new drugs in the pipeline, bought Marion Laboratories, a company with a lucrative heart drug. Bristol Myers, a company with significant marketing capabilities, bought E. R. Squibb & Sons, an ethical pharmaceutical company, to acquire strengths in research and development and a group of successful drugs. Genentech, a biotechnology company, recently sold the majority of its shares to Roche Holding to enhance the ability of both companies to develop new drugs and market them worldwide.

Still another perspective is offered by asking what is different about the pharmaceutical industry: what activities or concerns differentiate it from other industrial enterprises? Bert Spilker's *Multinational Drug Companies* contains the following list of differentiating factors. "1. the long period of time required to develop and market a newly discovered drug, 2. the high degree of financial risk and uncertainty of a drug's future, even after it is launched, 3. the large number of highly restrictive regulations that govern all aspects of a drug's development, production, and marketing, 4. the inability to predict when the next important drug discovery will occur, and 5. the large number of variables and factors that are involved in biological experiments, technical development and especially clinical studies."[19] The nature of the sales and marketing activities that are directed to professionals and not to the general public is an additional differentiating characteristic.[20]

With this basic understanding of the types of companies, their purposes, history, and differentiating characteristics, the documentation strategy team can look inside the companies to understand the nature of the functions or activities that are carried out and their documentary problems. Some of the relevant functions are research and development; testing of drugs; toxicology tests; clinical studies; regulatory submissions and approval; marketing and sales; and corporate management. A study of each function examines its purpose and the individuals within and outside the company who are involved (see Figure 2). For instance, clinical studies are now multinational activities

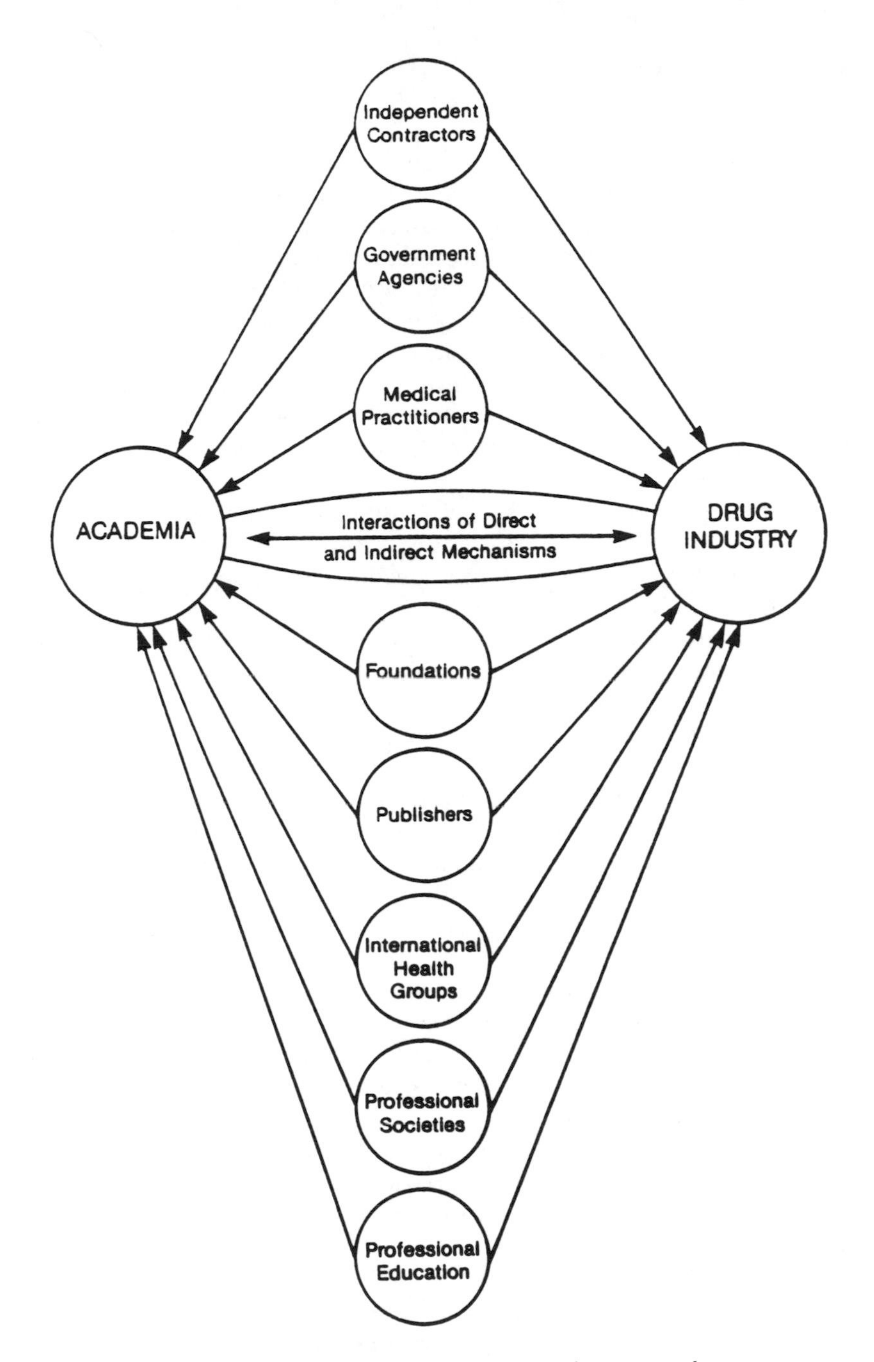

Figure 2. Selected areas where indirect interactions occur between academia and the drug industry, research and development. From Bert Spilker, *Multinational Drug Companies: Issues in Drug Discovery and Development* (New York: Raven Press, 1989). Reprinted by permission.

because drugs must be tested and approved in each country where the drug is to be marketed. The task of planning, coordinating, and analyzing the data assembled by the clinical trials rests with the staff within the company. Hospitals and physicians in each country administer the drug, evaluate its efficacy, and record any adverse reactions. Data on each test subject are entered in the individual patient records maintained by the doctors and the hospitals. These observations are then compiled and submitted to the company by the doctors. The company scientists then analyze and submit the data from the clinical studies in the New Drug Application submitted to the Food and Drug Administration (FDA). Only then can the drug be marketed and sold, and used by patients.

What does all of this analysis say about the documentary problems? How does the knowledge gained support the selection decisions that must be made? Both general and specific observations can be made about the documentation.

The examination of the pharmaceutical industry reveals some general conditions that encourage documentary activities. The industry values information and recognizes the need for long-term access for scientific, regulatory, and management purposes. "Pharmaceutical companies are extremely information-intensive enterprises. A large quantity of information is obtained and relied upon from external sources and is also generated internally throughout drug discovery and development processes. Information is systematically recorded for easy retrieval to support each product's survival, from its conception through its life in the market."[21] Archival activities are also supported by the age, and one would hope pride, of the companies that are nearly a century old. Merck's efforts to establish its archives and prepare a centennial celebration are indicative of the positive benefits that can be derived from a long history. The Miles Laboratory, Eli Lilly, and Squibb all have archival programs, and Upjohn and Abbott have both published centennial histories.

Unfortunately, there are also general issues that militate against documentary activities. The first problem is the size of the endeavor. Drug companies are billion-dollar enterprises that often occupy many facilities in different locations. Mergers and relocation of corporate headquarters to Europe or Asia further jeopardize the records. Each activity produces voluminous documentation, but only a small portion of these records has continuing value. The task is to locate and assure the preservation of just those records.

The other significant issue is the number of actors and institutions involved in these activities. The main focus of the strategy is the pharmaceutical companies. A full understanding of the industry, however, must also consider:

the government agencies such as the Food and Drug Administration, Federal Trade Commission, Department of Agriculture, Occupational Safety and Health Administration, and Environmental Protection Administration, etc., who regulate the industry;
the consultants from academic institutions and government who advise the industry;
the government, academic, and hospital laboratories that carry out part of the research efforts;
the hospitals, doctors, and patients that participate in the clinical studies;
the doctors, HMOs, pharmacies, and other buyers and distributors of the products;
the patients who use the drugs.

The documentation for any given drug is naturally dispersed among all of these individuals and institutions and cannot be gathered together, for two reasons. First, each portion of the record is part of the archival record of a separate institution and should be preserved as part of its history. Second, it is simply impractical to think of any institution that could handle the volume of records involved. What the documentation strategy provides, however, is the intellectual control that coordinates the placement and access to the dispersed record.

Specific documentary problems arise from the nature of the activities to be documented. Some functions are concentrated and controlled by the parent organization; others are highly dispersed. Some activities are easier to document because a record is created as a natural product of the activity; other activities generate no documentation. For instance, the problems associated with documenting marketing activities are different from those associated with documenting clinical studies. Marketing activities are carried out within the company. Records of the executives in charge document the decisions made on marketing strategies. Published literature on each product reveals how each drug is marketed, surveys and promotion plans document how the marketing strategy is developed, and correspondence and sales figures demonstrate the reactions to the marketing strategy. Clinical studies, on the other hand, are a highly dispersed activity. The company has the records of planning the study and a summary of the results, but not the individual case records. The government also receives large amounts of the analyzed data from the clinical studies.

Documentary analysis also suggests the less tangible aspects of the pharmaceutical industry: the decisions of the scientists on what compounds to examine, the management deliberations on corporate strategies, and the reactions of patients to the drugs. These and other intangible aspects can only be captured through deliberate documentary projects such as oral histories.

From all of these studies the documentation strategy team can determine the answers to the two key questions, namely, what is to be kept? and who will keep it? Addressing the latter question first, documentation strategies rest on the belief that creators of records, especially large corporations such as the pharmaceutical companies, must take responsibility for their own records. No university, state historical society, or research library can take on this task. The documentation strategy can assist these efforts in several ways. As the Beckman Center has done very effectively, the strategy team can educate, cajole, and pressure the companies to accept this responsibility. Information about the benefits of effective records management and archival activities encourages the companies. More specific advice on selection and retention of specific records is even more helpful to the companies and to the documentary project as well.

Another type of assistance could be provided if the documentation strategy team worked with the FDA historians and records managers, for instance, to issue records retention guidelines to facilitate the coordinated selection and preservation of records. It might also be possible to lobby the Congress and government agencies to use regulatory and tax policies to create incentives for the companies to retain their records and make them available to researchers.

And then, finally, what is to be preserved? Each company that has an archives collects material in response to its own vision and needs. The documentation strategy looks beyond the needs of any specific company and instead considers the required total documentation of the industry. The strategy must also pay particular attention to areas that are not well documented. Informing all of these selection activities is the knowledge of the history, trends, and unique characteristics of the industry.

For each of the four types of companies delineated in the typology (ethical, over-the-counter, generic, and start-up biotechnology) the team can identify the key companies and encourage the establishment of archival programs where none exists. Key drugs should be identified either because of their tremendous success (scientific, clinical, and/or market), failure, or because they are more typical of the industry. Working with all of the actors to coordinate the retention, a full record is identified for preservation: correspondence, laboratory notes, toxicology and clinical records, product samples, processing equipment, marketing strategies, and sales records. To assist the companies and future researchers, guidance should also be provided on the minimum information that should be retained for each drug once the product is no longer on the market.

Will future historical researchers know anything of the firms and drugs not included in this selection? Are all the other firms and their products to be neglected? No. While a documentation strategy for the pharmaceutical industry should assure the full documentation of at least the core companies and drugs, educational materials about the need for archival and records management programs, and guidance on the selection of records should be made available to all firms. And although not all firms will create and maintain archives, considerable documentation of the companies is available in published sources and records preserved by other institutions, including government repositories where environmental and tax information is available.

This brief summary describes the rationale and procedures of documentation strategies. It might sound utopian, Machiavellian, or just plain naive. The truth will be revealed as the few strategies that are under way are assessed. The ultimate test will come when future historical researchers use the documentation left to them. Will the documentation strategy team have made all the right choices? Probably not. But the analysis and planning that documentation strategies require seem to assure a fuller and a more coherent record than the one that is currently assembled by chance and passive acquisition.

The most important concepts embodied by the documentation strategy idea, I believe, are the following:

- The size and complexity of modern documentation require a planned and active approach by the archivist. It is no longer satisfactory to build collections by waiting for donors to offer material. Planned approaches are required to actively select adequate documentation.
- Historical study of the topic and analysis of the documentary problems are required to establish goals and shape the documentation.
- Assembling the historical record is not a task for the archivist alone. Creators, users, and other curators all have knowledge to contribute to this complex task.

At the Pennsylvania Avenue entrance of the National Archives are two pensive statues that advise the archivists and researchers who enter the building each day to "Study the Past" and remind us that "What Is Past Is Prologue." Studying the past informs the process of selecting the records that will be left to the future. In this way the past can continue to serve as prologue.

Notes

1. I thank Clark Elliott, associate curator of the Harvard University Archives, for suggesting Richards and Woodward as examples. The surviving records for both chemists are housed at the Harvard University Archives.

2. The most useful analysis of the evolution of the archival profession is Richard J. Cox, "Professionalism and Archivists in the United States," *American Archivist* 46 (1986): 229–247. The Cox article also touches on the continuing debate concerning the validity of historical vs. library training as the proper preparation for archivists, and cites the literature on this topic.

3. Joan Warnow-Blewett, "Documentation Strategy Process: A Case Study," *American Archivist* 50 (1987): 29–47, quote p. 33.

4. Joan K. Haas, Helen Willa Samuels, and Barbara Trippel Simmons, *Appraising the Records of Modern Science and Technology: A Guide* (Cambridge, Mass.: Massachusetts Institute of Technology, 1985; distributed by the Society of American Archivists).

5. David Bearman, *Archival Methods* (Pittsburgh: Archives & Museum Informatics, 1989), p. 9.

6. The combined impact of regulations and volume is demonstrated in the new drug applications (NDA) submitted to the Food and Drug Administration (FDA). Merck's NDA for Mevacor, submitted to the FDA in 1986, consisted of 190 400-page volumes.

7. This definition is based on the original version included in the Samuels and Hackman articles cited below as revised by Samuels, Richard J. Cox, and Tim Ericson in 1989.

8. Helen W. Samuels, "Who Controls the Past," *American Archivist* 49 (1986): 109–124; Larry J. Hackman and Joan Warnow-Blewett, "The Documentation Strategy Process: A Model and a Case Study," *American Archivist* 50 (1987): 12–47; Philip N. Alexander and Helen W. Samuels, "The Roots of 128: A Hypothetical Documentation Strategy," *American Archivist* 50 (1987): 518–531; and Richard J. Cox, "A Documentation Strategy Case Study: Western New York," *American Archivist* 52 (1989): 192–200.

9. *Proceedings of the Evangelical Archives Conference* (Wheaton, Ill.: Billy Graham Center, 1988).

10. Cox, "Documentation Strategy Case Study."

11. Beckman Center for the History of Chemistry, *Final Report: Polymers and People. NSF Project No. SES 84-21278* (Philadelphia: Beckman Center for the History of Chemistry, University of Pennsylvania, 1988).

12. Haas et al., *Appraising the Records.*

13. Bruce H. Bruemmer and Sheldon Hochheiser, *The High-Technology Company: A Historical Research and Archival Guide* (Minneapolis: Charles Babbage Institute, University of Minnesota, 1989).

14. Helen W. Samuels, *Varsity Letters: Documenting Modern Colleges & Universities* (Chicago: Society of American Archivists; Metuchen, N.J.: Scarecrow Press, 1992).

15. Administrative activities are discussed under the function Sustain the Institution of this book (note 14). The other six functions are Confer Creden-

tials (admit and graduate students), Convey Knowledge (teach and learn), Foster Socialization, Conduct Research, Provide Public Service, and Promote Culture.

16. Haas et al., *Appraising the Records,* pp. 69–76.

17. The current interest of historians and sociologists of science and technology in examining activities in the laboratory and the role of artifacts is explored in Jeffrey L. Sturchio, "Artifact and Experiment," *Isis* 79 (1988): 369–372.

18. I thank Jeffrey Sturchio for all of his advice about this example. The mistakes in interpretation, however, are all my own.

19. Bert Spilker, *Multinational Drug Companies: Issues in Drug Discovery and Development* (New York: Raven Press, 1989), p. 6.

20. The observation about sales and marketing was made by Jeffrey Sturchio.

21. Spilker, *Multinational Drug Companies,* p. 336.

Two Useful Tools for Documentation Strategies: Historical and Documentation Research

Joan Warnow-Blewett

I am in complete agreement with the preceding paper by Helen Samuels; here it is my intention to supplement her discussion of documentation strategies. Specifically, I hope to provide some characteristics of two different tools that can be employed in a documentation strategy by national discipline history centers for science and technology. These tools are historical documentation projects and documentation research on archival problems. Helen has blended these in her presentation; I will point out some differences.

One way a center carries out its charge is through historical documentation projects. Most of these projects focus on a particular field or subfield that has been little studied by historians or documented by archivists. The main project elements are historical research—including the identification of key individuals and institutions—the preservation of papers and institutional records, and oral history interviews. The oldest of these projects, the Sources for the History of Quantum Physics Project[1] and the American Institute of Physics (AIP) Project on Recent Physics in the United States,[2] go back to the summer of 1961. The AIP Project led to the establishment of the AIP Center for History of Physics—the original model for a discipline history center—and the Quantum Physics Project established a model for future historical documentation projects.

Historical documentation projects are also conducted by other institutions—for example, the University of California at Berkeley, Stanford University, and MIT—but the point I want to make here is that this kind of project is one of the chief tools that can be employed by national discipline centers to carry out their documentation strategies. In addition to projects of the AIP Center—for such fields as nuclear physics, astrophysics, and solid state physics—the Beckman Center has carried out its study of polymer chemistry[3] and the Charles Babbage Institute its investigation of Engineering Research Associates of St. Paul and its current project on the Advanced Research Projects Agency of the Department of Defense.[4] Because such historical documentation projects are relatively well known, let us move on to documentation research on archival problems, a newer and less familiar subject.

The objective of documentation research, as I define it, is to examine a broad range of functions and activities in a given area, find out what is needed to document the activity, and address policy and programmatic issues regarding the preservation of the documentation. There are times when a documentation research project is a more appropriate tool than a historical documentation project. The end product of a historical documentation project is good documentation; documentation research, on the other hand, is aimed at gaining knowledge about some problematical archival environment. A documentation research project is appropriate when a major area of contemporary activity is so fraught with unknowns that it is impossible for individual historians or archivists, let alone individuals who create the records, to state with any confidence how the area of activity ought to be documented.

Documentation research on archival problems is far broader in scope than historical documentation projects. In the latter, one knows how to assemble the record: key individuals and institutions are identified, oral histories with leading figures are conducted, and documentation is preserved. Documentation research also goes far beyond any research project in which a historian or other scholar seeks information on his topic of research in an archival repository; in fact, one of the major characteristics of documentation research projects is a focus on the activities and records in the contemporary scientific setting.

An early example—the first on a large scale—is a major documentation study of Department of Energy (DOE) National Laboratories conducted some years ago by the AIP Center. The overall purpose of the study was to learn how to document post-World War II Big Science, and the only way to do this was—and is—through understanding the process.[5] At the time, historians and archivists were preoccupied with academic science, and there was little knowledge of what went on in nonacademic settings. In addition, much of the documentation was new (such as machine-readable records) or altered in importance (for example, correspondence). The AIP study was called a documentation research project, because—although it included historical research, oral history interviewing, and preservation of records—its main objective was to look at the functions and activities of these postwar laboratories, seek answers as to what is needed to document the activity, and develop a methodology for saving the records. In short, the central purpose of the study was to bring a solid understanding of what was needed to document postwar physics research. The purpose was to put the Center and its allies in a position to provide guidance: first, to a wider group of archivists about appraising postwar scientific records, and second, to large research laboratories about initiating or improving archival programs.

Documentation research projects, although designed primarily by archivists, draw heavily on the participation of historians, sociologists, and—most important of all—the records creators themselves. This was particularly important in the DOE study when we needed to assess the potential research value of the records created by the postwar laboratories for a variety of academic disciplines.

A lot has happened since the DOE study. The report of the Joint Committee for Archives of Science and Technology, issued in 1983, assessed a broad spectrum of problems facing the documentation of postwar science and technology.[6] One of these problems was the need for appraisal guidelines beyond those put forth by the DOE study. Documentation research by Helen Samuels and her colleagues at the MIT Archives produced descriptions of research activity (particularly valuable in the university area) and provided guidelines for appraising the records produced.[7] Another documentation research project by the Charles Babbage Institute addressed difficulties associated with documenting large high-technology companies. This project investigated a broad range of industrial activity at the Control Data Corporation and elsewhere; it covered business functions—such as research planning and development, production and marketing—and support services.[8] The Babbage Institute was motivated to conduct the project by its overall concern about the state of historical documentation of the computer industry. It also needed to comprehend better the activities of high-technology industry, the value of the records produced by that activity, and the development of a methodology whereby important documentation can be identified and preserved. Yet another documentation research project is nearing completion—again by Helen Samuels. This study focuses on the basic functions of colleges and universities, the activities in place to carry out those functions, and the nature and quality of the documentation produced.[9]

At the AIP Center, the DOE study led directly to our present study of archival problems in the area of multi-institutional collaborations. To illustrate some of the issues related to documenting multi-institutional collaborations, here is what may happen in the field of high-energy physics. Members of a single collaboration may currently number in the hundreds; they come from a score of institutions based in several countries; they prepare for a decade and then conduct experiments at a unique accelerator facility. In many cases, collaborations may employ subcontractors in the design and fabrication of equipment. Although the collaborations themselves can be considered institutions, they are transitory; furthermore, even during their existence, team membership is fluid. Collaborations are not only multi-institutional but multidisciplinary and multileveled as well: in addition to physicists we find computer specialists and engineers; in addition to principal investigators and team leaders we find associate professors, postdoctoral fellows, graduate students, and technicians. One demonstration of the growth of multi-institutional collaboration in science and technology is the proposed National Collaboratory, a resource that would use networking and computer technology to support remote interaction.[10]

Once again, the first goal of our current documentation research project is to achieve an understanding of the process of scientific research and the records thus created. Only then will we be in a position to determine what is needed to document the work and to recommend possible solutions for saving the historical record. Because of the many complexities, the design of our project involved sociologists and outside archivists as well as historians. We are in the first year of a two-year study of high-energy physics research; this will be followed—funding permitting—by studies of space science and geophysics, some exploratory forays into other disciplines employing collaborative research, and—finally—workshops on project findings and archival and scientific policy issues. The period we are studying is from the early 1970s to the near present.[11]

To understand the community of collaborative research, we have set up a three-tiered research program. First, we are developing a census of all experiments carried out at major research facilities. Second, we have selected a sample of collaborations from this set and have a program of structured interviews with members of the selected collaborations. Finally, we have a program of "probes," or in-depth studies, of three highly significant collaborations.

The census draws on existing data bases developed by the scientific community. For high-energy physics, we are blessed with two detailed data bases: one tracks all experiments from the time they are approved by the accelerator facility, and the second tracks all reports and other publications as well as citations to these publications. We do not expect such a rich resource when we try to develop a census for other disciplines! However, even with high-energy physics, we are experiencing some difficulties—just what one would expect when using a scientific data base for historical and sociological purposes for the first time. For example, when using the computer to count numbers of physicists in collaborations, the most common individual we found was (Mr.?) "et al." More serious is the determination of the duration of each experiment, thanks to the fact that some data are given in terms of days, others in terms of date span or numbers of hours of accelerator beam time; this problem may be incapable of resolution. Finally, most laboratories have not completely linked publications with the number of the relevant experiment. Without such a link we cannot compute this measure of experiment productivity, nor can we analyze the numbers of citations of publications for experiments. Happily, we are getting help, for the laboratories are making special efforts to complete the linking of publications with specific experiments.

The program of structured interviews with members of selected collaborations is moving ahead. At an early meeting of our working group—composed of scientists, historians, sociologists and archivists[12]—we took our first steps in selecting the collaborations that would give us the broadest exposure to a range of scientific and social characteristics over our period of study. Also, we developed structured question sets addressing the major issues of the archival research project. I might mention that we have a main question set for physicists, a subset for graduate and postdoctoral students, additional questions for women and minority physicists, and still other separate question sets for engineers, computer scientists, and technicians. These have all been fully tested and are in use. They seem to be effective. We plan to conduct more than two hundred of these structured interviews. In all, we expect to uncover patterns of organizational and scientific activity, records creation, location of records, and any plans covering the retention of records.

Work is also underway on the probes to study in detail three significant collaborations: the discoveries of the J, the Psi, and the Upsilon particles. These collaborations ranged in size and complexity from the J team—simply MIT and Brookhaven National Laboratory—to the Upsilon collaboration, which is a series of seven experiments at Fermi National Accelerator Laboratory carried out by groups from as many as eight institutions, including four institutions abroad. On the level of the probes, the project will carry out more extensive historical research, interview many more individuals per collaboration, and make special efforts to save documentation. We expect in these cases not only to determine the documentation that should be saved but to take steps necessary to save the records. From this we hope not only to get a piece of our basic preservation mission accomplished but also to test in practice the opportunities and difficulties in documenting all multi-institutional collaborations.

At the time of writing we have been conducting interviews for only four months, but findings regarding records creation and retention are beginning to emerge. One particularly interesting example is a possible evolution toward more systematic record keeping by spokespersons—at least in the larger, more recent collaborations. This potential simplification contains another complexity in that the spokesperson can be in a surprising location; for example, a spokesperson for a collaboration carrying out an experiment at the Fermi National Accelerator Laboratory may be on the staff of the CERN laboratory in Geneva. The issues of who owns the records, and where should the records be preserved, are far from resolved. But so far, our findings are very preliminary and fragmentary; a more complete and formal report will be issued in about one year.[13]

Although much has been accomplished, there is a continuing need for documentation research. As in research of any kind, the resolution of one problem usually uncovers another, and the changing nature of science and technology presents us with ever new organizational arrangements, experimental techniques, and types of documents that need to be understood. Documentation research is ideally suited to this kind of problem solving.

Finally, I want to stress how each of these efforts contributes not only to the historical, archival, and scientific communities but also to the work of the other discipline history centers to fathom the complex issues underlying all modern science and technology. In multi-institutional collaborations, for example, the work of physicists and electrical engineers is so overlapping that the AIP Center has called for direct help from the Institute of Electrical and Electronic Engineers (IEEE) Center for the History of Electrical Engineering.

The Beckman Center is ideally suited to address several of the obstacles facing us today. For example, archivists and historians have only begun to investigate the process of postwar industrial science and technology. The Beckman Center might consider taking on some aspects of this perplexing area with a documentation research project as a follow-up to its polymer project. This new study of documentation problems might, as an example, focus on the industrial-academic and industrial-business relationships and include case studies tracking the activities and records creation from new discoveries to marketing products. It could draw on the methodology and findings of the CBI study of a high-technology corporation and the MIT study of colleges and universities. A second area ideally suited to the Beckman Center is that of internationalism in modern science and technology. One obvious target is the operations of those well-known far-flung chemical corporations with settings around the globe. Another target might be multi-institutional collaborations in chemistry—for example, the Human Genome Project; in this case, the methodology and findings of the AIP study could be useful as a starting point.

In the preceding essay Helen Samuels has illustrated, through the example of the pharmaceutical industry, the benefits of a fully-developed documentation strategy for chemistry. I endorse her description and want to emphasize that much work is needed to create a good strategy. I hope my discussion of historical documentation projects and documentation research projects has served as a useful supplement. Documenting modern science and technology is certainly a fascinating, ongoing process.

I am indebted to the following people who have read this paper and made helpful comments: William Aspray, John P. Blewett, Bruce H. Bruemmer, Frederik Nebeker, Jeffrey L. Sturchio, and Spencer R. Weart.

Notes

1. John A. Wheeler, preface, in Thomas S. Kuhn et al., *Sources for History of Quantum Physics: An Inventory and Report* (Philadelphia: American Philosophical Society, 1967), pp. v–ix.

2. W. James King, "The Project on the History of Recent Physics in the United States," *American Archivist* 27 (April 1964): 237–243.

3. Beckman Center for the History of Chemistry. *Final Report: Polymers and People. Final Report on NSF Project No. SES 84-21278* (Philadelphia: Beckman Center for the History of Chemistry, University of Pennsylvania, 1988).

4. For more information on the study of Engineering Research Associates, see "Origins of the Computer Industry: A Study of ERA," *Charles Babbage Institute Newsletter* 8, 3 (Spring 1986): 1, 5. Articles on the Advanced Research Projects Agency project are also in that *Newsletter:* "CBI Begins Study of the Impact of DARPA on Computing," 11, 2 (Winter 1989): 1, 3, and "Progress on DARPA Study," 12, 2 (Winter 1990): 1–2.

5. The Project was funded by the National Science Foundation and the Department of Energy. Field work began in 1977 and was completed in 1981. Project publications, including a final project report, appraisal guidelines, and a handbook for secretaries, are available from the AIP Center. I should mention that earlier archival work by Margaret Gowing and others at the United Kingdom Atomic Energy Authority provided at least a modest hope that our project would succeed.

6. Clark A. Elliott, ed., *Understanding Progress as Process: Documentation of the History of Postwar Science and Technology in the United States: Final Report of the Joint Committee on Archives of Science and Technology* (Chicago: Society of American Archivists, 1983).

7. Joan K. Haas, Helen Willa Samuels, and Barbara Trippel Simmons, *Appraising the Records of Modern Science and Technology: A Guide* (Cambridge, Mass.: Massachusetts Institute of Technology, 1985; distributed by the Society of American Archivists).

8. Bruce H. Bruemmer and Sheldon Hochheiser, *The High-technology Company: A Historical Research and Archival Guide* (Minneapolis: Charles Babbage Institute, University of Minnesota, 1989).

9. Helen W. Samuels, *Varsity Letters: Documenting Modern Colleges and Universities* (Chicago: Society of American Archivists; Metuchen, N.J.: Scarecrow Press, 1992).

10. Joshua Lederberg and Keith Uncapher, "Towards a National Collaboratory: Report of an Invitational Workshop at The Rockefeller University, March 17–18, 1989." I want to thank Helen Samuels for bringing this report to my attention.

11. There are a number of recent books and articles on multi-institutional collaborations. For example, on high-energy physics, Peter Galison, *How Experiments End* (Chicago: University of Chicago Press, 1987); Michael Riordan, *The Hunting of the Quark* (New York: Simon & Schuster, 1987); Gary Taubes, *Nobel Dreams: Power, Deceit, and the Ultimate Experiment* (New York: Random House, 1986); and Sharon Traweek, *Beamtimes and Lifetimes: The World of High Energy Physicists* (Cambridge, Mass.: Harvard University Press, 1988). And for other disciplines, Robert W. Smith, *The Space Telescope: A Study of NASA, Science, Technology, and Politics* (New York: Cambridge University Press, 1990) and two articles in *Science* 248 (6 April 1990); C. R. Cantor, "Orchestrating the Human Genome Project," pp. 49–51, and J. D. Watson, "The Human Genome Project: Past, Present, and Future," pp. 44–49.

12. "First Steps in AIP Study of Collaborations in High-Energy Physics," *AIP Center for History of Physics Newsletter* 21, 2 (Fall 1989): 1–2.

13. For further details on the AIP Study of Multi-Institutional Collaborations, copies of the project proposal and progress reports are available from the AIP Center, 335 East 45th Street, New York, NY 10017.

CHALLENGE TO PUBLIC OUTREACH

History of Chemistry and the Chemical Community: Bridging the Gap?

William B. Jensen

Introduction

When I and several other chemists in the Division for the History of Chemistry of the American Chemical Society (HIST) received a letter in October 1988 soliciting our opinions on a possible conference on "Issues and Challenges in the History of the Chemical Sciences and Technologies," a fair amount of enthusiasm was expressed for the third of the three potential emphases outlined in the initial draft of the accompanying proposal, namely, that the conference focus on "discussions between historians and chemical scientists about new initiatives which might be forthcoming for historical research and about ways and means of disseminating its fruits to wider audiences."

It was hoped that this would be an opportunity, similar to the symposium sponsored by HIST in 1968, to assess the current status of history of chemistry courses, both within chemistry departments and history of science departments; the availability of textbooks; publication opportunities; museum artifacts; and historical collections of monographs, photographs, prints, and documents relating to the history of chemistry.[1] It was also hoped that the conference might explore ways of improving communication between professional historians and the beleaguered few within the chemical community who still suffer from the illusion of being chemist-historians, as well as the role that such organizations as HIST and the Beckman Center might play in encouraging this dialogue. With this in mind, I indicated that I would be willing to contribute an essay assessing the current state of historical interest within the chemical community, based largely on my experiences in trying to develop a new historical journal for HIST directed primarily at practicing chemists and teaches of chemistry.

As events developed, the conference did not focus on the issues of pedagogy and resources but became instead a meeting for professional historians, with the emphasis on "accounts and discussions of current research directions" in the field, the few chemists and teachers of chemistry present being largely in the form of "invited guests" in order to imply some link, however tenuous, with the chemical community and to fulfill the requirements of one of the funding agencies. My reason for mentioning this is not "sour grapes" at having the conference develop in a direction contrary to the preferences of my HIST colleagues—for in fact the cast is truly stellar and the resulting proceedings will without a doubt become an important contribution to the field—but rather to indicate the origins of my embarrassment at being left on the program to deliver a paper originally intended for an alternative conference that failed to materialize.

Having done a great deal of work on the historical development of solid-state inorganic chemistry, a subject that has important consequences for both the conceptual structure of chemistry (in terms of nonmolecularity and nonstoichiometry and their bearing on traditional distinctions between solutions and compounds) and its disciplinary structure (as indicated by the recent revelation brought on by the excitement over superconductors that this field has largely been captured by departments of materials science rather than departments of chemistry), I would much rather have used this opportunity to provide, like the other conference participants, a substantive report of work in progress. Instead I appear to be saddled with what is essentially a policy paper and one which, owing to lack of hard data, is based largely on anecdotal information. Denied the protection of both historical substance and strong documentation, I will liberally interpret my function to be that of stimulating discussion on the touchy subject of interdisciplinary interactions and will come out with both guns blazing.

With this in mind, I will caution that some of following comments are intended to be provocative, though in all cases they represent opinions (however caricaturized) I have heard expressed by chemists and historians relative to each other's work. While some of them may be without foundation in fact, there is nevertheless a perception that they are true and, alas, in the field of public relations, perception is often more potent than fact. I might also add, in what I suspect is a futile hope that commentators will focus on the issues rather than attacking me personally, that I do not necessarily hold all of the opinions expressed in what follows.

The Nature of the Chemical Audience

When asked why chemists should study the history of chemistry, the doyen of American chemist-historians, Edgar Fahs Smith, once answered that it provided an excellent source of diversion and entertainment when the duties of the laboratory and classroom became too burdensome. "Tired," Smith wrote, "of pressing obligatory duties, one may well turn to the history of chemistry for the desired rest and recreation."[2] Like it or not, Smith's answer is still the most accurate characterization of most of the potential audience for history of chemistry in the chemical community. As holder of the Oesper Position in Chemical Education and History of Chemistry at the University of Cincinnati, I give about a dozen seminars a year to chemistry departments and local ACS Sections throughout the Midwest. Though a fair selection of historical topics is offered, ranging from talks on the origins of chemical thermodynamics to the state of American chemical education in the nineteenth century, the seminar most requested is entitled "Famous Chemists in Caricature and Anecdote." Efforts to talk the host into some alternative are usually met with the remark that the nature of the audience requires a large ratio of entertainment value to historical substance, and regrettably the same prescription also applies to any articles addressed to this audience. Historians can choose to ignore this group as unworthy of their professional efforts or cultivate it in the hope of seducing the occasional reader into a more serious study of the history of chemistry.

The second largest group of chemists interested in the history of chemistry are the teachers of introductory chemistry at both the secondary and college levels, who pursue it with the vague, but hopefully not vain, hope that it will provide a means of humanizing the subject for their students. This traditional alliance of interests finds its most obvious expression in the *Journal of Chemical Education,* which has, since its inception in 1924, been the journal of choice for the vast majority of chemists who have published historical articles, and which today, despite the rapid growth in the number of professional chemical educators clamoring for publication opportunities, still finds room to publish an average of twenty-four historical articles a year.[3] This group is potentially the most challenging for the historian hoping to reach a wider audience, as they are the people who write the textbooks. Regrettably, their attempts to inject history into the chemistry curriculum have also provided historians with their most fertile source of whiggism. Though this has proved to be an unending source of historiographical amusement, the novelty of thirty or more years of whig bashing is surely growing thin, and it is about time historians move from the critical mode to the constructive mode and begin to provide some positive suggestions as to how key historical events in chemistry can be presented in the introductory textbook without offending the historical sensibilities of the professional. The requirements of brevity and "linearized" context necessary to any textbook format whose primary objective is to teach chemistry, rather than a lesson in the methodology or the sociology of science, are far from easy and raise the question of whether history in a form acceptable to the professional historian is inherently incompatible with textbooks designed to teach the facts and theories of modern science.

Both of the above groups are generally passive in their historical interests and seldom, if ever, attempt to write historical articles beyond the "famous chemists on postage stamps" variety. They constitute a potential audience for, rather than a source of, books and articles on the history of chemistry, but a relatively unsophisticated audience for whom straightforward chronology must be maximized and heavily interlarded with entertaining anecdote, while issues of interpretation and context must be minimized.

Active involvement in the history of chemistry really begins with the third group. This generally consists of chemists who, as a result of their professional activities, have either developed a strong admiration for the work of one of their predecessors, an interest in the history of their department, or, less commonly, an interest in the history of their particular specialty or of one of its techniques. Thus most chemical biography, virtually all departmental histories, and most specialty histories have been, and continue to be, written by chemists rather than professional historians, who, in their search for more grandiose historical patterns, are often not interested in the morass of local, technical, and biographical detail that forms the bulk of these studies. On the other hand, chemist-historians, while usually willing to wade through, if not wallow in, this wealth of detail, often end up marring their biographical work with hero worship and their departmental and specialty histories with a lack of both context and an overall unifying theme. Indeed, it is not uncommon for the latter types of histories to degenerate into minibiographical dictionaries. Nevertheless, this group, by the very fact that it is active, rather than passive, forms a important source for the final group, which, though more sophisticated in its interests, is regrettably also the least numerous.

This fourth group consists of chemists who tend to view history of chemistry as an important adjunct to their understanding of current chemistry, as a way of assessing where we are and where we are going by understanding where we have been. For them the understanding of a chemical concept or technique is inseparable from an understanding of its origins and past historical permutations, and the idea that any scientific theory or fact can exist in some sort of pristine experimental purity, independent of its human origins, is a manifest absurdity. For this group the tools of scholarship and historical perspective are an important addition to the skills of the laboratory in assessing the worth of a concept.

In asserting the importance of historical, as well as experimental, skills in science, I do not mean to suggest that I believe that science is ultimately reducible to the transient whims of men, as has been claimed by certain writers in the sociology and philosophy of science. Like any practicing scientist, I operate on the belief that some form of physical reality exists and that science has become progressively more powerful in its ability to understand and manipulate that reality. What I am convinced of, however, is that this reality is so complex, at least by the time one reaches the level of chemical phenomena, that one must ultimately adopt a pluralistic attitude toward the use of theories and concepts. Even if a totally unified theory, à la Dirac, exists, as individuals we will always be limited by our inherent inability to combine detail with generality, by a sort of generalized uncertainty principle for knowledge in which the product of the intensivity of our knowledge times the extensivity of our knowledge becomes a constant. Consequently, I believe that there will always be room in chemistry for simpler incomplete theories, each of which provides a partial, but slightly different, insight into the phenomena at hand, and just as I increase my insight into the whole by viewing it from as many current partial viewpoints as possible, so is my insight further enriched by also looking at it from past historical viewpoints as well.

A critic's answer to these claims would be the obvious observation that many scientists, indeed the vast majority, operate quite successfully without such an historical perspective. In fact, examples can be cited of scientists who have actually impaired their productivity as a result of historical interests which led them to adopt an unfruitful viewpoint about the nature and function of their science. It would be foolish to deny that this is the case. If one's goals in science are research productivity in one's area of specialty, job security and peer prestige, then history is probably irrelevant. If, however, one also wants to know how it all fits together, if these career goals are also tied to the desire to gain as much conceptual insight as possible into the nature of chemical phenomena, then I would deny that this is possible without an historical perspective. The fact that so many chemists operate without such a perspective is merely a depressing commentary on why they became scientists in the first place.

It is the members of this last group who tend to be interested in the conceptual, rather than the biographical, aspects of chemical history and who have often authored either general histories of chemistry or studies of critical conceptual revolutions. Such names as Ostwald, Duhem, and Pattison Muir come immediately to mind as examples of chemists who seriously pursued history of chemistry in an effort to increase their understanding of current developments. Indeed, Colin Russell has recently reassessed many of the classic histories of chemistry written by chemist-historians in the nineteenth century and has shown that most of those histories fail to fit the caricature of them created by historians of science in the last thirty years as paradigms of whiggish self-indulgence written by chemists in their old age to justify the historical significance of their careers.[4] Many of the better histories, such as those of Kopp and Ladenburg, were in fact written at the beginning rather than the end of the author's chemical career and represent an obvious attempt to assess the current state of their science.

The possibility of adding the tools of historical perspective and scholarship to the laboratory skills of chemists graduating at either the B.S. or Ph.D. level is, in my opinion, the primary justification for teaching a specialized concept-oriented, internalized history of chemistry course at the senior-graduate level within the curriculum of the traditional chemistry department, in contrast to the broader-based general history of science courses offered at the freshman-sophomore level by most history and history of science departments.

None of the above motivations necessarily coincide with those of the professional historian of science. For the historian, topics in the history of chemistry—or, for that matter, the history of any science—are more likely to be pursued as "case studies" designed to illustrate some larger social or methodological issue, such as the evolution of scientific disciplines; the nature of scientific discovery or revolution; the impact of science on culture; the nature of the interaction between science and economics, science and government, and so on. Though chemist-historians often find this work of great interest, they are unlikely to become actively involved in historical work for the same reasons. Conversely, the probability is quite small that a professional historian would be interested in writing, let alone be qualified to write, the kind of detailed technique- and concept-oriented internalistic history of interest to the members of the fourth group. The key to bridging the gap between the historians and the above groups within the chemical community lies in developing a constructive educational attitude toward the members of the first three groups and a respect for the goals of the fourth.

The Size of the Chemical Audience

Having said something about the nature of the chemical audience, what can be said about its size? A surprising number of organizations exist for the encouragement of the study of the history of chemistry. The oldest of them, the Division of the History of Chemistry of the American Chemical Society (HIST), was founded in 1927 and in recent years has had a membership that fluctuates between 715 and 730.[5] Though it has sponsored roughly a dozen books based on special symposia organized by the division, it has traditionally relied on the *Journal of Chemical Education* as a publication outlet for individual papers. In 1988 the division finally began its own journal, *The Bulletin for the History of Chemistry,* which now appears three times a year. This is heavily subsidized by the Oesper Collection and also receives some support from the Chemical Heritage Foundation. Next in order is the Society for the History of Alchemy and Chemistry, founded in 1937, whose journal, *Ambix,* was for many years the only publication dealing exclusively with the history of chemistry. Like the *Bulletin, Ambix* appears three times a year and was also subsidized for many years. As of 1987, the Society reported a circulation of 570.[6] This figure apparently includes both libraries and individuals and does not give a breakdown as to how many of the individuals are practicing chemists rather than professional historians. The Fachgruppe Geschichte der Chemie of the Gesellschaft Deutscher Chemiker (GDCh) was begun in 1962 and currently has a membership of 233.[7] In 1988 it began publication of a small journal that reproduces key papers given at its biannual meetings, as well as providing news of symposia, displays, and other events of interest. The Japanese Society for the History of Chemistry was begun in 1973 and has published a quarterly journal, *Kagakushi,* since 1974. Its current membership is approximately 450.[8] More recent is the Historical Group of the Royal Society of Chemistry, which has a membership of approximately 229, of whom 204 are also members of the parent society.[9] It publishes a newsletter twice a year.

Liberally interpreted, this gives a worldwide audience of approximately 2,200 chemists whose interest in the history of chemistry is sufficient for them to fork out the annual membership fees for these organizations. When viewed against the total number of chemists belonging to the parent scientific organizations, this result is rather depressing. The ACS, for example, has a current membership of 140,089, which means that less than a half percent of the practicing chemists in the United States belong to HIST. If the makeup of the division is further analyzed in terms of the four audience groups outlined in the previous section, the situation moves from sad to worse.

Based on eight years of activity in the division, I would estimate that the number of members belonging to categories three and four, that is, those actively rather than passively interested in the history of chemistry, is less than 30, and that many of these are actually historians who are affiliate members of the division. This result has sobering consequences for those who would publish journals directed at this audience, in terms of the ratio of advanced scholarly versus introductory educational articles.

Assessing the audience within the chemical community for books and monographs on the history of chemistry is much more difficult. Based solely on personal experience in two large chemistry departments with over 30 faculty and one smaller department with 15 faculty, I would say that chemists in general do not buy books, historical or otherwise. Within the larger departments, fewer than 10 percent of the faculty had serious personal libraries of technical literature. In the vast majority of cases, only two kinds of books were found on their shelves: former textbooks, which they had used as students, and free sample textbooks given them by book representatives. The general attitude was that if they needed a book, they could get it from the library. Put more charitably, chemists tend to "look things up" in books rather than to collect and read them as a source of opinions and ideas.

There is also evidence suggesting that chemists, in contrast to biologists, for example, do not consider advanced monographs as a serious medium for presentation of scientific results. At Cincinnati the chemistry and biology libraries are combined, and purchase records show that while the number of journals purchased in the two fields is approximately the same, the ratio of biology books to chemistry books is roughly five to one, and that most of the chemistry purchases can be categorized as either textbooks or reference books.[10] The average purchase price for the chemistry versus the biology books of ninety-six versus sixty-seven dollars is also of interest, suggesting that the publishers have long since decided that the primary chemical market is with libraries, whereas a significantly larger individual market still exists in the field of biology. The bottom line, then, is the none-too-profound observation that the individual market within the chemical community for books and monographs on the history of chemistry is essentially nonexistent, and that marketing and pricing should be directed at the libraries instead.

The situation with respect to history of chemistry courses largely reinforces this picture. In a 1987 survey of 574 ACS-approved chemistry departments, Everett and DeLoach found that fewer than 10 percent offered history of chemistry in some form.[11] In no case was the course required, and of the courses surveyed, only 37 or about 3 percent were "designed primarily for chemistry majors and dealt exclusively with the history of chemistry." More than one third of these were also reported to be in danger of being dropped from the curriculum owing to "insufficient enrollment" or "retirement of the faculty member offering the course."

Barriers to Communication

It would be nice if simply stating that there should be an increased dialogue between the chemist-historian and the professional historian were equivalent to the act itself. Unfortunately, there are still significant barriers to this process. Despite grandiose philosophical rationales for these barriers, many of which will be touched on below, I feel that most of them, when shorn of their rhetoric, can be reduced to the simple fact that any activity directed primarily at another academic discipline, and especially any educational activity, necessarily involves commitments of time and energy that are seldom highly valued by the members of one's own academic discipline. In short, they contribute neither to tenure nor to peer group recognition. The best way of avoiding this, if the area of activity potentially overlaps with several other academic disciplines, is to shift the center of gravity for the field into one's own speciality, usually by denying the validity of opposing claims on the field, the relevance of their point of view, the quality of their scholarship, and so on. Regrettably, the rise of the history of science as a legitimized speciality within history departments has been accompanied by a great deal of nonconstructive fence building of this type, most of it directed at the professional scientist turned historian, who had previously dominated the field.

Sins of Omission: The Chemists

The most common criticisms leveled by historians at the work of the chemist-historian have to do with questions of internalist versus externalist history and the closely related question of context. Both of these are, in my opinion, red herrings that are impossible to justify on objective grounds, though a great deal of paper and ink has been expended in the attempt. One man's interest is another man's irrelevancy, and ultimately they all boil down to either the complaint that "you are not interested in the same kinds of historical questions that I am" or "this is superficial; tell me something I don't already know." Obviously, if I believe that the universe is already sufficiently complicated at the level of chemical phenomena to require the adoption of a pluralistic attitude toward our theories and interpretations, the same is even more true at the level of human activity. History is just too damn complicated, and each point of view, each cross-section—be it internalist concept-oriented, externalist culture-oriented, economic, sociological, or philosophical—provides a different and necessary insight into the whole.

There is, however, one area of criticism I feel is legitimate, namely, the frequent complaint from historians that chemists almost totally ignore the secondary historical literature when writing on the history of chemistry. My experience with both the *Journal of Chemical Education* and *The Bulletin for the History of Chemistry* bears this out. Of the fifteen or so articles written by chemists in the first two years of the *Bulletin*'s existence, not one in manuscript form made reference to a historical article written in either *Isis* or *Ambix,* and the few references appearing in the final printed form were added by either the editor or the reviewer. The same is largely true of historical articles appearing in the *Journal of Chemical Education.* Citation of full books, on the other hand, was more common, though still of sufficiently low frequency to be distressing. If it is any consolation, secondary articles on the history of chemistry appearing in the *Journal of Chemical Education* were often not cited as well, and in reviewing purely chemical papers for the same journal I have found the same sloppy scholarship with respect to the chemical literature itself. Indeed, the authors often do not even check back issues of the journal they are publishing in, and as a result the subject matter of the articles in the *Journal of Chemical Education* has a depressing tendency to repeat every ten years or so.

The reasons for this neglect are simple enough. Chemists are just not trained to the standard of historical scholarship found in the humanities. Although they generally take a superficial "chemical literature" course, most of this is devoted to the technique of locating specific chemical species in either *Chemical Abstracts* or *Beilstein* or to the technique of doing a computer search. Unless their work is being directly challenged, they seldom make use of a historiographical analysis of previous work in a field, seldom read footnotes, and frequently copy and reference literature citations they have never read. These distressing trends are reinforced by the fact that most libraries, for lack of space, are now beginning to put journals that are more than twenty years old in storage; that students are given the impression that computer searches, which seldom extend back more than thirty years; are a proper substitute for reading the literature; and that more and more departments are dropping their foreign-language requirements. Anyone who has worked or taught in a large chemistry department also knows that there is a kind of chemical macho bravado that is hostile to the concept of careful library research. Phrases such as "it is easier to measure it in the laboratory than to look it up in the literature" or "I never look at the literature before starting a project, as it inhibits my creativity" are frequently heard and held up to graduate students as acceptable role models for research.

In an effort to determine whether this lack of careful scholarship was due to a simple ignorance of how to locate materials in the historical literature or, as suggested above, to a general disregard for the standards of historical research, I had our librarian punch the name of Lavoisier into our computer search system in the chemistry library at Cincinnati. Out came 114 responses, covering the period from 1966–89, which referenced all of the relevant articles in *Ambix* and *Isis,* as well as in many other historical journals. Many of these were embarrassingly relevant to articles that had appeared in the *Bulletin,* though their authors had failed to reference them. Since this search was done with a data base specifically designed for chemists and available to virtually every chemistry department, one must regrettably conclude that sloppiness rather than ignorance lies at the heart of this problem.

A careful reader has probably noticed that I have said nothing about the charges of whiggism frequently leveled at the chemist-historian. This overworked phrase, which seems to have become the equivalent of calling someone an incompetent historical fascist, has been amply criticized by Hall.[12] One also cannot help but be amused by the fact that the most famous whiggism in the history of chemistry, Herbert Butterfield's infamous "postponed revolution," was committed by a professional historian rather than a chemist-historian, and by the very historian who had introduced the concept of whig history in the first place. On a more serious level, one must also have doubts about the sincerity of the profession on this issue, since it currently tolerates without a murmur a form of militant revisionist feminist history that specializes in passing historical judgments using a set of values that has been current only since the 1960s and which, in the righteous indignation of its historical conclusions, puts even the most extreme whiggism of the scientist-historian to shame.

Sins of Commission: The Historian

Having dumped on the chemists, it is now time to deal with the historians. Prior to the establishment of the *Bulletin,* I often heard the complaint from chemists interested in writing articles on the history of chemistry that there were too few publication opportunities available. By this they generally meant they wanted to publish in the *Journal of Chemical Education* but felt, in light of the fact that this journal has to reject, for lack of space, about 60 percent of the manuscripts submitted and has a publication backlog of almost two and a half years, that the probability of having a historical article rejected as "educationally irrelevant" was too high to make it worth the effort. When I brought up the possibility of publishing in either *Ambix* or *Isis,* it became apparent that they wanted to publish in a journal that was considered to be part of the chemical, rather than the historical, literature and would be found in their departmental library and be known by name to their chemical colleagues. In short, it was an example of the barriers produced by the pressures of tenure and peer group recognition.

A more serious problem, however, were those chemists for whom this was not an issue but who still did not want to publish in the historical literature because of their discomfort with what they saw as its "overly critical attitude" toward, if not outright denigration of, the work of others and especially that of outsiders. In contrast to the chemical literature, where one's results are usually presented as an extension or refinement of previous work in the field, current trends in the historical literature seem to require that one prove one's work totally invalidates the work of one's predecessors. Each new fact, however minor, is presented not as a further refinement of the existing picture but as the key missing element which demands a total revision of the field. It is difficult to take seriously the claims of the historian that there are no such things as good and bad guys in science or that scientific revolutions may not be as discontinuous or as common as scientists claim, when in the opening rhetoric of virtually every article we are told that there are not only good and bad historians but downright evil historians, and that a revisionist revolution lurks beneath each new piece of trivia.

Much of this critical rhetoric had to do with the growing pains of the history of science as a legitimate historical subdiscipline in the period between 1950 and 1980 and its attempts, as outlined earlier, to move the center of gravity of the field from science departments to history departments. However, no one doubts its legitimacy anymore, and if it is to have some real impact on science and scientists, the field needs to move from its current hypercritical historiographical mind-set, in which it appears to be obsessively stuck, into a more constructive educational phase.

As a case in point, after nearly forty years of activity, the professional historian of science, though producing many important specialist monographs, has failed to produce a single substantive general textbook or reference book on the history of chemistry. As far as I can see, though both works are nearly thirty years old, Partington has little chance of being displaced as the single most important reference work in the field nor Ihde as the most important textbook, and both will continue, for lack of any alternatives, to be the works of choice for those teaching history of chemistry courses within the context of traditional chemistry departments.

There are several probable reasons for this. As in all fields, the writing of textbooks is a relatively low-status occupation compared with the writing of specialist articles and monographs. Textbooks are usually generated in response to the individual's teaching needs, and because few historians are given the chance to teach a course in the history of chemistry targeted at chemistry majors and located within a chemistry department, their lack of enthusiasm for this type of project is perhaps understandable.

A more subtle reason, however, lies in a change in the style of historical writing. Older articles and books tended to place the primary emphasis on chronology and the effective organization of names and dates. Interpretative insights, when present, were subordinated to this factual organization. Thus, although a historian may disagree with the interpretations of Ihde or Partington or even consider them to be incorrect, and though he or she may find Partington impossibly boring to read as historical commentary, the factual content and efficient organization of these books gives them a value that transcends these supposed defects and has resulted in their continued use for nearly three decades, despite their having been the target of much historiographical criticism.

Most current historical writing, on the other hand, inverts this emphasis. The interpretative thesis is made the focus of the work, and the selection and presentation of facts, names, and dates are subordinated to this central thesis. Once this thesis ceases to provoke controversy, the work fades from view except perhaps as fodder for some future historiographical diatribe, since unlike the earlier works it has neither a factual content nor organization capable of transcending the passing interest in its interpretative thesis. In its most extreme form, this trend has resulted in historical articles that are little more than historiographical editorials, and books that are really extended historiographical "essays," which sketch out bold programs for the revisionist reconstruction of a field but stop short of actually putting the program into effect, the usual excuse being that "much work still remains to be done." In most cases we are still waiting for the work to be done, and this rhetoric has had little effect beyond discouraging chemists from entering the field. Using the field of literature as an analogy, we are, in effect, seeing an unhealthy confusion of literature (history) with literary criticism (historiography).

The Role of Organizations

In closing, I would like to briefly say something about the role of organizations in furthering the dialogue between historians and chemist-historians and in encouraging the cultivation of history of chemistry in general. As for the United States, I am unfortunately only able to speak authoritatively about the activities and intentions of HIST in this regard. Like all divisions of the ACS, it was founded to encourage the activities of chemists in the area of its specialty through the organization of symposia, the publication of books and journals, and the sponsoring of awards. When HIST encouraged the ACS to join in the founding of the Center for History of Chemistry (now the Chemical Heritage Foundation), it was with the understanding of the division that this organization would supplement the division's activities by acting as a permanent archive and reference center for historical documents relating to the history of the chemical community, a function which HIST, owing to its ever-changing administrative staff and lack of permanent physical location, was unable to perform. In recent years, however, the Center has significantly changed in direction and emphasis. Since HIST is not privy to these plans and has no input into them, I am unable to evaluate them in this essay.

As for HIST itself, it has made significant attempts, beginning with the symposia on Priestley, Dalton, and G. N. Lewis, organized several years ago by Derek Davenport, to include professional historians in its special symposia. In a more recent event, the symposium in honor of the "Bicentennial of the Chemical Revolution," held in Dallas, Texas, in April 1989, four out of nine, or nearly half, of the speakers were professional historians.[13] The division also attempts to maintain a similar mix of chemist-historians and professional historians in terms of both the authors and emphasis of the articles published in its journal.

As for pedagogy, the division hopes in the future to initiate a series of short informational bibliographies and pamphlets entitled *Data Sources in the History of Chemistry*. Most of these will be targeted at the first two of the four audience groups mentioned earlier and will include "A Guide to Chemical Museums," "A Handbook for Organizing a History of Chemistry Course," "A Bibliography of History of Chemistry Articles in the *Journal of Chemical Education*," "A Bibliography of Chemical Genealogies." If your eyebrows go up at some of these titles, it means that you are losing track of the nature of this audience. Though a historian may not consider the *Journal of Chemical Education* to be the premier choice as a reading resource for history of chemistry, this selection is dictated by the fact that it is, in contrast to *Ambix* and *Isis*, a source available to virtually every teacher of chemistry, whether at a junior college or a high school. Likewise, however questionable the historical value of chemical genealogies, the fact remains that they are still one of the most effective ways of getting chemists interested in the history of chemistry. Finally, the division also hopes to eventually get the ACS to officially recommend that a history of chemistry course be made available in all departments as one of the possible options for fulfilling a major's electives in advanced chemistry.

Notes

1. George B. Kauffman, ed., *Teaching the History of Chemistry: A Symposium* (Budapest: Akadémiai Kiadó, 1971).
2. Edgar F. Smith, "Observations on Teaching the History of Chemistry," *Journal of Chemical Education* 2 (1925): 533–555.
3. Based on "The 1988 Bibliography for the History of Chemistry," *Bulletin for the History of Chemistry* 4 (1989): 27–29; and "The 1989 Bibliography for the History of Chemistry," ibid., 6 (1990): 37–41.
4. Colin A. Russell, "'Rude and Disgraceful Beginnings': A View of History of Chemistry from the 19th Century," *British Journal for the History of Science* 21 (1988): 273–294.
5. "Aus dem Fachgebiet," *Mitteilungen Fachgruppe Geschichte der Chemie, Gesellschaft Deutscher Chemiker* 3 (1989): 83.
6. William A. Smeaton, "The Society's First Fifty Years," *Ambix* 34 (1987): 1–4 and 57–61.
7. "Events of Interest," *Bulletin for the History of Chemistry* 4 (1989): 32.
8. "Events of Interest," *Bulletin for the History of Chemistry* 7 (1990): 35.
9. "Aus dem Fachgebiet," 73–74.
10. *Management Information Report, July 1988–June 1989,* University of Cincinnati Libraries.
11. Kenneth G. Everett and Will S. DeLoach, "Who is Teaching the History of Chemistry?" *Journal of Chemical Education* 64 (1987): 991–993.
12. A. Rupert Hall, "On Whiggism," *History of Science* 21 (1983): 45–59.
13. Special Bicentennial Issue: "Lavoisier and the Chemical Revolution," *Bulletin for the History of Chemistry* 5 (1989): 1–52.

The Museum, Meaning, and History: The Case of Chemistry

Robert Bud

Introduction

"Chemistry is a French science and Lavoisier was its creator," boasted Wurtz in 1868.[1] No wonder the Paris Conservatoire des Arts et Métiers preserves Lavoisier's apparatus. The presentation of chemistry in the historical museum has its origins in the nineteenth-century confidence that chemical expertise equalled national virility. The Deutsches Museum recreates laboratories of the great German chemists and the Science Museum shows the atomic models of John Dalton, Britain's contender for the title of founder of modern chemistry. The combination of revered objects, national champions, and a message of progress harmonized so well that it spread to every such national museum.

The legacy of this model endures in many galleries of the world's historically oriented science museums. However, the inspiration no longer carries weight. Neither the message of progress nor the spirit of nationalism will suffice. Without them we either abandon the enterprise or radically reformulate its justification. Of course the Museum is not alone; academic history has gone through the same crisis and is potentially a great and powerful ally. It offers both information and methodology, but, tempting as these are, such a superpower tends to impose its own imperial values on over-reverential neighbors. Is the approach of an academic historian in discourse with a few hundred highly sensitized peers appropriate to the museum visited by several million visitors a year? A single gallery may address ten million people in its lifetime. The discrepancy between these audiences, and therefore functions, should give pause to those historians of science so ready to locate classically scientific knowledge in its cultural context. That note of caution does not mean that either the discipline of history or the skills of historians are irrelevant to the museum. Rather, it means that historical tools be applied with precision and awareness.

Instead of acting as a pale popular representation of real history, the museum can respond to needs of its visitors which are urgent and important. It can articulate and alleviate, I will argue, one of the profound problems of modern life: the confusion of categories created by the interpenetration of scientific and popular cultures. Humanity, technology, and the environment are examples of categories which are fundamental to our culture but whose distinct identities, so important to us as citizens, are challenged by science, perhaps chemistry particularly, thereby creating confusion and concern.

Historiography

The uses, interpretation, and communicative power of exhibitions have been debated intensively for years. This questioning is of course true in historical museums, among others. Yet, still, the factual content of history has appeared to be beyond dispute, and therefore the emphasis of the debate has been on how to communicate a given account rather than on how one can determine the story. This is perhaps strange because historians have shown themselves increasingly adept at relating purpose, context, and content in the works of those they study. Thus, while historians of science have explored the sociology of scientific knowledge, in their own work they have been more timorous and have tended to assume "progress" in historical methodology, and that there are "right" and "wrong" rather than merely "appropriate" styles and methods. Academic historians have even avoided examining the particular contexts within which their own trade is practiced.[2] This "black boxing" of history is neither necessary nor useful. It is not necessary because there is a newly available body of historiographic discussion that analyzes different rhetorical modes of history as alternatives rather than mistakes. The pioneer in this area is clearly Hayden White, the author of the pathbreaking text *Metahistory*.[3] Nor is the curator's self-denying ordinance useful: it prevents us from looking at the kind of rhetorical accounts that are appropriate for the complex context of the museum.

Accounts of the Past

To the professional historian, narrative may have allegorical significance. However, such significance is esoteric. Appreciation of it relies on a carefully cultivated understanding that in subtle ways attunes the professional to the ways in which "The Past is a Foreign Country." It also speaks, normally implicitly, to fundamental problems of which the public is unaware, and to many laymen narrative about the history of science is without apparent meaning. The segregation of history from its social base is of course not a problem that I alone have discovered. David Cannadine expressed one aspect, the crisis of British history, in his famous *Past and Present* lecture.[4] In it he pointed out the loss of meaning to be found in the once-great historiography of Great Britain. Instructive revolution had seemed to follow instructive revolution at convenient intervals of a century, culminating in the formation of the welfare state. Pedantic specialists had unfortunately undermined this majestic Fabian march of time. With the demolition of the components, the meaning of the whole had gone.

At the same time that academic political history has suffered a crisis of confidence, the past has come to be very popular. Museums dealing with historical themes have been doing very well, spurring the growth of the so-called heritage business.[5] Those museums and country houses may be persuasive arguments that it is social history whose roots are really in, and whose relationships are really with, popular culture. There domestic life of the recent and just-remembered past can be celebrated. Dramas such as "Upstairs Downstairs," television histories of the world wars, and agricultural folk life museums can each draw on rich wells of memory and education. Aspects of their economic and military heritage are also part of the common cultural property of many adults. Even if dim, memories may be reinforced by television programs and magazines. People's own memories of the news and their parents' news give poignancy to recent economic history, as in recall of the Great Depression. Lowenthal has explored the cultural uses of such kinds of history. There is a strange mixture of escapism and search for "reality."[6]

Certain aspects of technology may share this appeal—think of steam trains or historic cars and even cameras. However, these examples highlight the lack of popular concern in the past for a far wider range of science and technologies from quantum physics to ecology. At the erstwhile entitled Museum of History and Technology the most popular exhibit was for long the display of dresses of presidential wives through the ages. By contrast, the history of science and of most technology can draw on neither the education nor the memory. As remembered folk history these subjects relate to a tiny proportion of the population. Perhaps for chemists the history of chemistry can have the sort of nostalgic appeal that the rural life museum captures so well for hundreds of thousands. For them an antique beam balance or even some etched glass bottles can evoke a world we have lost. But there are relatively few chemists. No amount of being in a recreated laboratory can provide the layman with that mass of related memories that are evoked by artifacts from yesteryear. It is indeed dishonest to suggest to a visitor that he or she is seeing the world with the eyes of another. The historian is well aware that it is only with laborious difficulty that he or she will become equipped with that ability. Technology has had a wider direct impact. Still, in most cases, any particular technology affected only a small proportion of men and women. Fundamentally, whereas many museums deal with the history of the community from which the intended audience derives, the museum of science and technology generally addresses quite different communities to whose folk memory the history of science and even technology are as remote as the Trobriand Islanders.

The Past and the Present

So, the traditional narrative form of history of science has little appeal for many of the less scholarly audiences of the museum. In chemistry the "heritage" angle, however legitimate in other contexts, is not available to most audiences. At the same time there is public interest in science and technology themselves; an enormous infrastructure from school to the television series "Nova" and even much of current political debate sustains that interest. And indeed, the faith that it will enhance popular understanding and interest in science and technology is what sustains the science museum. It is much more reasonable to believe that the presentation of the scientific or technological past can be promoted by reason of its links to the present than in its own terms. Study of the past does not need to emphasize the separation from the present. Many societies have far less of a differentiation between past and present than we as historians find useful. Think of the religious person's use of the Bible to find lessons from ancient events as if they were occurring now. Tell your son about the departure from Egypt as if you yourself had been taken out, the Jewish father is enjoined on Passover. Even in academic contexts, sociologists turn to the past for case studies, as laboratories for the demonstration of timeless patterns. And of course one of the most influential models of social organization—that of Marx—was based on the mining of human history. Unlike the historian's, such approaches assume that for all the particular changes the past is like the present, subject to the same rules and constraints.

Before its recent professional transformation the history of science was determinedly ahistorical in this sense. Most accounts were designed to demonstrate the pattern of progress in science and indeed technology. This was true both of museums such as the Science Museum and of literary histories that had their heyday in the period between the World War I and the 1950s. That era has been dismissed as the prehistory of the history of science, but it did have its own great vibrancy and dynamics, as was pointed out recently by Reingold.[7] One of its leading exponents, and by virtue of his position one of the most powerful, was J. B. Conant, the chemist president of Harvard University from 1933 to 1953. Conant, whose university was to be the American pioneer of professional history of science in the immediate postwar years, was himself an expositor of the case study approach to this history of science. In a later book, *Science and Common Sense,* he explained the relevance of the study of the past to the layman. Conant was concerned to bridge the gap between those who understand science because science is their profession and intelligent citizens who have only studied the results of scientific enquiry—in short, the layman. He believed that

> the experience which would come from visiting various laboratories may be communicated by a discussion of the methods by which scientists have advanced knowledge in the past. . . . And if someone objects (as many already have) that my illustrations are all taken from those periods when a given branch of science was in its infancy that I am offering history to those hungry for current information, I must reply the methods of science were the same then as now.[8]

The Durkheimian Answer

The challenge for the Museum can be expressed as how to find a form which works for our public as well as Conant's models did in his day or as well as academic history of science works for the professional.[9] As Conant saw, chemistry is a wonderful subject to pick for troubling issues that can be illuminated historically. In part, our old established exhibits dealing with "industrial chemistry" have sustained the assurance of increasing physical and cultural synthesis made possible by science. The 1929 catalogue of the Science Museum's collection assured readers that while previously there had been

> few industrial processes for which it was thought necessary to employ chemical control, in recent years so great have been the benefits derived through replacing empirical working by accurate scientific methods, that there are now few large-scale chemical operations which have not been brought directly under the supervision of men who have been specifically trained for the work.[10]

Therefore the title "industrial chemistry" covered a program—a moral, if you will—as well as a subject. It was similar to the vision that Slosson conveyed in his classic works of chemical boosterism and Morrison in his *Man in a Chemical World.*[11] Even in the 1920s the message was not universally acceptable. Two years before Barclay's catalogue was published, the Bishop of Ripon had suggested a moratorium on research.[12]

Today it may be too late for reassurance. Not only is there doubt over the balance of progress in the consideration of particular developments, but there is a host of ambiguities about the propriety of industrial progress itself. Much of this uncertainty is expressed in terms of outrage over the confusion of our culture's categories occasioned by science. Thirty years ago Roland Barthes struggled with the concept of Plastics. "In the hierarchy of the major poetic substances, it figures as a disgraced material, lost between the effusiveness of rubber and the flat hardness of metal." On the other hand, he had to admit, "The fashion for plastic highlights an evolution in the myth of 'imitation materials'."[13]

We may have domesticated plastics, yet we are still troubled by chemicals. Is it proper to eat them? Whatever the real health issues, such a category anxiety pervades debates over the use of additives and of food cultivated with natural or chemical fertilizers. These need not just involve the dichotomy between the natural and the chemical. Science involves a radical reductionism such that materials with a host of sensory properties and symbolic transformations become merely weights and formulas with a very narrow range of defined properties. The classic battle along these lines was conducted by the German poet Goethe on Newton's concept of color in the early nineteenth century.[14] Whatever professional science's verdict, the unease is alive and well in popular culture. Recent political debates highlight this concern. Think of the difference between water as we know it, the staff of life, and H_2O with given boiling and freezing points and precious few other characteristics. Which of the two is in my glass? In Britain the ambiguity has recently been the basis of political debate as the water industry was privatized. Opponents argued that water was a basic necessity not a commodity. The government responded with a campaign based on the catchphrase H_2Ownership.

The boundaries of Life have become matters of international concern. When is a group of cells, clearly a scientific category, a human being? Is fourteen days the boundary? Is man playing God? David Channell's recent survey of the history of the concept of *The Vital Machine* chronicles three centuries of concern about the boundaries between life and machines. The concept of biotechnology itself, as I have recently argued, has been developed explicitly to describe that boundary.[15] Recently Jeremy Rifkin became a star with his opposition to biotechnology. His book *Algeny* unscrambles the complex issues by offering a clear choice between the natural ecological approach and the technological domination of nature.[16] If we are to expect members of the public to experience a more complex world than Rifkin offers, then we must help them to cope with ambiguity.

Myth and the Machine

Elsewhere I have described the museum gallery as a combination of myth and machine.[17] It is a piece of technology in its own right, with a significant cost, possibly complex working exhibits, a technical job of upgrading a visitor's knowledge, and an opening date. To many it is a giant teaching machine, and certainly its physical structure enhances the analogy. Children in particular are brought by institutions such as schools with questionnaires in hand, ready to come out armed with information. Every curator who creates a gallery knows how obsessed one can become with the successful engineering of this machine. The metaphor stretches out to each of the common antimonies of education and entertainment. Contrasting as those two are, success in each can bc measured. There is a third function, inspiration, which is much less easy to measure, and whose elusive quality the psychologists label "affective." J. H. Randall's *Our Changing Civilization* of 1929 illuminatingly used the Deutsches Museum as displaying the essence of the technocratic society. His work is worth quoting at length because it is a rare record of a visitor's response to a science museum at the time the genre was being developed.

> [The Museum] is in truth the garrison shrine of the new religion and a new civilization. . . . It is a great temple erected to the gods of the new age, Science and the Machine. Here are spread out the records of human industry and human invention, from the first crude devices of barbarian craftsmen to the latest products of the rolling mill and machine lathe. Here is the arsenal of the weapons man has built himself in the long war with nature. . . . On the upper floors resting on this industrial foundation are the chapels of pure science where the natural principles that have made possible such technical advances are illustrated by working models . . . we still peer in wonder and awe at the machines. We still gaze at the scientific instrument with dimly comprehending respect.[18]

More recently, Umberto Eco's *Foucault's Pendulum* begins with a rumination in the Conservatoire des Arts et Métiers, as the hero sees the divine, the *Ein Soph*, in that pendulum.[19] Walter Benjamin talked of the "aura" of the unique object.[20] The museum role as a secular religious center has often been remarked upon, and as the quotation from Randall shows, this has been clear since the birth of the idiom. The mediational analogies were also quite self-conscious. At the time London's Science Museum was being completed in 1926, a Keeper, H. W. Dickinson, explained to an American audience that the new organization was based on the belief that

> The idea [of development] is one that is easily grasped and he [the visitor] realises that he is facing not the dead past but something living, something that is part of himself, something that touches his "business and bosom."[21]

Myths and Truth

So museums do have an inspirational, almost religious, role in their ascription of meaning. If they are to address the problems of category confusion in that spirit, then they need to have the character of what anthropologists following Durkheim, such as Mary Douglas, suggest have been the typical characteristics of myths.[22] Myths need not only be consigned to primitive societies or castigated as erroneous. I shall suggest that using a precise meaning of the concept (rather than in its vulgar form as mere superstition), the myth speaks to the public need for interpretations of the past that address their concerns.[23] The problem then becomes: How can one relate two apparently distinct genres—the myth and the museum gallery—in the service of a better public appreciation of chemistry?

The idea of myth is a considerable challenge to academic cultural perceptions in science and technology. It has a taint of error and of a mysticism that might sit well in an art gallery but that apparently contradicts the empirically validated positive truths to which the objects in a science museum supposedly relate. The use of the word "myth" is perhaps disagreeable to the historian of science used to the vocation of "demythologizing." The historian has seen his or her role as "enlightening," liberating the soul from the shackles of received myth to enable it to find his way within the more stark constraints of truth. Here there seemed an epistemological safeguard against evil. Nazi sculptors, for instance, perpetrated myths with meanings that were simple, vulgar, or pernicious. Barthes attacked more banal vicissitudes of modern culture in his *Mythologies*.

Two arguments therefore need to be made: that this metaphor is appropriate to the message of science and to the medium of the science museum. As for the first, science itself is not composed of desiccated positive truths. To some physicists, and since Primo Levi's *Periodic Table*, to chemists, a mystical experience is not irrelevant to intuiting the nature of their subject. It may be argued, of course, that museums deal with more than cosmic constants, and even more than the universal chemical ingredients of the universe. There is the assembly of historic artifacts associated with the very contingent development of technology. That, surely, is an altogether more prosaic subject. Yet in making this assumption, the engineer ignores his audience. As Robyn Penrose, David Lodge's fictional English lecturer, observes of a factory: "You could represent the factory realistically by a set of metonymies—dirt, noise, heat and so on. But you can only grasp the *meaning* of the factory by metaphor. The place is like hell."[24]

As for the museum medium, there are three responses in defense of myth. First, if one wishes a museum to communicate, one must accept the possibility of communicating something with which the visitor may disagree. The alternative is the totally opaque institution that poses no danger and offers no meaning at all. As in all art, the guarantee is the judgment and sophistication of curators and the evaluation of critics. Today's curators must resign themselves to being timely rather than timeless. While the museum does aim to communicate what is believed to be true, it is necessarily subject to the subjectivity and time-boundedness of its "experts'" judgment. Moreover, visitors need to know what makes sense to them now, not merely eternal universals. Refuge from vulgarity cannot be assured by seeking the epistemological high ground of "reality." Nor is it necessary, for there are more and less beautiful myths. In his short story "The Shatterer," Amos Oren tells the story of a mythbreaker. He was the Commissioner of a nation's Department for the Destruction of Myths, who secretly sought for the fragments created by the destruction and created a new myth from them, as a road builder creates a road from the chips of the rocks he demolishes.[25]

The second response is that of course the visitors are not the passive victims of all-powerful curators. Unlike membership in a religious sect, visiting a museum gallery is not a total experience. It lasts a few minutes, is unlikely to be repeated, and has little or no support from a visitor's other experiences. The criticism that a myth about technology might render natural, necessary, and unproblematic what should be seen as constructed is misplaced. Not only do our myths not need to be like that, but what is more, a five-minute experience cannot render unproblematic that which has already been questioned. Possibly, it may do the inverse, that is, point out new ways of looking at what was previously taken for granted.

The third response is that the museum can make plain that it is offering only a partial insight. Rather like the supplier of breakfast cereals who suggests recipes for their use, however much we endorse their use, we would positively not like cooks to exclude other uses of our product. Moreover, as I shall argue later, however powerful the mythic interpretation, there is the possibility of the ironic response to the objects themselves. The authentic device has a presence that transcends any prearranged mythic interpretation. Unlike the theme park, the museum offers the visitor the possibility to create his own personal view, which may be visionary or cynical, all-inclusive or factitious.

It is therefore no business of this essay to suggest the treatment of the museum as myth to be merely a perpetuation of socially conventional solutions to fundamental problems, or to prevent the public from realizing that they are using mythical solutions. One may summarize this argument as emphasizing the fundamental distinction between the museum as myth and the museum as fetish.[26] Indeed, it is only through an articulation of myth that the prevailing social conventions can be recognized.

Myths and History

Even literary historians have begun to explore the relationship between their activities and mythmaking. In his elegant essays William McNeill has called for the judicious creation of myth. Although the term myth is normally associated with prehistory, or the erroneous, a more sophisticated historiography going back to Nietzsche has seen in it the integrative, meaningful quality all too often missing in academic historiography. "What historians ought to do," proclaims McNeill, "is to celebrate the grandeur of our calling as mythographers. For mythmaking is a high and serious business." In his 1985 presidential address to the American Historical Association, he called for the attainment of "mythistory."[27] Although McNeill does not examine in detail the properties of myth, his own historical work covering great sweeps of history does provide a model.[28] Equally, the great public success of Kennedy's all-encompassing *Rise and Fall of Great Powers* illustrates the inspirational power of such work.[29]

Certainly myth is different from conventional history. The philosopher Peter Munz argues that the fundamental difference is that myth lacks time and space indices.[30] Myths are to be distinguished from sociological law-based generalizations, which also lack such specificity. In terms of their narrative structure, myths are cast in the form of stories or concrete universals, as he calls them, rather than the abstract universals of the social scientist. Like McNeill, Munz sees the possibility of a compromise between myth and history. Myth, he argues, can provide form for the historical account, an underlying paradigmatic structure through which the particular facts of a historical account can be viewed.

Coming from a slightly different tradition than that of the historiographers, Roger Silverstone's formalist analysis of television as mythic form also provides interesting analogies to the museum curator. After all, television is another mass medium that purports to provide truth to the masses. Critically drawing upon the anthropological literature, Silverstone argues that television replicates the form and function of myth.[31] He carries out an elaborate analysis in terms of narration, tradition, motivation, function, framing, and mediation. I want to emphasize just the last. The mythic acts as a bridge between the everyday and the transcendent, he argues. The known and unknown and the sacred and the profane are linked.

Thus the literature on narratology, historiography, and media has suggested the potency of the idea of myth. Two characteristics differentiate the genre from history:

1. The myth identifies enduring patterns and events as examples, rather than specifications, of where phenomena apply.
2. The myth links the mundane ordinary experience of a layperson with the essence of its subject.

To these I would add a third feature:

3. The myth need not be internally consistent.

Television copes with the complexity of the world through the strange juxtaposition of programs with very different philosophical implications. Indeed, in an age without consensus about many aspects of science and technology it would be inappropriate to complement the simplicity and lack of nuance of individual stories with a coherence between them.

How

There is a specifically museological dimension to the myth. Its pursuit can help the curator resolve a fundamental tension characteristic of his medium: the tension between object and narrative. A museum gallery actually has little on its side in persuading the visitor to enter the transcendent world. It is visited in the daytime, not in the dreamy dark night of the theatre; it is frequently noisy and the visit to any exhibit will typically last only a few minutes. Moreover the visitor does not know quite what he or she is supposed to take away—is it information they are supposed to gather? The one great ally in obtaining transcendence is the authentic unique artifact. Whereas this sits uneasily with a bounded narrative, it goes well with myth, which also applies across time. It assures us for a moment of the truth of the story, as we wonder about its implications or outcome today.

How is this wonder to be incorporated in the museum? Different as institutions are, we can identify three common exhibition components, which lie alongside each other. The success of the myth may be defined in terms of the integration of these components. One is the objects themselves, the second is the technical description that lies alongside them, and the third is the description of the context. At the Science Museum we have traditionally used two different sorts of labels: object labels and gallery labels. Corresponding conventions are widespread. Schematically, one can assert that these three sequences of interpretation acquire their inspiration from different sources. The object labels—whether they tell the visitor how the thing works or who used it, whether the labels are long or short—are by definition object oriented. The artifact is, apparently, the "signified." The artifacts themselves are self-referential; one looks at them for their own sake. By contrast, the contextual accounts are often inspired by the message the curator wishes to teach. The content, it often seems, is centered on academic history and not upon the object at all. So there is a tension between object and context orientation. This could be described in terms of content; however, it is more fundamental.

Hayden White has provided us with a language with which we can begin to describe the rhetorical and genre characteristics of each as they have been inherited, and which perhaps provides a way of thinking about them to facilitate convergence. I shall begin by describing the archetypical. The objects as they have often been arranged at the Science Museum or the Deutsches Museum go from the remote past to the present. The traditional sequence of objects is presented as an approximation of the entire progression, unfortunately not available. The object labels make this plain, telling visitors not only what they are seeing, but also, through reference to other items, telling them the ideal entire sequence. By their classic linear arrangement the objects could be compared to what White has called *annals*: in 1895 a laboratory looked like this, in 1965 like that, and here is a "modern" laboratory. The gaps are presumably unimportant; one can assume nothing much happened in between. It is true that, generally, the laboratories are presented as "typical" of their period rather than as breakthroughs. Yet these objects are presented episodically, as if they were the abstract of slightly fuller annals. Since the sequence goes to today or even tomorrow from the distant past, it is unbounded and the narrative is implicit.

The traditional structure of the exhibit, as caricatured here, is apparently crude, and to the cynical visitor it may constitute nothing more than rows of dead and dusty objects with no power at all to enliven any year. The detailed technical information provided by a precocious schoolboy in Eco's fable appeared superficial and beside the point. On the other hand, the traditional structure has great strengths: it is not separate from today or from the visitor. Implicitly, the visitor is part of that display at the current stage, and the very lack of narrative allows the visitor to impose his or her own fantasy. Again, we come back to Eco's portrait of Casaubon dreaming of the *Ein Soph* as he looked at Foucault's pendulum.

The explicitly narrative gallery-style labels providing interpretation have their own problems. Whereas the pendulum evokes the *Ein Soph,* literally the "without end," the narrative text of the museum message is bounded. It tells us what happened in eighteenth-century chemistry, or a panel may be devoted to the early nineteenth century. We are therefore in the world of bounded narrative separated off from us.

The tension in terms of narrative principles and emotional response, created by separate structures of academic narratives and object-related annals, can, this paper has argued, be replaced by a coherent myth. However, this does require great self-consciousness on the part of the curator. No longer can he or she rely on a belief that the story is determined solely by truth.

Are we therefore at the mercy of personal whimsy? I would argue not. Hayden White, through his study of *Metahistory,* provides us with a way of thinking about the kind of decisions that will need to be made about the underlying structure of the plot. He proposes a fourfold division between romance, in which the hero transcends the demands of physical reality; comedy, in which man is at least temporarily reconciled with his world; tragedy, in which such reconciliations proves impossible; and irony, in which the literal meaning of words is suborned by the underlying message.

In these terms the battles in our culture about the place of "chemicals" concern the underlying plot. Are the dilemmas challenges to be overcome or fundamental paradoxes? In White's terms, is the story a comedy or a tragedy? Are we to present the questions of safety as fundamentally, tragic, in which no ultimate resolutions are possible? Clearly we do not wish to decide this for the visitor. It is for each visitor to decide the relationship between Life and the genome.

In assuring the visitor's freedom to choose, the curator can be helped by two more devices: irony and the real lack of closure. Whatever we say, the public should know that there are alternative possibilities, or indeed that the ending has not yet occurred. The mentality being explored is not that of a safely dead culture but our own. This device of suddenly making the audience aware of where they really are is of course what Brecht did in his "alienation" theatre—*Verfremdungseffekt.*[32] He saw the role of art as offering alternative perspectives on our world by means of such effects. "To see one's mother as a man's wife one needs an A-effect [Alienation-effect]; this is provided, for instance, when one acquires a stepfather." This highlights the importance of rhetorically linking narratives and the ironic A-effect.

The power of the object can provide that powerful effect. The original artifact transcends any explanation. As anthropologists have shown, artifacts have meaning in society and themselves constitute a language. The key injunction should perhaps be that of Mary Douglas in *World of Goods*:

> Forget that commodities are good for eating, clothing and shelter; forget their usefulness and try instead the idea that commodities are good for thinking; treat them as a nonverbal medium for the human creative faculty.[33]

It is a property of innovative objects that they challenge the categories: saucepans are metal and clearly not wood; but how does one respond to a plastic frying pan? Its ambiguity therefore enables the visitor to make up his or her own mind. However powerful the mythic interpretation, there is the possibility of the ironic response to the objects themselves. The authentic device has a presence that transcends any prearranged mythic interpretation.[34]

A limited sequence of scientific instruments or pieces of equipment can be used, therefore, to address great issues. Eco describes a washing machine as the alchemistic transformation of black into whiter than white. As soap powder manufacturers have long found, that myth does have a way of raising interest.

Less facetiously, Erika Hickel and I have independently compared the ambitions of biotechnology with alchemy.[35] From this perspective, the fermenter still has the mystic and symbolic potential that alchemists found in the alembic. Constructed out of stainless steel, those devices seem a natural-born member of that most unnatural class, the chemical plant. Yet they are also the nests of living creatures. Thus, although apparently uninspiring, they could be used to address the fundamental anxiety of the natural and the artificial and the roles of God and humanity.

The significance of objects such as a fermenter need not be expressed in complex terms. Indeed it is the argument of this essay that

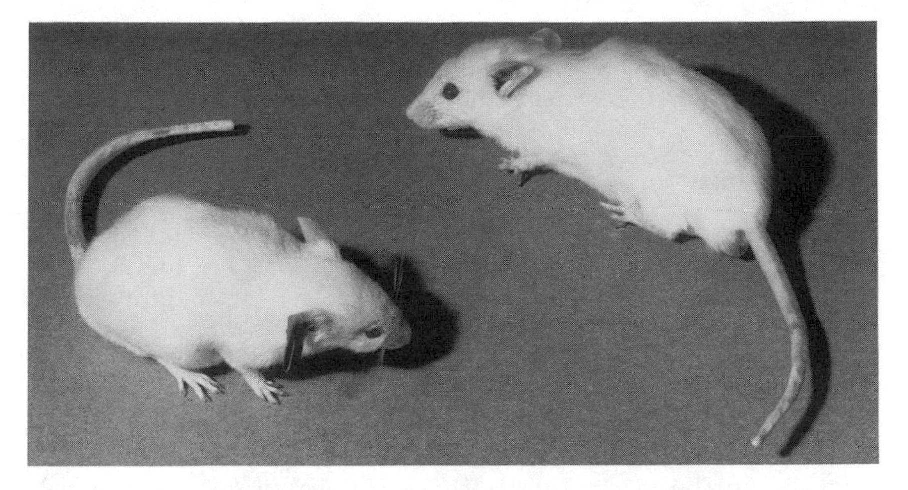

Figure 1. Patenting and life-mythography at the Science Museum. Two lyophilized transgenic mice. Photograph courtesy of the Science Museum.

the public are quite ready to recognize such significance, more so perhaps than the blasé professionals. The questions about the relationship between life and chemicals are far more salient in popular than in scientific culture. What is needed for the former is the account that expresses this ambiguity. That focus on this ambiguity and the fermenter itself as the emblem of the ambiguity is the hallmark of my mythical approach. Within this one could include reference to the Pasteur-Pouchet debate, to debates over design and materials at the beginning of this century as engineers wrestled with unfamiliar problems. After all, fermenters and stirrers have to be specially designed to preserve the lives of the tender living and breathing occupants. We can point out that similar equipment can be used to grow both the familiar yeast and the very problematic human cells.

Equally, the transgenic mouse incorporating DNA from a virus, which Harvard University developed as a research tool, raises a host of questions. Is it a biochemical test-bed or is it a living sentient being? To many the categories are exclusive, and scientists have been accused of playing God. The mouse was granted a patent in the United States in 1988 but was far more controversial in Europe. Only exceptionally, and after several years' delay, was the mouse classified as a proper subject for patenting. Gazing upon such an object is therefore a truly museological experience. Two freeze-dried specimens stare back at visitors to the Science Museum's Chemical Industry Gallery (Figure 1).

They will help members of the public prepare their own minds for the complex question as to whether the human genome itself is a proper subject for patenting. Is it just another useful chemical?

Conclusion

This paper has argued that the museum gallery is potentially a powerful and popular form for addressing chemical issues. The confusion of categories that has so often attended chemistry's impact on popular cultures is an ideal subject for the curator's mythography. Moreover, artifacts, far from being an irritating irrelevance to the curator, can be key components of the portrayal of myth in the museum.

Academic historians have not been concerned with providing the historical case studies for such an approach. This paper therefore ends with a research program for the scholar in the museum, whose special province has traditionally been a connoisseurial object-oriented research. This paper suggests, however, that there is a special category of story that needs researching. Focused on category confusion rather than lawlike generalizations, the curator can find his or her own territory appropriate to the millions in the audience.

Notes

1. C. A. Wurtz, *Histoire des doctrines chimiques depuis Lavoisier jusqu'à nos jours* (Paris: L. Hachette, 1869), p. 1.

2. See, however, the critical work of Theodore S. Hamerow, *Reflections on History and Historians* (Madison: University of Wisconsin Press, 1987).

3. See Hayden White, *Metahistory: The Historical Imagination in Nineteenth-Century Europe* (Baltimore: Johns Hopkins University Press, 1973).

4. David Cannadine, "The State of British History," *Times Literary Supplement,* 10 October 1986, 1139–1140. For commentaries see "Debate. British History: Past, Present—and Future?" *Past and Present* 119 (1980): 169–203.

5. This literature and debates over it have recently been reviewed by David Lowenthal in "The Timeless Past: Some Anglo-American Historical Preconceptions," *Journal of American History* 75 (1989): 1263–1280.

6. David Lowenthal, *The Past is a Foreign Country* (Cambridge: Cambridge University Press, 1985).

7. N. Reingold, "History of Science Today, 1. Uniformity as Hidden Diversity. History of Science in the United States, 1920–1940," *British Journal of the History of Science* 19 (1986): 243–262.

8. James B. Conant, *Science and Common Sense* (London: Oxford University Press, 1951), p. 6.

9. Academic historians of science have also been concerned with the problem of present-orientation. Loren Graham has argued that history and sociology of science legitimates the established priorities of historians. See Loren

Graham, "Epilogue," in *Functions and Uses of Disciplinary Histories,* Sociology of the Sciences Yearbook (Dordrecht: Reidel, 1983), pp. 291–295. See also his "Why Can't History Dance Contemporary Ballet, or Whig History and the Evils of Contemporary Dance," *Science, Technology and Human Values* 6 (1981): 3–6. See also David L. Hull, "In Defense of Presentism," *History and Theory* 18 (1979): 1–15.

10. A. Barclay, *Science Museum. Handbook of the Collections Illustrating Industrial Chemistry* (London: His Majesty's Stationery Office, 1929), p. 51.

11. A. Cressy Morrison, *Man in a Chemical World: The Service of Chemical Industry* (New York: Charles Scribner's Sons, 1937).

12. Carroll Pursell, "'A Savage Struck by Lightning': The Idea of a Research Moratorium 1927–1937," *Lex et Scientia* 10 (1974): 146–167.

13. Roland Barthes, "Plastic," in *Mythologies,* trans. Annette Lavers (London: Granada Publishing, 1972), pp. 97–99.

14. See Dennis Sepper, *Goethe Contra Newton: Polemics and the Project for a New Science of Color* (Cambridge: Cambridge University Press, 1988).

15. David Channell, *The Vital Machine: A Study of Technology and Organic Life* (Oxford: Oxford University Press, 1991). See my "Biotechnology in the Twentieth Century," *Social Studies in Science* 21(1991): 415–457; and *The Uses of Life: A History of Biotechnology* (Cambridge: Cambridge University Press, 1993).

16. Jeremy Rifkin, *Algeny* (New York: Viking Press, 1983).

17. Robert Bud, "The Myth and the Machine: Seeing Science through Museum Eyes," in *Picturing Power: Visual Depictions and Social Relations,* ed. Gordon Fyfe and John Law, Sociological Review Monographs 35 (London: Routledge & Kegan Paul, 1988), pp. 134–159.

18. J. H. Randall, *Our Changing Civilisation: How Science and the Machine are Reconstructing Modern Life* (London: George Allen & Unwin, 1929), pp. 4–5.

19. Umberto Eco, *Foucault's Pendulum,* trans. William Weaver (London: Secker & Warburg, 1989), pp. 5–6.

20. Walter Benjamin, "The Work of Art in the Age of Mechanical Reproduction," in *Illuminations,* ed. Hannah Arendt, trans. Harry Zohn (London: Fontana, 1973), pp. 219–254.

21. H. W. Dickinson, "The Science Museum, South Kensington, London," *Mechanical Engineering* 48 (1926): 104–108.

22. The classic work is Emile Durkheim and Marcel Mauss, *Primitive Classification,* trans. Rodney Needham (Chicago: University of Chicago Press, 1963); see also Mary Douglas, *Natural Symbols* (New York: Vintage, 1973).

23. For a synopsis of this approach, see Edmund Leach, ed., *The Structural Study of Myth and Totemism* (London: Tavistock Publications, 1967).

24. David Lodge, *Nice Work* (London: Penguin, 1978), p. 178.

25. Yizhak Oren, "The Shatterer," in *The Imaginary Number,* trans. Max Knight (Berkeley, Ca.: Benmir Books, 1986), pp. 35–42.

26. For an analysis of the museum as a classification system of its own in danger of fetishizing its objects, see Ludmilla Jordanova, "Objects of Knowledge: A Historical Perspective on Museums," in *The New Museology,* ed. Peter Vergo (London: Reaktion Books, 1989), pp. 22–40.

27. William H. McNeill, *Mythistory and Other Essays* (Chicago: University of Chicago Press, 1986). See especially the essay entitled "Mythistory or Truth, Myth, History and Historians," Presidential Address to the AHA, December 1985, pp. 3–22.

28. See, e.g., William H. McNeill, *The Pursuit of Power: Technology, Armed Force and Society since A.D. 1000* (Chicago: University of Chicago Press, 1982).

29. Paul Kennedy, *The Rise and Fall of the Great Powers: Economic Change and Military Conflict from 1500 to 2000* (London: Unwin Hyman, 1988).

30. Peter Munz, *The Shapes of Time: A New Look at the Philosophy of History* (Middletown, Conn.: Wesleyan University Press, 1977).

31. Roger Silverstone, *The Message of Television: Myth and Narrative in Contemporary Culture* (London: Heinemann, 1981). Although there are parallels between Silverstone's treatment of television and this analysis of the museum, I am proposing that we go beyond anthropological detachment toward using the self-awareness to develop the medium.

32. See John Willett, *Brecht in Context: Comparative Approaches* (London: Methuen, 1984), pp. 218–221. Willett points out that Brecht was inspired by the Russian formalist literary theorists, whose legacy has underlain many writers in this tradition. See, e.g., Silverstone, *The Message of Television.*

33. Mary Douglas and Baron Isherwood, *The World of Goods: Towards an Anthropology of Consumption* (London: Allen Lane, 1979), p. 62; see also Mary Douglas, *Risk Acceptability According to the Social Sciences* (London: Routledge, 1985), in which she examines whether risk is a natural or a social category.

34. The artifact can of course be used less ambiguously. Structuralist ethnologists have provided an ahistorical structure to give meaning to phenomena that happen to have been of the past. The Paris Musée des Arts et Traditions Populaires is an heroic attempt to exploit this approach within a museum setting. Punctuated with large panels each containing a quotation from Lévi-Strauss, it decodes the artifacts of French life. This museum is a remarkable attempt at pure structuralism and eschews any narrative structure. The French Revolution does not figure in this interpretation of the fundamentals of French rural life.

35. Erika Hickel, "Die neue Alchemisten-(Un)kultur," *Forum Wissenschaft* 6, 3 (1989): 28–31; see also R. F. Bud, "Great Expectations: The Tale of Biotechnology," *Chemistry in Britain* 24 (1988): 441–444, 466.

CHALLENGE TO PUBLIC POLICY

Between Knowledge and Action: Themes in the History of Environmental Chemistry

Christopher Hamlin

"Environmental chemistry" is a perilously vague term, for all chemistry is ultimately concerned with the workings of the environment and much of it is eventually applied to altering the environment to serve human purposes. Nonetheless, we are apt to agree on several characteristic features of such a field of study: inquiry into the chemical systems of the environment, a concern with the effects of the environment on human health and the effects of human activities on environmental systems, and a concern with the institutions through which chemistry is applied to environmental policies, and with whether these can guarantee sound knowledge and good policy.

In the last few decades we have become increasingly aware of the issues environmental chemistry encompasses and of its remarkable power to treat those issues. Environmental toxicologists and atmospheric and oceanic modelers are discovering crises at a prodigious rate, solutions are being suggested, and governments and international agencies are taking steps to implement them. It is tempting to see these changes as products of modern science, particularly modern chemistry: to believe that science has finally progressed sufficiently to allow us to run the world properly. We can claim that our ability to model has allowed us to predict a set of photochemical reactions in the upper atmosphere and that our analytical abilities have enabled us to document, with startling sensitivity, that a number of these predictions are accurate.

In this perspective environmental chemistry is likely to be seen as an outcome of the accumulation of chemical knowledge, the development of statistical research methods, and much better understandings of transport processes in soils, bodies of water, and the atmosphere. Here it will be the product of a mature science where a certain level of knowledge has generated new research possibilities whose exploration leads to a recognition of needs for policy, to policies themselves, and to further research problems. It would seem to follow that there could have been little environmental chemistry prior to that maturity, simply because there would have been no grounds for concern nor any sound basis for policy. In such a view, then, environmental chemistry is a product of knowledge and leads to action.

A quite different view is suggested in recent work on the social construction of science. It suggests that the environmental policies we undertake (or the crises we worry about) are better explained as generated by social needs of some sort, with knowledge being created or called up to legitimate what is being undertaken on other grounds. Certainly in the cases both of predictions of future environmental conditions and in the recognition of contamination we have superb examples of the ways in which underdetermined theories and value-laden observations allow extra-scientific factors to affect what is being presented (and what the public would like to regard) as scientific certainty.[1] Both Alvin Weinberg in his exposition of trans-science and Jerome Ravetz in his discussion of large technological innovations suggest that environmental predictions are commonly unfalsifiable since manifold changes are being made in the system while the experiment is going on. Thus our science must function less in telling us what will happen or is happening than in giving us confidence to face the future.[2] A similar view appears in the work of the well-known cultural anthropologist Mary Douglas on the culture-boundedness of environmental perception. Douglas has argued that judgments of pollution, as well as predictions of imminent environmental catastrophe, are generated by the imperative to maintain social stability or to achieve (or retain) social power.[3] In these views we do not act when we know, we create knowledge to justify our actions.

The kinds of histories we write will depend on which of these views we take. To take the former would entail starting from the present and working back, recognizing the enormous twentieth-century advance in analytical capabilities[4] and the striking conceptual advances—for example, in the understanding of the nature of soil fertility or of the oxidation of organic wastes—that took place in the nineteenth century.[5] To take the latter would be to start in the past and watch how people tried to derive imperatives for social action through the development of concepts and standards, visions of natural order, and explanations of causation and responsibility.

Is one of these views right? It is true that two of the central concerns of environmental chemistry—pollution and sustainability (or what might better be called cyclic renewal) have long been matters of concern in our own culture and in other cultures, and have been endowed with moral and sacred significance as well as being subjects of scientific inquiry. We could see the "chemical philosophy" of the sixteenth and seventeenth centuries as a proto-environmental chemistry in this regard: an attempt, heavily laden with religious and social concerns, to understand how the totality of quasi-chemical entities in the environment affected the well-being of the individual human.[6] It is also true that as Douglas would lead us to suspect, environmental chemistry is still regularly appealed to, to prove that the righteous will triumph, that sin will be punished, that revolution is imperative, or that those out of power must now be put in charge.[7] But even if we admit all this, to adopt the relativism of Douglas and to see both problems and solutions as products of temporary social strains and cultural discourses seems an unsatisfactory attitude with which to approach contemporary policy-making. It is striking that although a number of science studies scholars have been attracted to Douglas's perspective, none of them, so far as I know, has systematically applied it to the explanation of the full range of concerns in modern environmental chemistry.[8] Sociological or anthropological analyses of the claims of environmental scientists have become part of the ammunition with which modern environmental controversies are fought, used to expose as mere social constructions some conceptions of what nature is or will do, so that the correct conceptions (those held by the debunkers) can be recognized.[9] Yet we run a great risk of wholly undermining our ability to act if we undertake a systematic debunking, imposing a relativistic framework on all contemporary perceptions of the environment.[10] It may be that we do face grave changes in the earth's systems from factors potentially in our control, and if that is true it seems wiser to take our knowledge seriously, all the while recognizing its tentativeness and insufficiency.

I am not sure it is possible to reconcile these two polar conceptions of environmental chemistry. But it does seem possible to explore a middle ground, to find more practical ways of resolving environmental controversies by understanding both that people in the past tried to manufacture certainty (just as we do now) and also that they occasionally found ways of negotiating through uncertainty (as we perhaps need better ways of doing).[11] Such history will be unabashedly policy-oriented and correspondingly normative, but it will look to the history of environmental chemistry not for the pioneers of a correct understanding of nature (whatever one takes this to be) but better to recognize problems and discover options (certainly not lessons) for environmental management.[12]

What I shall do here is to take three case studies of features or issues common to modern environmental chemistry, but which probably exist as far back in the history of science as anyone would care to go. I label these "expert disagreement," "glorious vistas and dire threats," and "satisfactory authority." I will focus on their appearance between 1780 and 1900 in Britain, both a pioneer in pollution as the first industrial nation and a pioneer in the guarantee of environmental quality as the birthplace of the modern public health movement.

Expert Disagreement

In the last edition of his noted textbook on medical jurisprudence, the forensic chemist Alfred Swaine Taylor, king of the profession in mid-Victorian London, devoted a great deal of attention to a problem that had become increasingly prominent during his lifetime, that of the disagreement of paid experts in courts of law and other forums of policy-making. "No difficulty has ever been found in obtaining any amount of evidence . . . on either side of any point in issue. There is a contest as to whether a vitriol or a gas manufactory is a nuisance. Twenty chemists of fair character and scientific acquirements come forward to swear that the effluvia . . . are producing the most deadly fevers, and twenty others equally eminent will give just as positive testimony that the gases are absolutely wholesome, and rather fattening than otherwise. These things are of every day experience."[13] Dorothy Nelkin has made much the same point more succinctly: "The extent to which technical advice is accepted depends less on its validity and the competence of the expert, than on the extent to which it reinforces existing positions."[14] This is a troubling prospect, for it would seem that one of the reasons we look to experts is to find a way beyond present perplexity. Yet most of us are familiar with a variety of cases in which experts clash: on the fates and effects of toxins in the environment, on the availabilities of resources, and even on the quality of feedstuffs and fertilizers.

Expert disagreement is not new: the long tradition of mineral water analysis illustrates how old are problems of expert disagreement among chemists. Even though this example takes us back before the origins of modern chemistry, it seems a valuable one for highlighting continuities. As with much modern environmental chemistry, the concern in the mineral water literature was with the effects on health, sometimes quite subtle effects, of some quasi-chemical entity in the environment. The mineral water tradition combined philosophical and practical aspects of chemistry. Many believed that springs, like animals, plants, metals, and celestial bodies, possessed a unique spirit, which interacted with the microcosm, the particular human, in a unique way. During the Middle Ages, religion had provided the social authority to support the belief that particular springs had real powers; each tended to be associated with a saint.[15] But increasingly during the early modern period the proprietors of springs commissioned chemists to do analyses. As early as the fifteenth century in southern and central Europe, a need was felt for a level of authority beyond the legend of a saint or the testimony of the patient. This was provided by a rudimentary chemical analysis, such as the behavior when heated of the crystals produced upon evaporation.[16] It is important to note that this chemistry was rarely the means by which mineral springs were discovered, nor was it the means by which their medicinal powers were determined; instead it was an appeal for social sanction to certify that the claimed effects were real.

In the succeeding centuries literally thousands of tracts and treatises were written, most of them including analyses chemists had been commissioned to make to show the powers of some springs and the lack of powers of others.[17] The size of this literature, and the fact that much of it was directed to the informed invalid, suggests that for much of early modern history a large public was quite as much exposed as we are to ongoing debates about the effects of the environment, and that chemistry had become the principal idiom in which this debate would occur by the end of the seventeenth century.[18] Most of these tracts were written by chemists and physicians now unknown, but some famous names—Priestley, Lavoisier, Berzelius—were contributors to this literature, and for others—Bergman, F. Hoffmann, Boyle, Fourcroy—it occupied a significant role in their careers. A chemist wishing to participate in this market for authority need not sully his reputation by making great claims about what his results indicated; indeed his authority still could be valuable even if he gave no interpretation at all; he could, as did A. W. Hoffman, simply carry out an analysis and let the spring's promoter interpret it, only allowing his name to be attached to it.[19] These controversies were interminable; there were many more springs to be promoted than wealthy patrons to patronize them.[20]

Eighteenth- and nineteenth-century chemists recognized that they were engaged in expert disagreement; there was the regular exchange of condemnations of the motives, morals, and methods of one's opponents. Writing in 1717 on behalf of the spring at Pyrmont and against that at Spa, Frederick Slare, FRCP, FRS, suggested that his rivals were utterly arbitrary in their interpretations of analytical results: "or else his Fancy shall run upon another notion, and shall call it a Sulphur of

Mars; and not rest here, but endeavour to make up the three Chymical Principles, namely Salt, Sulphur, and Mercury; he will call it Sulphur, though it be uncombustible, it must be a salt, though it be undissolvable, and a Mercury, although it look like a gravel stone."[21] Charles Lucas, perhaps the most eloquent of the mineral water chemists, characterized the literature similarly: "Most of the voluminous and numerous tracts," he wrote, "and of these the most pompous, [that] we have, upon mineral waters, have been published by men living and practicing upon the spot, . . . always interested in the fame of the particular water, which was their idol." He added that "such a man's evidence must therefore be deemed as doubtful . . . as that of any other priest touching the miracles of the shrine, by which he gets his daily bread."[22] He complained that on the continent "a mercenary slave of the profession [of medicine] can hardly be wanting to perpetrate the foulest deed" and not only individuals, but "formal associations have been formed, . . . that call themselves the College of Physicians at Liège, to pronounce the solemn condemnation of this water [of waters in competition with Spa]."[23]

But to recognize and deplore the partisan character of the mineral water literature in this way was usually simply one of the techniques of controversy, an attempt to place one's own efforts in a different category of disinterested inquiries into the truth of the matter. The rejection of rivals was always to be founded on scientific rather than commercial grounds. In representing a vista of corruption and incompetence against which he stood as an isolated island of virtue, Lucas was presenting himself as the only trustworthy advisor available to the bewildered invalid.

Doubtless the rejection of some springs and the endorsement of others was most effective when the writer could, like Lucas and Slare, extricate himself from the context of controversy. But this could be done in much more subtle ways than simply by pronouncing oneself a disinterested authority, in ways that indeed raise very troublesome questions about what stance we should take toward expert testimony. One could appeal to the philosophy of science or to various images of scientific progress to contrast one's own work with the backwardness of one's opponent. Such images could do exactly the same work as the broadsides of Lucas or Slare in upholding the virtues of certain springs at the expense of others. But they could do so without for the most part suggesting that the profession as a whole was corrupt and therefore inviting the reader's suspicions. One was saying not that "I am truthful and he is a liar," but that "I am part of progress, while he, though perhaps well intentioned, is stuck in antiquated and stagnant tradition." The dirty work was thus done with bits of ideology. We can see how this was done by contrasting two views from the 1770s and 1780s, those of the Swedish chemist Torbern Bergman, whose approach to mineral water analysis would remain the dominant one until roughly 1830, and John Barker, a Cheltenham physician active in promoting that town's mineral waters.

Even more than now, prevailing medical theory in the seventeenth and eighteenth centuries emphasized the interaction between the interior of the human and his external environment in explaining one's state of health and seeking to change it, but there were great problems in demonstrating that changes in one's state of health were associated with certain properties of environment. Bergman was a pharmacological reductionist. He would not admit that an environment could cause real changes in health unless these could be explained by the demonstration that active substances were present in that environment. With so many factors in the environment, the problem was not one susceptible to controlled clinical trials, even had such methods been available at the time. Among the corollaries of Bergman's view were the beliefs that one knew, and could identify the presence of, all the substances in the environment that could possess pharmacological activity, and therefore that artificial mineral waters could be made with powers equal or even superior to those of natural waters.[24] In a lengthy historical introduction to onc of his papers on mineral waters, Bergman made it abundantly clear that he was on the side of social and scientific progress. There had been a time, he noted, when the active properties of mineral waters had been attributed to a "spiritus mundi [that] . . . constitutes the spirit and life of medical waters."[25] At the time Bergman was writing, this mysterious substance was being shown to be in many cases nothing more exotic than carbon dioxide, but as we shall see, that recognition did not stop chemists and medical men from continuing to claim that certain springs had unique and irreducible medical properties.[26]

Full of confidence in his analysis, Bergman envisioned a world in which patients would no longer be fleeced by quacks trying to treat them with waters devoid of properties, in which patients would truly be sent to the waters that demonstrably had contents that would relieve their symptoms, and in which what had been exotic medicines could be synthesized and made available to the poor. His proposal involved a concept of cosmic order in which nature was susceptible to human reason and in which science led to social justice. His proposals had plausible and orthodox philosophical foundations. There could be no results without causes, he was arguing; the medical effects of a water could be due only to the physical and chemical properties it possessed; by measuring those properties one could determine the range of possible effects.

Others, like Friedrich Hoffmann, well known as one of the developers of phlogistic chemistry, and his follower the English doctor John Barker, were self-proclaimed holists.[27] Barker, writing a few years later than Bergman in defense of the mineral water at Cheltenham, argued that it was dangerous to be guided by the imperfect science of chemistry; much better to put one's trust in "experience," by which he meant medical testimony: "We must . . . take nature as she is: not compare her easy, yet powerful methods of preparation too much to . . . a process chemical." It was impossible—even impious—to think that we could imitate natural springs. Each one of them had "specific properties . . . wherein it differs from every other of the same class." These differences he ascribed to a standby explanation of eighteenth-century obscurantists, the subtle fluid, in this case a peculiar volatile sulphur. Chemistry might "assist us indeed in forming some, though imperfect notions, of the nature of the grosser contents of waters; but their spirit is too fine and volatile in its nature, subtlety, and powers, to lie sufficiently open to such proofs."[28]

Barker too was able to appeal to the philosophy and sociology of contemporary science. In effect he was arguing that if there were real medical results from taking a certain water, the fact that chemists could find no explanations for those results ought to be seen as a scientific anomaly indicating where further research was needed. He depicted contemporary chemists as aggressively pursuing social authority, plunging into areas about which they knew nothing, and deducing conclusions willy-nilly according to a priori systems. However difficult to withstand the pressure to theorize and draw premature conclusions, it was necessary for the public good that scientists patiently stick to the facts and allow experience to indicate the true properties of waters. He too regarded nature as susceptible to human reason and full of undiscovered bounty, but presented an image of the humble observer to whom truth would eventually come.

In this controversy, as with modern controversies that involve environmental chemistry, both men claimed to speak on behalf of progress and justice, and each drew upon contemporary philosophies of science and concepts of nature. In both cases their positions can also be explained in terms of professional and financial interests: Bergman was a pioneer in an artificial mineral waters industry (he produced imitations of famous springs, though it is not clear he made much money on them); and he saw his work in the Swedish national interest and in the interests of chemistry. Barker was upholding the independence of medicine and urging invalids to spend their money on Cheltenham water wisely prescribed by experienced Cheltenham physicians. While Bergman's perspective is more in keeping with the way chemistry was developing at the end of the century—his assumptions that analysis could be exhaustive and his implicit use of gravimetric reasoning (that is, the only things that are there are what I find in my analysis) being among the most important methodological conventions that would characterize postwatershed chemistry—the case Barker made was valid. There was great risk that clinical knowledge would fall victim to the onslaught of a chemistry primitive enough for the purposes of analysis and completely incapable of offering any useful guidance as to the pharmacological action of medicines within the body. With the rapid discovery of new elements, some of them, like iodine, possessing powerful medical properties, and later with the discovery of radioactivity in many of the spring waters, Barker has in a sense been shown to have been right: waters could have important properties undreamt of by the chemists of his day.

Were this a modern environmental controversy, I suspect we would want to ask two questions. First, we would want to know which side was right, and second, whether those who were wrong were the makers of honest mistakes or the minions of vested interests—to put it bluntly, paid liars. Was it a pathological condition of chemistry that allowed this controversy to exist, or was that condition of chemistry in itself nothing more than a manifestation of these differences of opinion? In a practical sense the dispute was, at least given the methods available at the time, transscientific; the issues were statable as scientific problems but not soluble.[29] It is possible to imagine the dispute conducted more rigorously than it was—the Barkers of the day had no means of making controlled clinical trials to help them distinguish which were the active elements of the therapeutic environment—but it is difficult to claim that even much improved science could solve the problem once and for all. A right answer was not available.

The second question is more troublesome. We regularly do make assessments about the motives of contemporary scientists in various sorts of environmental controversies simply on the basis of the interests they appear to serve. But if the questions are transscientific, and a right answer is unavailable, we are unwarranted in assuming that all participants were consciously and in the most venal form selling their authority, though it is exceedingly likely that some of them were. And we are also unwarranted in assuming, as is often done, that it is only the scientists on the side opposite the one we agree with who are engaging in this practice.

One way to try to get beyond this dilemma is to ask why in this controversy, as in so many modern environmental controversies, the participants were so intransigent. Surely Bergman and Barker recognized the assumptions on which their claims depended at least as well as we do. Yet they acknowledged little middle ground and refused to admit that their positions were based on best guesses and had to be much qualified.[30] One possibility (insufficient, I think) is that they were simply representing the intransigent positions of the vested interests they served. Another is to explain their intransigence as an indication of incommensurably different views of competing professions (medicine and chemistry) with distinct methods, social identities, and ways of looking at the world. Still another possibility, not incompatible with either of the former, is that there was nothing to be gained from acknowledging uncertainty: patients needed certainty to go with confidence to take the waters, proprietors needed certainty to advance the claims of their own springs over those of others, and for chemists and physicians the claim of certainty was equally a claim of the independence and priority of their professions.

It is noteworthy that in the early nineteenth century a middle-of-the-road approach to the study of mineral waters did develop in Britain. Writers like Daubeny, Gairdner, and Saunders rejected the contentiousness of the existing literature and urged the need for an enormous research project of disinterested analyses to determine once and for all what the properties of the mineral waters were.[31] It was an early indication of the pretension of chemistry to a status in which its research was presumed to be relevant to all matter of social issues, though in fact it was fast becoming an academic enterprise.[32] This research program did not catch on, in part because there was insufficient financial support for disinterested knowledge on the issue, but in part because partial knowledge, tentative knowledge, sober and qualified assessments, did not meet the needs of a public that desired from its scientists a warrant for action, not an exercise in iconoclasm. Significantly, A. W. Hofmann, professor of the Royal College of Chemistry and a shrewd judge of the social support of chemistry, refused to cast his lot with those who would purify mineral water chemistry. In the late 1840s he established a program of water analyses by RCC students that was rigorous, academic, served the needs of clients, yet avoided the central philosophical ambiguities involved in making claims about the contents of a mineral water, which at that period concerned the nature of salts in solution.[33]

Glorious Vistas and Dire Prospects

One of the ways most of us are exposed to environmental chemistry is through reports that reach us in a variety of ways of a future of peace and plenty or, alternatively, one of imminent doom, all depending on how well we understand the chemical systems of the environment and manage the world in accord with them.

Arguably, the current crop of such visions is a product of two conceptual achievements of the chemical revolution. The first was the conviction of the conservation of elemental substances and, what may be seen as one of its extensions into a particular chemical problem, the overthrow of the humus theory of plant nutrition by those who, like Liebig, Sprengel, and Boussingault, would insist that the elements that made up plants or animals must come from the environment and could not be produced for the occasion from whole cloth, as it were.[34] The second was the recognition by Priestley, de Saussure, Ingen-housz, Dumas, and others of some aspects of the chemical complementarity or symbioses between plants and animals, the most important of these being the complementarity between photosynthesis in plants and respiration in animals (plant respiration was often conveniently forgotten).[35]

By 1840 both these recognitions had filtered into European culture, where they had several effects. The second discovery, of complementarity, was understood in terms of natural theology. Just as Newton had been taken to have shown how clever was the mechanism by which bodies on earth and in the heavens were moved, so the chemists were seen to have demonstrated how marvelously complex was the mechanism that sustained life.[36] It was as if a flood of purposes and explanations had been unleashed; a new level of design, in which a great many more observations were shown to be consequences of a necessary system, had been detected. J. F. W. Johnston, professor of chemistry at Durham and one of the most articulate exponents of this view, wrote in 1853 that "The Deity will that this corner of His great work should be the theater of new displays of wisdom, of consummate contrivance, of a wonderful fitting in of means to the accomplishment of beneficent ends." This, along with the discoveries in agriculture, led to a perspective in which every atom of every element was valued and regarded as having a destined role in the system of the world. It led to a belief that damage would result from managing things so that atoms did not go where they were supposed to, and to a view that through the complete understanding of the courses of these atoms, requisite knowledge could be obtained to manage the world properly, in the way God had intended us to. The whole chemical system of the earth and the universe was extraordinarily fragile, Johnston wrote; by a change no larger than "the simple turning of a screw . . . an alteration in the natural constitution of things of so small a kind as to be inappreciable to our senses would at once insure the certain extinction of animal and vegetable life."[37]

But in the nineteenth century these were not merely matters for pious reflection. With industrialization and urbanization, the fates and courses of atoms were changing markedly. For reasons that have in part to do with real changes in physical conditions, but also to do with changes in medical theories and views of the most feasible of alternative approaches to social reform, these natural processes became matters of political significance in mid-nineteenth-century Britain, and to a less dramatic extent also in France, Germany, and the U.S. The focus of this concern was with the recycling of waste products, chiefly, though not exclusively, the organic wastes generated in cities.

Edwin Chadwick's 1842 *Report on the Sanitary Condition of the Labouring Population of Great Britain* is the best known of a series of reports depicting the conditions of the mostly urban poor. It depicted a population unhealthy, immoral, dangerously revolutionary, barely able to scrape a living wage, and frequently out of work. Chadwick and his followers attributed all these problems to improper management of atoms in the social body. These were not metaphorical atoms, but the real things; indeed one might go so far as to say that the dominant view of the sanitarians was that all social problems were due to the mismanagement of fixed nitrogen.[38] For the early sanitarians the main problem was the retention of wastes within the city; despite the growing realization that plants took in nutrients in an inorganic form, sanitarians were preoccupied with hastily removing wastes to farmland before decomposition had a chance to take place; the fevers, the cholera, even the general debility that blighted the life of the city dweller was due, they insisted, to the presence of products of the decomposition of matter, which ideally should have quickly been applied to growing plants. These physical causes were deemed responsible for an amazing range of problems. Among the most serious causes of death that the chemist (and Health of Towns Commissioner) Lyon Playfair found in his tour of Lancashire cities in 1844 was the overdosing of hungry and querulous infants with opiates. Yet the cause of the problem was lack of sewers, Playfair insisted; if the environment were freed from the pollution that surrounded it, children would not cry and there would be no reason for parents or child-minders to dope them.[39] It is not surprising that as working class socialism developed toward the end of the century there was real resentment that demands for more money and shorter hours were still being met by the construction of new sewerage works.

Chadwick himself was only secondarily concerned with theological or ecological reasons for maintaining a balanced budget of fertilizing elements. In his view the main incentive for recycling sewage was economic: its worth as fertilizer could pay the full cost of sanitary infrastructure. But other sanitarians and reformers, like Playfair and Charles Mansfield, a Christian socialist chemist who died at an early age in 1853 in a laboratory accident, developed the argument that recycling was an ecological necessity, a part of the divine pattern of creation (as well as being financially advantageous). They sought to apply the perspective they found in the works of Johnston and Liebig about the importance of recycling the elements of fertility to practical matters of sanitary engineering and administration.

For example, inspired by Liebig's conception of limiting mineral constituents, Playfair argued that recycling ought to be done within a tight circle, since a population's "effete matter" was "of all things just the thing to supply as manure for the growth of the food of that population. . . . The Irish fed very much upon potatoes, and potatoes grow very well in Ireland, because the effete matter of the potato feeder was exactly suited for the growth of that crop."[40] It is clear that Playfair was conceiving of a virtually closed system: all the fixed nitrogen, phosphorus, and potassium in the potatoes getting back into the ground through human wastes or human corpses.

In Mansfield's pioneering sanitary utopia, "Hints from Hygea," the earthborn visitor to the truly ecological planet of Hygea explains the ruling conception of nature, and of humanity's place within it, that prevails there:

> They never willingly waste anything. They hold, that all the chemical and mechanical combinations that occur in nature are a constant production of reservoirs of force available for human use; that the conditions thus wrought abide for a certain time, as if to give man a chance of seizing an opportunity of power, which, once neglected, is gone for ever; that the compounds formed, if not used within the allotted period, split up by rapid self-destruction into their elements, again to recommence their cycle, again to culminate in a proffer of utility to man. They hold, that every substance within man's reach—from his own brain to his cabbage leaves—that is resolved into formlessness, without having done for the common good of our kind an amount of service proportional to its complexity, is wasted. And they hold waste to be a sin.[41]

The linked ideas that chemical elements were conserved, that they could assume more or less useful forms, that they were the source of wealth and survival, combined therefore to underwrite what might be called a bookkeeping mentality in which civilized humanity was to take responsibility for conserving natural capital.[42]

Chadwick's economics, together with a chemico-theology developed by Johnston, Playfair, and Liebig, led to a Victorian recycling movement concerned mainly with sewage recycling. In Britain there was great enthusiasm for sewage recycling in the late forties and early fifties, and again from about 1862 to 1875. As late as 1871 William Corfield, author of one of the better texts on sewage treatment, was declaring that "No scheme which does not remove ALL refuse matter . . . and utilise it so as to 'make it pay', can be accepted as anything like the final solution . . . nor can such a scheme be recommended to towns as a feasible solution of their difficulties."[43] By the late 1870s the realization had set in that recycling would rarely be profitable, but it was still being recommended as the best available technology for keeping rivers pure and the wisest in terms of long-term ecological constraints. Also, by this time, both the techniques of recycling and the idea that it was part of technological progress had spread from Britain to France and Germany (both Paris and Berlin built extensive sewage recycling works) and to a lesser degree, the U.S.[44]

It would be wrong to think that this recycling perspective was a simple and direct deduction from the discoveries of Priestley, Dumas, and their followers. It is better to see these discoveries as key parts of a conceptual background that could be (and was) exploited for all sorts of purposes. Images of a harmonious world in which all good things were recycled appeared in multiple contexts in Victorian Britain and were variously invoked to reinforce notions of God and of the social order, to guide development of technology and justify the setting of environmental standards, and simply to sell things. But even recognizing this, it still seems important to ask, as in the mineral waters case, how far knowledge constrained the uses that were made of it, how far it implied technical action, or was, on the contrary, a means of giving the actors an apparent post hoc justification for their actions.

A key figure here is William Crookes, whose career brings together perhaps better than anyone else's the multiple contexts that had come to characterize professionalized science, and particularly chemistry, by the last third of the nineteenth century. Crookes was (1) a serious researcher making valuable contributions in a specialized area of chemistry; (2) an explorer, through his interest in spiritualism, of the bounds of science and of its place in human affairs; (3) a promoter of chemistry, taking, as editor of the *Chemical News* and the *Quarterly Journal of Science,* an active role in all controversies that concerned the training and employment of chemists and the utilization of chemistry; (4) an environmental activist, using his professional knowledge and public position to call public attention to an environmental crisis, the wasting of fixed nitrogen; and (5) a chemical entrepreneur, serving for a number of years as one of the chief promoters, and for a time as chairman, of the most notorious of the sewage recycling schemes, the ABC process of the Native Guano Company.

What the relationship among these activities was is not at all clear, and here I shall be concerned only with numbers 3 to 5. Crookes became publicly involved in issues of sewage treatment, river pollution, and the waste of fixed nitrogen in the late sixties when, owing to prolonged agitation of the fisheries lobbies, royal commissions were studying the problems of sewage treatment and river pollution, and national legislation was contemplated. To Crookes as editorial spokesman for the profession of chemistry these were matters of concern: the viability of chemistry-based industries such as sulfuric acid manufacture and calico printing and the prospects of new chemical methods of sewage treatment were at stake, as was the employment of chemists as certifiers of pollution and purity. The positions he took in regard to these interests can be seen as straightforward defenses of the interests of chemistry: he opposed extreme standards of water quality and formulaic and simplistic definitions of water quality, and promoted the development of chemical means of sewage treatment (and recycling) in place of the practice of irrigating farmland with sewage.

Among chemists Crookes's principal antagonist on all these issues was Edward Frankland, the chief member of a Royal Commission on Rivers Pollution that sat from 1868 to 1873. In 1868 Frankland's commission had devoted the entirety of one of its reports to exposing as fallacious and fraudulent the claims of the Native Guano Company, one of the many companies that sought to purify sewage (and extract valuable fertilizer from it) by adding a mixture of precipitating chemicals to it. The firm had been launched in 1867 by two London financiers, William and Robert Sillar, and the young chemist G. W. Wigner.[45] Despite failures of many such processes, Native Guano had acquired a certain notoriety owing in part to the audacity of the Sillars' claims of purity and profit and in part to the unusual origins of their "ABC" mixture: "A" was alum, "C" was clay, but "B" was bullock's blood, authorized by the biblical instruction that "all things are by law purified by blood."[46] Frankland's investigators were able to find strong evidence of a willful attempt to mislead through the doctoring of samples going on at the company's Leamington works.[47]

Despite this exposure, Native Guano persisted, the claim being made that what might once have been true was true no longer. By 1871 Crookes had joined the firm, first as a director and chemical advisor at two hundred pounds a year, eventually receiving one thousand pounds a year and making numerous sales trips to the continent. He wrote (or lent his name to) numerous pamphlets of testimonials and explanations and only ended his formal relationship with the company (he may well have continued to hold shares) in 1880.[48] In these pamphlets he went so far as to deny some of the principal assumptions of chemistry—that, for example, the worth of a fertilizer could be established by analysis. When analyses of "Native Guano" were repeatedly inconsistent with the powers attributed to it in the testimonials of practical farmers, it was chemistry that had to be rejected.[49]

At about the same time that he became mixed up with the Native Guano Company, Crookes began a campaign to publicize "the nitrogen problem" as the "king-problem of practical chemistry." (He did, however, continue to be concerned with fixed-nitrogen recycling long after his separation from the company, his best-known treatment of the topic being his presidential address to the British Association for the Advancement of Science [BAAS] in 1898.[50]) There were certainly good reasons at the time to be quite as concerned about the transformations of nitrogen as we are now about atmospheric carbon dioxide. By the second half of the century it had become clear to most chemists that plants acquired their nitrogen in fixed form (nitrates, nitrites, or ammonia) from the soil. Even with the discovery toward the end of the century of nitrogen-fixing bacteria, there was no known natural mechanism that fixed atmospheric nitrogen at anything like the rate at which it was being lost (and along with discovery of nitrogen-fixing bacteria came the discovery of denitrifying bacteria). The problem only disappeared with development of synthetic nitrogen fixation processes (energy-intensive processes, no less) in the first two decades of the twentieth century.[51] Thus in the late nineteenth century the nitrogen problem could be seen as a threat to national survival, and even to human existence. It lent itself to striking imagery: the fixed nitrogen cast off by the British population each year was the equivalent of 4,380 million four-pound loaves of bread; the great threat posed by modern warfare was not in the killing of human beings but in the combustion of nitrate explosives, every ignited munition representing food lost forever—a "crime against humanity," as J. W. Slater, one of Crookes's associates, put it.[52]

What is unclear is the relation between Crookes's involvement with Native Guano and his evocation of the nitrogen crisis. Did he become involved with a slightly shady operation out of conviction that it was the closest thing to a technical solution to a crisis that threatened survival? Or did he champion the nitrogen problem in an effort to promote the company's welfare, or were the two activities mutually reinforcing in some more complicated way? Crookes, as would any good promoter, excelled at disguising the occasions in which he was being merely commercial from those in which he was speaking honestly as a scientist of what he thought to be true.[53] We face this kind of question regularly in trying to ascertain how seriously we are to take the various visions of environmental calamity, and it is by no means clear how we are to answer such questions, either in Crookes's case or that of others.

Satisfactory Authority

In the early 1970s J. Primack and F. von Hippel brought up the concept of "public interest science." They, like many others at the time, were concerned with what might best be called the democratization of expertise: "public interest science" was to be the means of democratization.[54] But it was never wholly clear how "public interest" science was to be distinguished; the title was open to be claimed by any group wishing to make the point that its interest was also the public interest. But could there really be something like public interest science, of expertise truly in service of democratic institutions?

My third case takes up this question by exploring the issue of where within a society environmental problems are brought up for solution and, correspondingly, how that location affects the ways problems are defined and resolved and the sorts of authorities that are found to be useful. The subject is a change in the context of the analysis of public water supplies in the 1880s and 1890s.[55]

In Britain from the late 1820s to about 1880 the question of whether a water was safe to drink or good to use for domestic supply was a matter of forensic combat. As with the mineral water analysis considered earlier, chemistry was harnessed to the defense of interests—vested in the ownership of existing supplies, or in land from which an urban water supply might be taken, or even in abstract principles, such as the right of the people to own a public water supply or the right to be protected absolutely from the threat of dangerous contamination.

Working in such a context where the investment of great capital and political power were at stake and where social tensions ran high, chemists found themselves pressed for extreme statements; they were expected, through their analyses, to guarantee not only that a water would do no harm at the time of analysis, but that it would always remain safe. There was no consensus at the time about the specific nature of the materials suspected of being harmful, great disagreement about whether any reliable indicators of danger might exist, and little in the way of satisfactory epidemiological evidence linking particular measurable contaminations with particular diseases. In such circumstances one may wonder why chemists were being consulted at all.

In effect, what had developed in the early nineteenth century was a crisis in social authority governing the allocation of water supplies. As cities grew and canal proprietors and textile manufacturers sought supplies of water, there was great pressure on legal and legislative institutions to decide the merits of rival claims over how much should be paid for the transfer of the rights to water sources or what qualities of water should be enforced for different purposes. These institutions were unprepared to make these kinds of decisions, especially in the burgeoning climate of utilitarianism, with its notions that the "greatest good" ought to be secured in the utilization of resources. Chemistry (and secondarily, geology, meteorology, civil engineering, and medicine) was looked to for answers to these questions, and chemists, full of confidence and members of a rapidly growing profession, were willing enough to accept the call. We may recognize that many of the key questions were bacteriological rather than chemical, and a number of contemporaries, chemists and others, likewise recognized, at least on occasion, the inadequacy of chemistry to deal with the questions they were pressed to answer. But two factors ensured that their expertise would dominate. The first was the widespread perception that chemists were closer to providing adequate answers than any other group of professionals, and the second was the recognition by all parties that decisions had to be made promptly.

The provision of this kind of authority called for a number of features on the part of the chemist who provided it. It called for confidence in public and coolness under cross-examination; for familiarity with the professional literature, and therefore with the details of any case one's opponent might bring up; for a stockpile of plausible theory, and for some long years of ineffable experience. Such a context led chemists into what may be styled a battle for the decimal points, a struggle to be seen to be in control of techniques of analysis more sensitive than those of one's rival, regardless of the doubtful relevance of the entity being measured (which for most of this period was organic matter or something presumed to be an element or aspect of it). The most ferocious of these disputes began in 1867 and preoccupied British water analysts for the next decade. The protagonists were James Alfred Wanklyn, who had many followers and was founder of the ammonia process, which determined, within a few hundredths of a part per million, the concentration of "albuminoid ammonia," an artifact of the analysis; and Edward Frankland, who had few followers and was founder of the combustion process, which determined with even greater sensitivity the concentrations of nitrogen and carbon in the organic matter in a water. Primarily what these men fought over was credibility: the privilege to be deemed the world's greatest authority on water quality on the grounds of being able to outdo one's rival in the laboratory and therefore to be able to assert before a judge, umpire, or parliamentary select committee that one's words reflected the state of the art, that one had gone further and was closer to truth than the opponent, who was still foundering with his primitive methods.

It was a mode of expertise adapted to its purpose, namely, securing political and economic control of water, not monitoring its quality or safety. Satisfactory results, in the form of forensic victory, did come to some parties through this clash of experts. Yet with advances in epidemiology, and particularly with recognition that lethal waters were only occasionally lethal owing to accidental contamination, there came a need for quite another sort of authority, an authority demanded by the users of water that their water would not harm them. It was not initially clear that there were two kinds of problems here, requiring two kinds of authority. The question "was the water safe?" was after all phrased in the same way in both contexts, the persons to whom it was asked were the same persons, but the issues in fact were enormously different. Previously, the issue had been control, and the authoritative sanction of a chemist had been the means to obtaining control. Had chemists been asked in a neutral forum what precisely their sanction meant, they might have said that the sample in question seemed in general a suitable sort of water for domestic use—that is, that it was not outrageously hard or that it did not contain blatant signs of sewage contamination—but they would probably not have wished to go on record as certifying that it would never become contaminated.[56]

Throughout the eighties samples continued to be sent by local authorities to the elite London consultants, the Franklands and Wanklyns, with requests for a kind of authority that was unavailable from them. The results were sometimes tragic. The most poignant case is that of the Stockton, Darlington, and Middlesborough water supplies. During the mid-1870s these towns had begun an arduous but eventually successful campaign to acquire control of their water supplies by obtaining parliamentary sanction to purchase the privately held waterworks at an arbitrated price. Frankland, a champion of public control, was one of the experts to testify on their behalf. In the summer and autumn of 1890 these towns were devastated by typhoid, suspected at the time and later shown by exhaustive epidemiological investigation to have been waterborne. During the epidemic local officials had sent samples to Frankland (and other chemists) for evaluation. They had all pronounced it "harmless" and "excellent."[57] It should be pointed out here that the issue was not simply one of bacteriological versus chemical methods of analysis: the waters in question passed bacteriological as well as chemical tests. The water had also been filtered according, as far as anyone could tell, to proper procedures.

Events like these, and the remarkably similar Hamburg cholera epidemic three years later, confirmed British public health authorities in their distrust of chemistry. But the problem, as some chemists had begun to recognize, lay less in the imperfection of their science than in the circumstances of its application. It was ludicrous, they began to suggest, to base management of a water supply on a single or an infrequent analysis made by a distant metropolitan chemist who might know nothing of the geographic circumstances of the source from which the sample had been taken, and who might not even get around to analyzing it for weeks. What was necessary was a continuous on-site management by someone skilled in a number of techniques of environmental assessment, including the use of chemical analyses both rudimentary and sophisticated; someone competent also in bacteriology and epidemiology, and alert as well to changes in weather or human activity that might temporarily threaten a usually safe water supply.

The developers of this outlook were not the Franklands or Wanklyns but a new group of regionally based chemists with public health concerns, the public analysts appointed under various antiadulteration acts to oversee the quality of foods and drugs on the retail market. The most important figure here is John Clough Thresh (1850–1932), whose *Examination of Water and Water Supplies* was the principal English text on the subject in the first decades of the twentieth century. Thresh was unusually well trained for the role he would play: he held a doctorate in chemistry, an M.D., and the new diploma in public health.[58] Speaking to the Society of Public Analysts in 1895, he pushed hard the theme that those in positions of authority must recognize that what mattered was not whether analyses were done or what their results were, but what actions the public took in using certain waters. He presented a list of some thirty cases from the previous eight years in which typhoid-causing water had been judged acceptable by chemists. Bacteriological techniques were little better: at the time, searches for the relatively rare typhoid organism were characterized both by frequent false negatives (among those few bacteriologists who claimed to be able to distinguish the typhoid organism from similar organisms) or frequent false positives (among those who could not).

Under such circumstances it was extraordinarily dangerous to trust one's analysis, and since one could not trust one's analysis it became an arguable moral question whether there was ground for doing the analysis at all. The possibility that a misleading analysis could lead to an unwise recommendation and consequently to the loss of life had to be taken seriously, Thresh argued. In this light the bold statements of the elite London consultants were morally outrageous: these analysts were actively and unconscionably inciting public action with complete disregard of the consequences. Whereas the London consultants had worried that chemistry would cease to be taken seriously if their disputes over analytical processes and verdicts on water quality were made too public,[59] Thresh was horrified that chemistry was taken as seriously as it was. He told of a case at Worthing in which a London analyst had certified on the basis of a water analysis that a typhoid epidemic was over. The populace stopped boiling its water and the epidemic promptly reappeared.[60]

It is important to stress that Thresh and his followers were not trying simply to expose the inadequacy of analysis or to replace inferior methods with superior. The very idea of a formulaic solution, of a table indicating that a certain set of findings dictated a certain set of conclusions and recommendations, was precisely what they opposed. Instead they focused on the character, experience, and prudence of the person who was to have authority over water safety. Such a person had to be locally situated, had to be continually (though not exclusively) engaged in monitoring the quality of a certain supply, had to be adept in a great many means of assessing environmental quality and able to assimilate lots of different kinds of evidence to test hypotheses, and had to apply to public recommendations at least as much caution as he applied to matters affecting his own well-being.

These themes were echoed by contemporary chemists unassociated with the British public analysts. In his textbook on *Public Health Laboratory Work,* Henry Kenwood urged students always "to keep in view the fact that our work is essentially in the cause of disease prevention, . . . our duties and responsibilities do not end in returning an analytical report."[61] Cornelius Fox, in a manual of analytical processes for the use of medical officers of health, described his intent "to avoid a consideration . . . solely after the manner of an *analyst* who mechanically deals with chemical operations and arithmetical calculations, but [rather] to treat them as a physician who studies them in connection with health and disease."[62] In 1893, *The American Journal of Medical Sciences* published a paper by A. R. Leeds, a New Jersey chemist with long experience in water matters, on a "Question of Water, Ethics, and Bacteria" in which he directly addressed the moral dilemma that went with being uncertain yet in a position of authority. After twenty years as an analyst, Leeds wrote that he now felt "a more painful sense of this responsibility" than ever, "and am at times more embarrassed in arriving at a decision than when I made my first analysis."[63]

Conclusion

In these examples I have meant to raise policy issues, complicated and troublesome issues, that arise regularly in the historical record as well as in contemporary environmental issues where chemistry and other sciences are appealed to. The advantages of historical exploration of these sorts of issues are twofold: first, they make it clear that the sorts of problems that seem critical today are instances of general problems of making social decisions in inevitable conditions of epistemic uncertainty and of ignorance about the future. In a second-guessing and relativistic society like our own, we must get used to regarding uncertainty not as a pathological condition of decision making, to be patched up with more science as quickly as possible, but as the normal condition of decision making, for precisely the reasons outlined by Weinberg and Ravetz. History helps here simply by making this condition of existence more familiar.

Second, as we go back and encounter unfamiliar environmental disputes, it becomes harder to impose our own prejudices on the participants. Most of us, I suspect, see contemporary environmental disputes in terms of good guys and bad guys, those with vested interests to protect and those of us who are but the disinterested representatives of truth and justice. Had I lived in the 1780s, doubtless I would have sympathized either with Barker or with Bergman, depending on my cultural situation. But from a distance of two centuries one sees only that both had a moral and a rational case to plead, and likewise that neither was wholly disinterested in the outcome of the controversy. Recognizing that, the problems of contemporary policy-making look quite different. No longer should we assume that all scientists with any moral integrity at all will take the same position on an issue. If expert disagreement is a conflict of competing rationalities, each one truly rational in some strong sense, or of social and institutional settings and goals, or even of personalities, the problems of finding satisfactory solutions look much different, and there is a much greater burden on participants for respect, comprehension, discussion, and persuasion, rather than for forensic victory. This approach need not leave out the exposure of hidden biases or the need to redistribute power, but it need not be limited to these concerns either.

In terms of the history of chemistry, the field of what may loosely be called environmental chemistry is rich with important and largely unstudied questions. It is perhaps the key area for understanding the interaction between chemistry on the one hand and society and culture on the other. Historians of physics and biology have paid considerable attention to such questions—to the responses to Newton, Darwin, Einstein, and Bohr, and to the social origins and impacts of eugenics or nuclear energy. But in the history of chemistry the gulf between the internal agenda of discovery and conceptual change and the external agenda of institutional growth appears to have been broader. Yet I suspect that we will yet find significant and deeply seated cultural responses to Lavoisier, Liebig, and Pasteur, and manifold (and fascinating) instances of social tensions reflected in the work of modern chemists on such matters as nutrition and other aspects of agricultural and environmental chemistry.

Notes

1. Giandominico Majone, "Science and Trans-science in Standard Setting," *Science, Technology, and Human Values* 9, 1 (Winter 1984): 15–22; Brendan Gillespie, Dave Eva, and Ron Johnston, "Carcinogenic Risk Assessment in the USA and the UK: the Case of Aldrin/Dieldrin," in *Science in Context: Readings in the Sociology of Science,* ed. B. Barnes and D. Edge (Cambridge, Mass.: MIT Press, 1982), 303–335.

2. Alvin Weinberg, "Science and Transcience," *Minerva* 10 (1972): 209–222; J. Ravetz, "Scientific Knowledge and Expert Advice in Debates about Large Technological Innovations," *Minerva* 16 (1978): 273–277.

3. Mary Douglas, "Environments at Risk," in *Science in Context,* pp. 260–275; Douglas, *Purity and Danger: An Analysis of the Concepts of Pollution and Taboo* (London: Routledge & Kegan Paul, 1966); Mary Douglas and Aaron Wildavsky, *Risk and Culture: An Essay on the Selection of Technical and Environmental Dangers* (Berkeley: University of California Press, 1982), pp. 85–95.

4. Cf. Ferenc Szabadvary, *History of Analytical Chemistry,* trans. Gyula Svehla (Oxford: Oxford University Press, 1966).

5. Christopher Hamlin, *What Becomes of Pollution: Adversary Science and the Controversy on the Self-Purification of Rivers in Britain, 1850–1900* (New York: Garland, 1987); Hamlin, "Providence and Putrefaction: Victorian Sanitarians and the Natural Theology of Health and Disease," *Victorian Studies* 28 (1985): 381–411; Hamlin, "William Dibdin and the Idea of Biological Sewage Treatment," *Technology and Culture* 29 (1988): 189–218.

6. Mircea Eliade, *The Forge and the Crucible,* 2d ed. (Chicago: University of Chicago Press, 1978); Allen Debus, *The Chemical Philosophy,* 2 vols. (New York: Science History, 1977); Carolyn Merchant, *The Death of Nature: Women, Ecology, and the Scientific Revolution* (London: Wildwood House, 1982).

7. See Charles Walters, Jr., "A Message to and from NORM," *Acres USA* (March 1981): 20, 23–26.

8. See, however, Bill Luckin, *Pollution and Control: A Social History of the Thames in the Nineteenth Century* (Bristol: Adam Hilger, 1986); and Alain Corbin, *The Foul and the Fragrant: Odor and the French Social Imagination* (New York: Berg, 1986).

9. See Eliot Coleman, "Impediments to the Adoption of an Ecological System of Agriculture," in *Agriculture, Change, and Human Values, Proceedings of an Interdisciplinary Conference, 1983,* ed. R. Haynes and R. Lanier, 2 vols. (Gainesville: University of Florida Press, 1983), 2:584, where conventional agricultural science is compared to Ptolemaic astronomy; and Stan Rowe, "Summing it Up," in *Planet Under Stress: The Challenge of Global Change,* ed. C. Mungall and Digby J. McLaren (Toronto: Oxford University Press, 1990), pp. 331–332.

10. Douglas and Wildavsky, *Risk and Culture,* pp. 85–95; Paul B. Thompson, "Ethics, Risk, and Agriculture," in *Agriculture, Change, and Human Values,* ed. Haynes and Lanier, 2:533–535. Some of the more radical sociologists of science have begun to recognize the problem. See H. Rose, "Hyper-Reflexivity: A New Danger for the Counter-Movements," in *Counter-Movements in the Sciences: The Sociology of Alternatives to Big Science,* ed. H. Nowotny and H. Rose (Dor-

drecht: D. Reidel, 1979), pp. 277–289; and S. Woolgar, "Irony in the Social Study of Science," in *Science Observed: Perspectives in the Social Study of Science,* ed. K. Knorr-Cetina and Michael Mulkay (London: Sage, 1983), pp. 239–266; and S. Woolgar, ed., *Knowledge and Reflexivity* (Beverly Hills: Sage, 1989).

11. Interesting in this regard are several papers in Thomas L. Haskell, ed., *The Authority of Experts: Studies in History and Theory* (Bloomington: Indiana University Press, 1984), particularly Haskell's "Introduction," pp. ix–xxxix; and Magali Sarfatti Larson, "The Production of Expertise and the Constitution of Expert Power," pp. 28–80. Both pieces recognize that expertise disguises uncertainty and supersedes independent rational decision making. Both authors try, without much success, to imagine an expertise that would not do this. I consider this issue in *A Science of Impurity: Water Analysis in Nineteenth Century Britain* (Berkeley: University of California Press, 1990).

12. I seek here a historiographical analogue to S. Restivo, "Critical Sociology of Science," in *Science off the Pedestal: Social Perspectives in Science and Technology,* ed. D. Chubin and E. Chu (Belmont, Calif.: Wadsworth, 1989), pp. 57–70.

13. A. S. Taylor, *The Principles and Practice of Medical Jurisprudence* 2d ed., 2 vols. (Philadelphia: Henry Lea, 1873), 1:32.

14. In Barnes and Edge, *Science in Context,* p. 239. See also C. Hamlin, "Scientific Method and Expert Witnessing: Victorian Perspectives on a Modern Problem," *Social Studies in Science* 16 (1986): 485–513.

15. William Addison, *English Spas* (London: Batsford, 1951), pp. 137–144; Andre Guillerme, *The Age of Water: The Urban Environment in the North of France, AD 300–1800* (College Station: Texas A & M University Press, 1988); E. H. Guitard, *Le prestigieux passé des eaux minérales: Histoire du thermalisme et de l'hydrologie des origines à 1950* (Paris: Societé d'Histoire de la Pharmacie, 1951). Issues in this section are more fully considered in Hamlin, *A Science of Impurity* (note 11), chaps. 1–2; Roy Porter ed., *The Medical History of Waters and Spas, Medical History* supplement no. 10 (London: Wellcome Institute for the History of Medicine, 1990)

16. A. Debus, "Solution Analyses prior to Robert Boyle," *Chymia* 8 (1962): 41–61.

17. There literally are thousands. See W. H. Dalton, "A list of works referring to British mineral and thermal waters," *Report of the 52nd meeting of the BAAS, 1888* (London: John Murray, 1889), pp. 858–897; F. R. Peddie, *A Catalogue of Books to 1880,* s.v. "mineral waters"; Guitard, *Le prestigieux passé des eaux minérales*; H. Bolton, *A Select Bibliography of Chemistry, 1492–1892,* Smithsonian Miscellaneous Collections, no. 850 (Washington, D.C.: Smithsonian Institution, 1893).

18. D. Hartley makes the point that the rhetoric of spa literature changed in the mid-seventeenth century, with appeals to religion and politics giving way to appeals to chemistry. D. Hartley, "Religious and Professional Interests in the Northern Spa Literature, 1625–1775," *Bulletin of the Society for Social History of Medicine* 35 (1984): 16–17 .

19. George West Pigott, *On the Harrogate Spas and Change of Air: Exhibiting a Medical Commentary on the Waters founded on Professor Hofmann's Analysis,* rev. and enl. ed. (London: Churchill, 1856), p. 280. Hofmann's analysis occupies only one page of this 280-page book.

20. Addison, *English Spas,* pp. 106, 121.

21. F. Slare, *An Account of Pyrmont Waters,* dedicated to Sir Isaac Newton (London, 1717).

22. Charles Lucas, *An Essay on Waters* (London: A. Millar, 1756), 1:126.
23. Ibid., 2, pp. 217–218.
24. T. Bergman, "Of the Artificial Production of Cold Medicated Waters," in his *Physical and Chemical Essays,* trans. E. Cullen, 2 vols. (London: Murray, 1784), 2:263–264.
25. Bergman, "Of the Analysis of Waters," in *Physical and Chemical Essays,* 1:101.
26. Jon B. Eklund, "Chemical Analysis and the Phlogiston Theory, 1738–1772: Prelude to Revolution" (Ph.D. diss., Yale University, 1971).
27. John Barker, *Treatise on Cheltenham Water and its Great Use in the Present Pestilential Constitution* (Birmingham: Pearson, 1782), p. 2.
28. Ibid., pp. 7–9, 75–76.
29. The concept is developed in Weinberg, "Science and Trans-science."
30. See Morris Berman's description of the contrasting expert-witnessing practices of Faraday and Brande in *Social Change and Scientific Organization: The Royal Institution, 1700–1844* (Ithaca, N.Y.: Cornell University Press, 1978), pp. 152–155.
31. Charles Daubeny, "Report on the Present State of Our Knowledge with Respect to Mineral and Thermal Waters," *6th Report of the BAAS, Bristol 1836* (London: John Murray, 1837), pp. 45–46; Meredith Gairdner, *Essay on the Natural History, Origin, Composition, and Medical Effects, of Mineral and Thermal Springs* (Edinburgh: Blackwood, 1832), pp. 356–357; William Saunders, *A Treatise on the Chemical History and Medical Powers of some of the Most Celebrated Mineral Waters* (London: Phillips, 1800), p. 477.
32. R. Bud and G. K. Roberts, *Science versus Practice: Chemistry in Victorian Britain* (Manchester: Manchester University Press, 1984).
33. Key papers here are George Merck and Robert Galloway, "Analysis of the Water of the Thermal Spring of Bath," *Philosophical Magazine,* 3d ser., 31 (1847): 56–67; J. H. Gladstone, "On the Salts actually present in the Cheltenham and other Mineral Waters," *Proceedings of the BAAS for 1856,* secs. 51–52.
34. See Richard P. Aulie, "Boussingault and the Nitrogen Cycle" (Ph.D. diss., Yale University, 1968).
35. Cf. D. C. Goodman, "Chemistry and the Two Organic Kingdoms of Nature in the Nineteenth Century," *Medical History* 16 (1972): 113–130; J. Reynolds Green, *A History of Botany, 1860–1900* (Oxford: Clarendon Press, 1909), pp. 419–420.
36. C. Hamlin, "Robert Warington and the Moral Economy of the Aquarium," *Journal of the History of Biology* 16 (1986): 131–145.
37. J. F. W. Johnston, "The Circulation of Matter," *Blackwoods Magazine* 73 (1853): 560. In general on these issues, see Hamlin, "Providence and Putrefaction"; and Hamlin, *What Becomes of Pollution* (note 5).
38. F. Krepp, *The Sewage Question* (London: Longmans, 1867) attributed to this factor all wars, prostitution, intemperance, revolution, and the fall of the Roman empire. See esp. pp. 42, 50, 178, 203, 205. See also Graeme Davison, "The City as a Natural System: Theories of Urban Society in Early Nineteenth Century Britain," in *The Pursuit of Urban History,* ed. D. Fraser and A. Sutcliffe (London: Edwin Arnold, 1983), 349–370.
39. *Second Report of the Health of Towns Commission,* 1845, 18 [602], app. II, 67.
40. In disc. of W. Fothergill Cooke, "On the Utilisation of the Sewage of Towns by the Deodorization Process Established at Leicester, and in the Economical Application of it to the Metropolis," *Journal of the Royal Society of Arts*

5(1856–57): 56; cf. Justus Liebig, *The Natural Laws of Husbandry* (1863; reprint, New York: Arno, 1972), p. 180.

41. Ithi Kefalende [Charles Mansfield], "Hints from Hygea," *Fraser's Magazine* 41 (1850): 304.

42. Cf. "Living Beyond our Income," "The Economy of Nitrogen," *Quarterly Journal of Science,* n.s., 8 (1878): 146; W. Crookes in disc. of C. N. Bazalgette, "The Sewage Question," *Minutes of Proceedings, Institute of Civil Engineers* 48 (1876–77): 173. It is interesting that writers like Mansfield were also often advocates of alternative energy sources: tidal power, wind power, hydrogen fuel, etc.

43. William Corfield, *A Digest of Facts on Sewage Treatment* (London: Spon, 1871), p. 2.

44. G. Rafter, "Sewage Irrigation," United States Geological Survey Water Supply Paper, no. 3 (Washington, D.C.: Government Printing Office, 1897); N. Goddard, "Nineteenth Century Recycling: The Victorians and the Agricultural Use of Sewage," *History Today* (June 1981): 32–36; G. E. Fussell, "Sewage Irrigation Farms in the Nineteenth Century," *Agriculture* 64 (1957–58): 138–147; H. A. Roechling, "The Sewage Farms of Berlin," *Minutes of Proceedings, Institute of Civil Engineers* 109 (1892): 197–228.

45. J. W. Slater, *Sewage Treatment, Purification, and Utilisation* (London: Whittaker, 1888), p. 119. On Wigner, who was one of the founders of the Society of Public Analysts, see DNB 21, 197–198.

46. William Shelford, "The Treatment of Sewage by Precipitation," *Minutes of Proceedings, Institute of Civil Engineers* 45 (1875–76): 144.

47. Royal Commission on Rivers Pollution, Second Report. *The ABC Process of Sewage Disposal PP,* 1870, 33, [c.-180.].

48. E. A. Fournier d' Albe, *The Life of Sir William Crookes* (London: Macmillan, 1923), 257–270; William Crookes, *Twelve Months' Experience with the ABC Process of Purifying Sewage. A Letter addressed to a Shareholder in the Native Guano Co., Ltd.* (London: Chemical News Office, 1872); William Crookes, *The Profitable Disposal of Sewage. A Paper Read before the Congrès International d'Hygiene . . ., Brussels, 4 October 1876* (London, 1876); Crookes, *Report on Recent Trials of the ABC Process made by the Leeds Corporation* (London, 1876).

49. Crookes, *Twelve Months' Experience,* p. 5.

50. William Crookes, "Address of the President," *68th Annual Report of the BAAS, Bristol, 1898* (London: John Murray, 1899), pp. 3–38. It is intriguing the way in which the address combines three of Crookes's interests: the nitrogen problem, the latest research on radiation (i.e., x-rays), and spiritualism. Crookes found a greater unity among these topics than would most modern readers, I suspect.

51. Green, *History of Botany* (note 35), pp. 320–350.

52. "The Economy of Nitrogen," *Quarterly Journal of Science,* n.s. 8 (1878): 142–144; J. W. Slater, *Sewage Treatment* (note 45), p. 126.

53. See his participation in the discussion of C. N. Bazalgette, "The Sewage Question" (note 42), pp. 164–174.

54. J. R. Primack and F. von Hippel, *Advise and Dissent: Scientists in the Political Arena* (New York: Basic Books, 1974). For more recent treatments, see Haskell, *The Authority of Experts* (note 11); and Malcolm L. Goggin, ed., *Governing Science and Technology in a Democracy* (Knoxville: University of Tennessee Press, 1986).

55. These issues are explored more fully in Hamlin, *A Science of Impurity* (note 11), chaps. 8–10, and conclusion.

56. Major Charles Smart, part of a U.S. National Board of Health team that compiled the most searching report on the capabilities of British analytical processes, somewhat ingenuously assumed that when a British chemist pronounced a water "good," he was simply translating a number obtained in analysis into a relative standard; that is, that the sample scored better on some test than most other samples. But Smart worried, and quite rightly, that "adjectives [might] become in many instances the expression of an actual opinion." Charles Smart, "On the Present and Future of Sanitary Water Analysis," *Reports and Papers of the American Public Health Association* 10 (1884): 82–83.

57. F. D. Barry, "Enteric Fever in the Tees Valley," in *21st Annual Report of the Local Government Board. Supplement Containing the Report of the Medical Officer for 1891, PP,* 1893–94, 42, [c.-7054], vii, 58–59, 119–136.

58. "John Clough Thresh," *The Analyst* 57 (1932): 549–550.

59. Charles Meymott Tidy, "Processes for determining the organic purity of Potable Waters," *Journal of the Chemical Society* 35 (1879): 46.

60. J. C. Thresh, "The Interpretation of the Results Obtained upon the Chemical and Bacteriological Examination of Potable Waters," *The Analyst* 20 (1895): 80–91, 97–111.

61. Henry R. Kenwood, *Public Health Laboratory Work* (London: H. K. Lewis, 1896), p. 25.

62. C. B. Fox, *Sanitary Examinations of Water, Air, and Food: A Vade-Mecum for the Medical Officer of Health,* 2d ed. (London: Churchill, 1886), p. x.

63. A. R. Leeds, "A Question of Water, Ethics, and Bacteria," *American Journal of Medical Sciences* 105 (1893): 259. See also Emile DuClaux, "Les microbes des eaux," *Annales de l'Institut Pasteur* 3 (1889): 569.

The Chemogastric Revolution and the Regulation of Food Chemicals

Suzanne White

Following World War II, America's food and chemical industries expanded greatly in search of new customers and sustained profits as part of a much larger and broader pattern of postwar changes. Accompanying the chemotherapeutic revolution in medicine that accelerated with the discovery of penicillin and other antibiotics, a parallel revolution took place during this period in the nation's food supply. New agricultural practices, changing food manufacturing processes, new food additives, potent new pesticides, and chemically sophisticated packaging all converged, sparking what one historian has termed a "chemogastric" revolution.[1]

These widespread changes in the American food supply were indicative of some of the profound societal changes that were taking place simultaneously. Grounded in the growing prosperity that followed the Depression and World War II, and fed by the postwar "baby boom," this chemogastric revolution was created by chemical and technological innovation on the one hand and consumer demand and advertising on the other, and was closely associated with an evolving suburban lifestyle. It was in the broad context of this midcentury revolution that a new public policy toward food additives was forged and that controversy over the regulation of chemical carcinogens, which has dominated the second half of the twentieth century, emerged. This new policy reflected a profound change from the food policies forged by Harvey Wiley and the Bureau of Chemistry in the early twentieth century following enactment of the Progressive era's Pure Food and Drugs Act. The 1906 law had drawn clear distinctions between natural and nonnatural food ingredients, holding added ingredients to a more stringent standard of safety. In doing so, the law reflected society's traditional antipathy toward food adulteration and commercial deception. The new legislation enacted in the midst of the chemogastric revolution seemed to reflect a new acceptance, or at least tolerance, of nonnatural, "chemical" ingredients, in choosing to regulate rather than to restrict their use in foods.

Historians of chemistry looking at environmental science and public policy issues in the twentieth century have been faced, as Christopher Hamlin has pointed out, with adopting one of two opposing perspectives. One viewpoint posits scientific knowledge, and chemistry in particular, as a guide to correct environmental action, while the other maintains that science plays a smaller role in regulation and policymaking than do conflicting political and social values, interests, and power. Hamlin wisely argues for a broader, long-range perspective on environmental issues that encompasses both perspectives and that assumes that "negotiating through uncertainty" is the norm rather than the exception, historically as well as currently.[2] To this wider perspective I would add that negotiating through change—industrial, social, governmental, and scientific—also is the norm rather than the exception. Indeed, in individuals no less than in society, it is during periods of rapid change that tensions mount, perceptions of power become heightened, and conflicting values are more obvious, while political solutions to vexing public-policy issues may reflect the influence of scientific uncertainty more than scientific knowledge.

The chemogastric revolution was predicated on a new association between elements of the chemical industry and elements of the foods industry. The nature of these associations, however, has not yet been studied in depth by twentieth-century historians. In his latest book, *Scale and Scope: The Dynamics of Industrial Capitalism,* Alfred Chandler, Jr. does suggest that both the chemical and the food industries shared an extensive diversification into branded consumer goods, which accelerated following World War I. During the 1920s, the major food and chemical companies all invested heavily in research and development laboratories. Chandler indicates, however, that food companies continued to market established products, while the consumer chemical manufacturers were more innovative and concentrated on the development of new items. Proctor and Gamble, for example, developed washing soaps (Ivory Flakes) and cooking oil (Crisco). Chandler concludes though that "neither Procter & Gamble nor its competitors attempted to move beyond soap and cooking oil. Their substantial investments in high-volume, low-unit-cost systems of production and distribution for these related products deterred them from going further afield, at least until well after World War II."[3] At least some of the new chemicals sold for use in food products from the 1930s on were developed from chemicals pioneered not by the consumer chemical companies but by industrial chemical and munitions manufacturers.

From an economic perspective, the boom in food chemicals following World War II was the result of a profitable engagement between the nation's largest and most important industry—foods—and its most technologically sophisticated and fastest growing industry—chemicals.[4] Both industries were also positioned to profit from the postwar demographic profile. Between 1938 and 1952, the U.S. population grew by 25 percent, and between 1945 and 1960 it increased by almost 40 million, the largest growth spurt since the peak years of immigration in the first decade of the twentieth century.[5]

In part because of population growth and urbanization, but also because of a change in the tastes and purchasing patterns of consumers, the American processed-food industry, already well established at the turn of the century, experienced a sustained period of growth following World War II. Processed-food sales increased from $63.4 billion to $82.1 billion between 1951 and 1960, while the total assets of food manufacturing corporations more than doubled between 1947 and 1962, from $10.8 billion to $22.3 billion.[6] The American chemical industry had not come into its own until the 1920s, but it was stimulated and sustained by war throughout the twentieth century. During World War II, chemical technology made possible new products such as light metals, synthetic rubber, anti-malarial drugs, jet fuels, nuclear isotopes, and synthetic fats. After the war, ongoing nuclear research and the new space exploration continued to feed this new technology, but domestic uses also burgeoned for synthetic fibers such as nylon, rayon, and dacron; new plastics, dyes, cosmetics, drugs, fertilizers, and pesticides, all of which ushered in a new chemical age for consumers epitomized in Dupont's classic slogan "Better Living Through Chemistry."[7] The American chemical industry was in a state approaching pure competition following World War II, with no one company maintaining a clear advantage over its competitors.[8] Research and development therefore became the key to continuing success.

No one displayed "better living" better in the postwar period than the new suburbanites. Middle income families with rising expectations began an exodus to the suburbs in ever increasing numbers, reaching a peak between 1948 and 1952.[9] The postwar American dream centered around the home in an era described as the most domesticated period in American history.[10] Although thoughtful critiques of American suburban life in the 1950s alternately idealize and vilify the changes it wrought, almost everyone writing about the suburban phenomenon agrees that consumption was both a change and a cornerstone of the new lifestyle.[11] From station wagons to Volkswagens, telephones to televisions, and power mowers "to give crew cuts to handkerchief-sized lawns" to power tools, the suburbs were materialism's mecca. Suburban kitchens, especially those "built in," sported a more affluent look, evident not only in new labor-saving appliances such as freezers and dishwashers but also in new foods.[12]

Food shopping practices changed dramatically as well. Middle class housewives of the 1930s had often telephoned their orders to grocery clerks, who filled orders largely with their own choices of brands and grades and delivered them to the customer's home.[13] Suburban shoppers, in contrast, did their own shopping and went in person to select goods. Food producers, seeking their fortune in the suburbs, carefully studied and surveyed their new consumers. Frustrated in its attempts to draw simple conclusions, one packaging company characterized the "average" American housewife of the period as simply unpredictable, noting: "She's a woman with $20, 20 minutes, a second child waiting in the car and a third in escrow. Her shopping list is incomplete at best and many of her buying decisions are made in the seven or eight seconds it takes her to pass the counter where the items are displayed. Some call it *impulse buying*."[14]

"Convenience foods" for both humans and their pets, prepared mixes for cakes and other baked goods, instant coffee, and frozen foods all gained popularity throughout the 1950s in spite of the higher prices charged for them.[15] In the early years after the war, however, the term "convenience foods" encompassed a limited variety of products. During the late forties and early fifties, marketing experts used the term "convenience foods" primarily to distinguish them from "shopping goods." According to the American Marketing Association in 1948, convenience foods referred merely to goods that the consumer bought by description, including brand and grade, rather than by inspection. Shopping goods were bought by examining the quality of the products themselves. Where convenience foods at that time involved "only a small amount of money," and brands were less important than availability, shoppers carefully compared prices among different shopping goods.[16]

By the late sixties and early seventies, however, the term "convenience food" had become as much a popular term as a marketing term. Acceptance of many of the new precooked and prepackaged products had engendered new ideas about food preparation.[17] Like the proverbial blind man and the elephant, however, consumers most often appreciated only part of the elephant rather than the whole. "Convenience foods" connoted abundance and greater ease of preparation to some consumers, including the growing number of women who worked outside the home, while to others they denoted overpriced, inferior products, stripped of flavor and food value.[18] Consumption of such foods nonetheless rose from seventeen pounds per person in 1949 to forty-five pounds per person in 1955, as differences between the prewar and postwar pantry became more striking to producers and consumers alike.[19]

Food forms and food packaging changed with the appearance of "heat and eat" foods, "boil-in-bags," and ready-to-serve foods, many of which had been developed during the war for the armed forces.[20] New flavors were introduced into processed foods, especially desserts, and during this era the standard package shrank in size to accommodate the buying practices of smaller suburban families.[21] Foods in pills and capsules, instant foods, and push-button foods were popular novelties, widely believed to be the harbingers of the future. Meanwhile, modified and easy-to-prepare versions of rice and tapioca, such as Wonder-Rice, and Minute Rice, and Minute Tapioca, were beginning to create national culinary interest in regional staples whose popularity had been on the wane.[22] The apparent yet unpredictable willingness on the part of consumers to try new products and pay more for foods that had undergone a great deal more processing than in the past spurred the food industry to new levels of innovation, both in the making and in the marketing of new, chemically sophisticated products.

At the heart of the chemogastric revolution was the coming of age of the frozen foods industry, which emerged as one of the leading food innovators.[23] Schoolboy Johnny no longer came home from school asking "What's cooking?" but rather "What's thawing?" according to a popular anecdote of the period.[24] Until World War II, frozen foods had been a luxury item, available only in stores serving wealthier patrons.[25] Quality was uneven, and the industry had been plagued by persistent rumors that frozen products lacked the vitamins contained in fresh products.[26] Viewed largely as an aristocratic offshoot of the canning industry, the frozen foods business did not begin to gain ground in middle-class markets until World War II, when, in the words of one promoter, "Tin went to war; frozen foods stayed at home."[27] Boasting that frozen foods could help win the war by eliminating waste, conserving not only steel and tin but precious space in railroad cars as well, industry leaders convinced Washington officials that theirs was not a "silk-stocking" product, but rather an underappreciated one. Steel freezers, constructed with increased metal allocations, began to appear in large chain grocery stores. Recruiting distributors from the swelling ranks of unemployed automobile salesmen, the frozen foods industry began to gain a national market. Corn, spinach, soups, apricots, carrots, and coffee all became casualties of tin and a boon to frozen foods during the war. The point rationing system, moreover, provided free advertising for the fledgling industry because posters listing the point values for frozen foods were on display whether or not the store actually carried such wares.[28]

During the war only a fraction of the frozen foods packed reached home markets. Most of these were simple fruits and vegetables, although some more complicated prepared foods, not subject to rationing, were also marketed. Promoters anticipated a postwar frozen foods bonanza.[29] To some extent their optimism was well founded. Farmers, for example, profited enormously from the increased demand for Southern vegetables (yellow squash, whole okra, turnip greens, and black-eyed peas), California strawberries, lima beans, and frozen poultry in the national market.[30] Overproduction, however, soon became a recurrent problem in the new industry. In 1946 alone, more than 860 million packages remained unsold, and of every ten plants opened, only three to four stayed in business.[31] This enormous competitive pressure stimulated a fierce battle among the brand names—Libby, Snow Crop, Stokely's, and others—for freezer space in retail groceries. By 1948, the vast majority of fly-by-night operators and small packers had been forced out of the frozen food business.[32]

The Florida "gold rush" in orange juice, launched in 1948 when Bing Crosby signed his first Minute Maid contract, rendered all other industry successes pale by comparison.[33] Initially advertised as being cheaper than whole oranges and better for one's health, frozen orange juice was an immediate success, and more than any other single product it helped establish and stabilize the market for frozen foods. Frozen juice concentrates not only eliminated the "summer slump" in frozen foods, they boosted the price of Florida real estate sky-high as well. As early as 1950, orange juice concentrate absorbed 25 percent of the Florida orange crop, a figure that grew to 70 percent by the end of the decade.[34] Orange juice became one of the first truly national convenience foods.

Processed-food manufacturers also anticipated a postwar boom in frozen foods. They discovered quickly, however, as they began to experiment with products in frozen format, that not every food could be successfully frozen. Simple vegetables and fruits proved far easier to freeze than more exotic and complicated dishes. There were serious problems in developing quality products, and consumer demand proved to be capricious: not so high as anticipated for some products, and inconceivably high for others.[35] Experiments with Neapolitan cakes, cherry flips, biscuit tortes, lemon and chocolate pies, Scotch scones, pecan muffins, crepes suzette, apple brown betty, and other fancy baked goods were often short-lived, in spite of a barrage of coupons and sales promotions.[36] Sara Lee cheesecakes, on the other hand, were popular from the start, and their producer soon expanded production into other high-quality desserts.[37] Some companies began to experiment with ethnic specialties such as Mexican hot tamales, chili con carne, and beef enchiladas, along with Chinese, Italian, and kosher foods. The baking industry in particular experimented heavily with frozen goods, hoping that freezing would allow holiday baking to be completed on straight time rather than overtime and hoping to counter the industry-old problem of stale merchandise as well.[38] Frozen waffles and French toast were notable successes. So was frozen breaded shrimp. William Millis, who reportedly "liked fried shrimp more often than his wife wanted to prepare it," experimented with breading and freezing, and after investing $5,000 in capital was able to launch a stable $10 million business.[39] Breaded (and often deep-fried) shrimp, onion rings, and fish all became popular frozen foods, in turn stimulating the development of commercial breading mixes. The variety of prepared foods available was remarkable by almost anyone's estimation.

Successful freezing often required altering recipes. Although food manufacturers would later stand accused of using chemicals merely to increase production, reduce cost, retard spoilage, and increase shelf life, chemicals were also used to improve the appearance, taste, texture, and nutritional value of foods during this period of mass experimentation.[40] Prepared foods, whether packaged or frozen, often required a variety of alterations in order to make them acceptable to consumers. Packaged cake and pudding mixtures and frozen dinners, to cite one of the most persistent problems, suffered a loss of flavor during processing, which a later generation would refer to as "blandness." Frozen foods were subject not only to the drying effects of freezing, but also to the temperature fluctuations common in early food freezers.[41] Food scientists, seeking to compensate and overcome these and innumerable other technical problems in food production, experimented with a variety of new ingredients, new forms of packaging, and the use of various chemical additives. In frozen prepared foods, for example, flavor enhancers such as monosodium glutamate became nearly ubiquitous.[42] In packaged pudding and cake mixes where volatile flavor oils quickly evaporated, a solution was achieved using gums from plant cellulose—methylcellulose. The gums contributed no flavor or nutritive value to the food but were used to coat flavoring ingredients, preserving them intact until the heat employed in cooking and baking dissolved the coatings.[43]

However successful they may have been in solving technical problems in food production, unfamiliar ingredients, with chemical-sounding names such as monosodium glutamate and methylcellulose, were largely enigmatic to consumers. Consumers, commercial competitors, and regulators alike, moreover, had begun to pay more attention to ingredients following enactment of the 1938 Food, Drug, and Cosmetic Act, which required that all ingredients in a nonstandardized food product be listed on the label of the product along with any preservatives, artificial colors, and flavors.[44] The new law had also provided the U.S. Food and Drug Administration (FDA) with the authority it had sought since the turn of the century to establish standards of quality and identity for foods.[45]

Although the nature of standard setting changed over time, standards were historically sought for large-volume staple foods, such as canned fruits, jams and jellies, ice cream, flour, bread, and canned and condensed milk.[46] Consumers benefited from standardization because with contents and quality assured, products could compete on the basis of price. Prohibiting the use of cheap substitutes protected both larger and smaller producers from unfair competition. Standards hearings had begun to address questions about what new ingredients were to be called; how they were to be listed on the product label; their chemical composition; and appropriate and inappropriate uses, when World War II interrupted the standards-setting process. Many food standards were left unformulated or incomplete, while final standards for others, such as those for commercial white bread, were held in abeyance during the war. After the war, as new products such as frozen orange juice and frozen breaded shrimp successfully entered the market, standards were established for them, but the backlog of foods for which standards would be required increased dramatically as a result of the chemogastric revolution, and foods with unfamiliar ingredients appeared with increasing frequency on supermarket shelves.

The chemogastric revolution in foods, as well as the chemotherapeutic revolution in drugs, forced the FDA, whose scientific roots lay in analytical rather than biological chemistry, to confront a brave new regulatory world after the war. The most immediate goal was to reestablish customary peacetime vigilance over the food supply. Competition and innovation in the food field presented one set of scientific challenges at war's end, while inexperience and carelessness by food producers created a different set of regulatory concerns. As new processors and packers entered the food business, problems with misbranding, adulteration, and the use of poisonous and deleterious ingredients multiplied. Typical of some of the new problems in regulating food chemicals were those encountered at the end of World War II with monochloracetic acid and thiourea.

Monochloracetic acid was a European discovery, patented by the French in 1933. A small amount had been found useful in preparing fruit juices, beer, wine, and soft drinks, because it permitted pasteurization at lower temperatures, which, in turn, preserved the clarity of the drinks. Originally, pharmacologists had concluded that monochloracetic acid was harmless. Later studies, however, showed that the substance was irritating to the gastrointestinal tract.[47] In spite of the conflicting pharmacological evidence, the chemical was offered for sale, and the manufacturer of a synthetic orange beverage used it as a "stabilizing" agent. During World War II, soldiers in a southern training camp were seized with an epidemic of nausea and vomiting that was attributed to the orange drink. Stabilizers, unlike preservatives, did not have to be identified on the product label. FDA officials suspected that the manufacturer valued its preserving characteristics more than its stabilizing properties, but the acid was a difficult ingredient to identify chemically in any case.[48] Yet it was not difficult to convince the courts that a problem existed with the chemical, and between 1945 and 1947 hundreds of seizures were instituted, not only of the particular orange drink in question but also of other products that used monochloracetic acid as an ingredient, including other fruit beverages, pancake syrups, chocolate flavoring syrups, salad dressings, and chowchow relishes.[49] The seizures hit the beer and wine industries hardest, and some smaller wineries, which lost an entire year's production, were reportedly put out of business.[50]

Thiourea presented a more complicated chemical problem. It too had been discovered earlier in the century and was patented in 1937 and 1943 for the prevention of browning in cut apples, peaches, and pears before ascorbic acid was found to be effective for the same purpose.[51] A chemical cousin, thiouracil, moreover, had been investigated as a drug. In 1941, Johns Hopkins University researchers experimenting with it in the treatment of thyroid disorders concluded that it inhibited the body's production of thyroxin. But in clinical trials for hyperthyroidism the drug proved unpredictable, and experimental animals were found to vary greatly in their sensitivity to it.[52] Promoters and chemical advertisers, meanwhile, continued to tout the virtues of "the literally magic chemical thiourea," as an antioxidant in cut fruit. In 1946 the FDA instituted a seizure against frozen peaches that had been treated with excessive amounts of thiourea and had reportedly caused illness, and found that portions of the frozen peaches fed to laboratory rats killed them. FDA scientists concluded that the substance was poisonous and unpredictable even in small amounts. More alarmingly, it produced liver tumors and enlarged thyroids in rats fed the substance in small amounts over an extended period: its adverse effects seemed to bear little relationship to the dose.[53] The Florida Citrus Commission, responding to the booming demand by the frozen foods industry for juice oranges, experimented with thiourea and discovered that it was very effective in preventing stem end rot and effectively inhibited blue and green mold in oranges. In response, FDA food chemists and pharmacologists jointly developed a method to determine whether the substance as used commercially would penetrate the skin of the orange. They documented that thiourea indeed penetrated into the juice of the orange, because they found that when the juice was fed to laboratory rats a single dose of orange juice containing 45 parts per million of thiourea killed two out of six adult rats. The citrus industry immediately abandoned use of the chemical on fruit.[54]

The Florida Citrus Commission was just one of a growing number of groups, trade associations, companies, toxicologists, and academic scientists that had begun to seek advice from and consultations with the Food and Drug Administration regarding new chemicals and toxicological testing methods after World War II. Conferences between regulated industry and the FDA increased exponentially between 1945 and 1955.[55] Food chemists continued to test products and devise new analytical testing methods for many of the new chemicals, but the biological chemists in the FDA's Division of Pharmacology bore the responsibility for passing scientific judgment on the suitability of new chemicals for use in and around foods. The division pioneered the development of regulatory methods for appraising the relative toxicity of new chemicals and performed the earliest evaluations of consumer chemical risk in the immediate postwar period.

The FDA's Division of Pharmacology had been established in 1935. Erwin E. Nelson, a prominent University of Michigan pharmacologist, was loaned to the FDA to recruit promising researchers and later joined the staff himself. A former pharmacology professor, Herbert Calvery, headed the new division staffed by biochemists, pharmacologists, and pathologists. Described by scientific colleagues as a "gung-ho outfit" of excited people working in a frontier area, the division was originally charged with devising bioassays to assure the proper strength and purity for newer biological products, especially glandular preparations, and with conducting studies of the chronic health effects of lead and arsenic insecticide residues.[56] Animal studies were a crucial component of both studies.

The division's initial tasks, in seemingly unrelated ways, focused on two key types of biological testing—acute and chronic toxicity. Seeking methods that they could use to establish standards for glandular preparations, the scientists refined and statistically adapted a biological method used first by Alfred Trevan, a British biologist in the 1920s.

Trevan used a measure of the LD-50 (Lethal Dose—50 percent) to standardize biological potency in digitalis, insulin, and diphtheria antitoxin.[57] Preparations that met the established standard would result in the death of 50 percent of the test animals at the designated dose.[58] The scientists also gained valuable experience in conducting low dose, long term toxicity studies through their insecticide work.

In 1937, after funding for the insecticide work had been cut, the Elixir Sulfanilamide crisis changed the focus of the division's work. What happened, in brief, was that a liquid dosage form of sulfanilamide was hastily marketed by a southern drug firm. The company had employed a new solvent in the drug's formulation, the industrial chemical diethylene glycol. Neither the solvent nor the drug was tested in animals or humans before the product was sold. The preparation killed over one hundred people, mostly children, and public reaction to the disaster was singlehandedly responsible for the requirement, enacted in 1938, that the FDA oversee the safety of all new drugs prior to marketing.[59] In the midst of the crisis, it was clearly only that the preparation had caused renal failure in its victims. Several scientific groups around the country scrambled to establish that the diethylene glycol was actually the chemical culprit, since sulfanilamide itself could also affect the kidneys. The Division of Pharmacology used its LD-50 statistical methods to plot comparisons between the toxicity of sulfanilamide and that of diethylene glycol, showing conclusively that the solvent was the deadly ingredient. Other laboratories corroborated the results of their testing.[60]

After the immediate crisis had passed and all the Elixir Sulfanilamide had been recovered and destroyed, FDA inspectors looked more closely into the use of diethylene glycol, discovering that it had also been promoted as "ideal" for household food flavorings and extracts, and had been advertised to keep the flavors from "freezing out" of ice cream and "baking out" of cakes.[61] As the Division of Pharmacology continued its research into the glycols, the scientists reinforced their growing conviction as to the importance of lifetime animal studies in identifying long term problems with a substance.[62] Small amounts of both ethylene and diethylene glycol given over a rat's lifetime (about two years) did not kill quickly, as did larger doses, but nonetheless produced identifiable adverse effects including what were considered to be characteristic lesions, kidney and bladder stones, and tumors.[63] In a pioneering article published in 1939, Edward Laug, Herbert Calvery, Herman Morris, and Geoffrey Woodard of the Division of Pharmacology summarized their glycol work and their use of the LD-50 statistical method to compare the acute toxicities of different substances.[64]

During the 1940s, the group continued to expand their use of the

LD-50 as a measure of acute toxicity, and to refine their methodology for studying chronic toxicity, concentrating on studies of preservatives, artificial sweeteners, and surface-active agents.[65] They also began to explore the toxicity of chemicals that came into contact with the skin and mucous membranes. In 1939, John Henry Draize joined the Division of Pharmacology to help implement the FDA's new responsibilities for cosmetics under the 1938 Act, and address scientifically growing concerns about the safety of new ingredients in hair dyes, shampoos, and soaps. Draize, a pharmacologist who had done pioneering work in dermal toxicity at the chemical warfare center at Edgewood Arsenal in Maryland, worked with others throughout the forties in developing animal testing procedures, primarily with rabbits, to study skin and eye sensitivity to new cosmetic ingredients. The testing methods they developed, the so-called Draize tests, were valuable not only in evaluating the potential harmfulness of a new substance but also in helping to identify sensitizing chemicals—chemicals that seemed harmless upon first application but could cause severe allergic reactions after a later application.[66]

As a result of its pioneering methodologies and added drug and cosmetic responsibilities under the new law, the Division of Pharmacology was well positioned during the 1940s: publishing, consulting with industrial and academic colleagues, and providing guidance and training to toxicologists setting up university, private, and corporate testing laboratories.[67] Neither an industrial nor an academic laboratory, the Division of Pharmacology established itself in its regulatory capacity as an adviser to both and as an acute observer of the chemical landscape. A mandatory clearinghouse for new drugs, it soon became an unofficial clearinghouse for other new chemicals as well, including food additives, new packaging materials, pesticides, drugs, and new cosmetic ingredients.

The scientific and regulatory challenges created by the chemogastric revolution beginning in the late 1940s are evident in the swelling correspondence and chemical files from the Division of Pharmacology, and echoed in the oral histories of FDA pharmacologists.[68] O. Garth Fitzhugh, for example, recalled sitting down with the head of the division in its early years and quickly coming up with a list of the new chemicals deserving careful study. The list contained less than a dozen items, all important, but after these had been studied they had no others in mind. "After the upsurge came on," however, Fitzhugh recalled, "I began to make cards of all the materials that came to the Food and Drug Administration. I'm sorry I stopped it, but I did stop it when it got up to 10,000."[69]

From large chemical companies who consulted with FDA pharmacologists in the final stages of multigenerational testing, accompanied by staffs of research scientists and attorneys, to budding entrepreneurs who had heard that the FDA might be interested in their new chemical ventures and had performed no tests at all, an element of enlightened self-interest did lead many to seek advice and assistance during this period. In particular, they sought assurances that the government would not seize their products as being "poisonous or deleterious to health." Acute toxicity testing, therefore, was widely recognized as both necessary and prudent. Adhering to a "hundredfold margin of safety" in evaluating new chemicals, the pharmacologists considered an additive "safe" if it produced no observable adverse effects or pathology when fed in daily doses a hundred times larger than the largest daily dose expected for humans.[70] Companies were less convinced of the need for chronic toxicity testing, especially in the competitive climate which prevailed in both the food and the chemical industry at war's end. Because testing was expensive and competitors were not required by law to do likewise, lengthy long term studies were often considered a tedious luxury.[71]

FDA scientists also found it increasingly difficult to encourage restraint and promote adequate testing given the sheer volume of new chemicals available to food producers. The Division of Pharmacology itself often attempted to dissuade correspondents from the use of one chemical ingredient by suggesting another, albeit one with a lower LD-50 calculation. Indeed, during this new chemical bonanza the LD-50, which had been developed for biological standardization, became routinized as a mass screening tool; first by an agency swamped with new substances requiring testing, and more and more by the regulated industries themselves, once both drugs and food additives were subjected to premarketing approval. The LD-50 constituted a new shorthand scientific method for separating more from less hazardous materials, and its use spread to regulatory bodies around the world during the next decade.[72] Its utility in identifying immediate hazards and in facilitating comparisons among chemical choices was undeniable.[73] For a subsequent generation, however, the LD-50 became a controversial symbol of science gone amuck assuming that every new chemical was a potential penicillin.

Even during the late 1940s, dissatisfaction was voiced about the growing proliferation of chemicals in the food supply. Concerns about the safety and overuse of many of the new organic pesticides, insecticides, and rodenticides were among the most serious.[74] The Division of Pharmacology was particularly outspoken about the peacetime use of DDT as an insecticide, and the runaway popularity of the more acutely toxic persistent pesticides aldrin and dieldrin.[75] The division had tested DDT for the Army and concluded that, although it indeed had a low order of toxicity, its long term effects were uncertain, particularly because it was stored in human fat. Expediency in wartime, the scientists warned, was no justification for its widespread postwar use, for during an illness dangerous effects might appear when the body drew on its fat reserves and DDT stores were released.[76] At least a quarter of the Division of Pharmacology's correspondence during the late forties warned of insecticide dangers.

In 1947 the *Food, Drug, and Cosmetic Law Quarterly* carried an article by W. B. White, the Chief of FDA's Food Division. White called attention to the growing problem of chemicals in foods and observed that the food industry, like the pharmaceutical drug industry, seemed to be moving in the direction of specialization and increasing complexity. Foods, he warned, should not be permitted to pose the same calculated risk as some beneficial drugs. White advocated an "ecological approach" to the complex problem of multiplying uses for the multitude of new chemicals. He insisted above all that the job was too big for either government or science to solve on its own, and that industry would have to assume its share of the burden.[77] At its December 1949 meeting, the Food and Nutrition Board of the National Research Council agreed to establish a committee to offer scientific advice on the newer chemicals in foods.

Congressman Frank Bateman Keefe, a Wisconsin Republican, also became interested in a growing list of food chemical problems in the late forties. His work with agency appropriations and his oversight responsibilities in the House of Representatives put him in close touch with the FDA and its leaders, particularly Commissioner Paul B. Dunbar, with whom he shared a mutual interest in public health issues. A strong supporter of public health initiatives during his service in Congress from 1939 to 1951, Keefe was one of a group of Congressional leaders who helped secure funding for scientific research projects to investigate the causes and treatment of cancer, arthritis, and heart disease. Keefe also worked toward the expansion of the National Institutes of Health in the 1940s.[78] Keefe's initial interest in the chemogastric revolution and its attendant problems, however, began closer to home: the hearings to set standards for white bread.

In 1941, the FDA held its first hearings on the subject of bread, attempting to set national standards of identity for white bread, which according to regulatory observers had become "the most popular target for modern chemical experiments."[79] Issues raised at this hearing centered around the approval or disapproval of a number of substances that had begun to be employed in commercial bread baking operations. Basically, the bread makers argued that theirs was an ancient and honorable art, and that they should therefore have all the ingredients they wished at their disposal. The government countered that whereas bread making might once have been an art, modern mechanization had brought it much closer to a science by 1941, and that in the interests of consumers bakers should be able to specify what was being used in their products, in what amounts, and for what purposes.[80] Government witnesses expressed concern that bread makers had been sold ingredients that were not only unnecessary but wholly foreign to the very definition of bread.[81] In the end, many of the so-called chemical "bread improvers" and "dough conditioners" were disallowed.

Two kinds of emulsifiers were permitted in the final standards. One was soy lecithin, also used in chocolate, which had been manufactured in the United States since 1935 by the American Lecithin Company.[82] Favorable testimony declared that it had been found to "promote better shortening dispersion, increase fermentation tolerance and gas retention properties as well as improve the internal characteristics of bread."[83] The second emulsifying ingredient was more controversial. Procter and Gamble, manufacturer of Crisco to the household trade and Sweetex to the baking trade, wanted to be certain that their shortenings to which mono- and diglycerides had been added were included in the standards.[84] These special shortenings had been on the market since 1933 and were also manufactured by Swift, Armour, Lever Brothers, Wilson, Durkee Famous Foods, and others, including the South Texas Oil Co.[85] Witnesses testified that through their emulsifying action the shortenings imparted tenderness and increased volume to most yeast-raised products. The government's attorney tried to argue that the product's approval was not in the interests of consumers because the product provided relatively little advantage in bread as opposed to other baked goods and that it cost more than ordinary shortening.[86] Procter and Gamble's attorney countered that mandating oleomargarine and excluding butter from the standards would not hurt the public interest either but would be a "rank injustice" to the dairy industry.[87] There were also disputes as to the proper labeling of the enhanced shortenings. Scientists testified that mono- and diglycerides were natural components of fat, which had merely been concentrated in the final product. At the suggestion that the word "glycerinated" be used, Procter and Gamble produced a survey showing that consumers understood the term to mean that the "product contained hand lotion or explosives."[88] In the end, both lecithin and the hydrogenated shortenings were allowed as optional ingredients in white bread. The hearings themselves took 26 days and generated a transcript of

4,162 mimeographed pages. Then, to promote flexible war management, the final bread standards were not issued.

During World War II, artificial enrichment of both flour and bread began, so that standards for both white bread and enriched white bread were needed after the war.[89] In 1948 the hearings were reopened to consider revisions in the original proposed standards, including the addition of several new optional ingredients. When the hearings ended in September 1949, the published transcript exceeded 17,000 pages with 400 exhibits.[90] Unlike the 1941 hearings, which were dominated by the government's attorney, the 1948 hearings were so crowded with industry lawyers that it became difficult at times to separate substance from semantics. Even before the hearings formally reconvened, government and industry representatives were voicing concerns about the sheer number and variety of chemical additives being proposed as optional ingredients in the bread standards. The counselor for the American Baking Association warned its members prior to the hearings that the government officials with whom he had spoken were concerned about the "impressive list" of chemicals used by bakers. He suggested that they might reconsider asking for admission of some of the chemical ingredients, which included twenty-four different bread softeners and a flavoring agent called "pick up" or "hypo" that contained sodium thiosulfate and in a different context was used to develop photographs.[91]

The hearings, as predicted, became polarized over the use of surfactants (also called surface-active agents), emulsifying agents (also called bread softeners), and shortening extenders. The major manufacturers were Atlas Powder Company and Glyco Products, Inc. Atlas Company witnesses testified that the company had been making sorbitol, hexitol, and mannitol sweeteners for use in gum and candy since 1935 and had developed its line of emulsifiers from the stearate portion of the sorbitol, hexitol, and mannitol molecules.[92] Atlas was interested in having three types of surfactants, Spans, Tweens, and Myrjs, recognized as optional ingredients in bread. POEMS, manufactured by Atlas, was a mixture resulting from the direct reaction of ethylene oxide with commercial stearic acid in the proportion of eight molecules of ethylene oxide to one molecule of stearic acid. Polyoxyethylene (8) monostearate (POEMS) was given the trade name Myrj-45 and had been sold to the baking industry beginning in 1947. Glyco Products manufactured its competing S-541, trade-named StaSoft, by reacting ethylene oxide with water to form a glycol and then reacting this glycol with stearic acid. Clearly, however, neither Myrj-45 nor StaSoft was a single specified chemical entity. Rather, the two products were a "reproducible reaction product" and "a mixture of very similar chemical compounds." Adding to the apparent chemical mystery surrounding these substances, Glyco's chemist stated that he did not know the chemical composition of Sta-Soft or even how to find it, since the ultimate composition would depend in part on the temperature of the reaction and the catalytic agents.[93]

Furthermore, neither Atlas nor Glyco marketed products directly to the baking industry. C. J. Patterson Company distributed Glyco's S-541 to bakers as Sta-Soft in direct competition with the R. J. Vanderbilt Company, distributor of Myrj-45. Although Atlas's biggest sellers were Span 60, Tween 60, and Myrj-45, part of the confusion during the hearings came from Atlas's determination to propose its entire line of emulsifiers for inclusion in the standards for white flour, some twenty-seven different products. It was soon evident that not all of them had been subjected to the same levels of scientific scrutiny, and Myrj-45 in particular had only been on the market for two years. Atlas appeared to be scrambling to market a product that was still undergoing testing. Experienced observers felt that if the company had limited its proposal to the more established and thoroughly tested products, it might have been more successful.[94] Indeed, mono- and diglyceride manufacturers, including Swift and Company and Procter and Gamble (represented by a young attorney named Potter Stewart), and their trade association, the Institute of Shortening Manufacturers and Edible Oils, turned the hearings into a full-fledged trade war. The bread softeners had taken some of their business during World War II, and they were anxious to remove the foothold POEMS had established in the lucrative breadbaking industry.

Bread softeners, it is clear in retrospect, were largely a convenience for large bakeries, which found them advantageous because they yielded a larger loaf, allowed them to use flours of borderline quality, contributed to the keeping quality of bread, and gave them, in particular, more margin for error in meeting mechanized production schedules.[95] Their proponents, therefore, were chemical manufacturers aligned with large bakers, and they were opposed by smaller bakers supported by the mono- and diglyceride and lecithin manufacturers. Several witnesses were called to testify to the fact that the bread softeners had been introduced during the war and sold to bakers as a replacement for shortening, which had been in short supply. Most bakers testified that they had simply added the softeners to their recipe without cutting down on milk or shortening; only a few admitted to having altered their recipe somewhat during the war.[96] The statistics presented showed a very modest drop in fat consumption by the baking industry over prewar figures. The softeners also stood accused of deceiving customers as to the freshness of a loaf of bread.[97]

Effective organized opposition to bread softeners came as well from the Council of Foods and Nutrition of the American Medical Association and the Committee on Cereals of the Food and Nutrition Board of the National Research Council.[98] The former, represented by William J. Darby, professor in the Nutrition Division of the Vanderbilt Medical School, expressed his committee's twofold concern over the bread softeners. First, nutritionists regretted the lack of detailed information about the toxicity of the proposed substances, especially chronic toxicity; and second, they opposed the dilution of the natural nutritive value contributed by fats and milk in baked goods resulting from use of the shortening extenders. R. R. Williams, representing the National Research Council, expressed similar concerns, concluding that the NRC intended to "work for the exclusion of ingredients of non-biological origin which are not fully tested as to their toxicity."[99]

Ultimately, it was scientific testimony on the bread softeners that led to the downfall of the POEMS as an ingredient in bread. POEM opponents went so far as to bring in René Dubos from the Rockefeller Institute to testify that Tween 80 was used in Institute labs as a surface-active agent that made it easier to inject mice with the tubercle bacillus, though he refused to speculate on any possible relevance to humans.[100] The most important testimony was that of John C. Krantz, Jr., of the University of Maryland. In what the trade press reported as a turning point in the scientific testimony, Krantz, who had shown a "brilliant grasp of the various scientific fields, especially pharmacology," was led by the government's attorney in the course of a week-long cross-examination to agree that more information was needed to establish beyond doubt that certain of the softeners were harmless. On the one hand, Krantz, citing his own work, disputed an article by a Cornell research group linking monostearates with elevated serum cholesterol and arteriosclerosis, claiming that the work on Myrj-45 and some of the Tweens was conclusive. On the other hand, he did admit that studies on Tween 60 were still underway and that other studies were still in the planning phase.[101]

The hearings to set standards for white bread gave substance to the accusations of pharmacologists concerned that companies were cutting long term studies short and marketing products before testing was completed and properly interpreted.[102] The bread hearings also drew attention to the aggressive marketing techniques of chemical ingredient manufacturers. In an effort to maximize profits, chemical companies had routinely developed an entire line of chemical derivatives from a single basic chemical. Desiring a market for the derivatives that would stimulate demand for the basic chemical, yet not finding it profitable to conduct the research themselves, companies often resorted to what a marketing consultant of the period referred to as the "we-have-these-chemicals—but-we-don't-know-what-they're-good-for method" of advertising.[103] Listing outstanding properties as well as chemical and structural formulas in trade journal advertisements, the advertisers encouraged other companies to perform the research needed to see how these chemical derivatives might serve their own commercial needs. Food processors and other chemical companies took this bait. Atlas Powder Company, for example, had rewarded R. T. Vanderbilt, who pioneered the use of POEMS in baked goods, with an arrangement granting the company exclusive marketing rights to the baking industry.[104] Although this approach was not illegal, the testimony clearly showed that there was no system of industry accountability for new chemicals. Indeed, while the public made no fine distinctions among chemical competitors, holding the industry as a whole accountable for the actions of its members, the chemical industry's fragmented and competitive nature during this period contributed to the overall impression of a cavalier and irresponsible attitude toward food safety.

The final standards for bread disallowed POEMS as an optional ingredient, claiming that its use served to mislead consumers into thinking that the softened loaves were actually fresher. During a break in the bread hearings in May 1949, Representative Frank B. Keefe seized the moment to indulge in a bit of grandstanding and to introduce a resolution calling for a Select Committee of the House to investigate the use of chemicals in foods and agricultural products, including fertilizers and insecticides.[105] Keefe represented Wisconsin, a key dairy state, in which the safety of agricultural chemicals was a particular concern, and growing questions about the use of DDT around milk cows were particularly compelling; but it was allegations that synthetic ingredients such as POEMS were being substituted for milk and shortening in America's bread that moved him to action in defense of Wisconsin's chief agricultural commodity. Dramatically depicting the vulnerability of small bakeshops at the hands of chemical salesmen, he argued that the competitive nature of the baking industry forced individual bakers to use ingredients that they neither needed nor desired. The manufacturers of the softeners could not be trusted either, he suggested, citing test data "full of irregularities." Claiming that 25 percent of Americans were already eating chemicals in their baked goods, he warned that the battle over the use of chemicals in staple products was being waged at the front door of the baking industry.

The House Select Committee to Investigate the Use of Chemicals in

Foods received its charges from Congress in House Resolution 323, in June 1950. A committee of six was instructed to look at three issues. First, they were to examine the "nature, extent, and effect" of "chemicals, compounds, and synthetics" on the production, processing, preparation, and packaging of food to determine their effects on health and on the agricultural economy. Second, they were to look at the effects of agricultural pesticides and insecticides on consumer health. Finally, they were to consider the effects of chemicals on the soil, the quality and quantity of food produced, and their overall effect on the health of the nation. These charges clearly reflect the priorities of Representative Keefe. When Keefe suffered a heart attack before the committee began meeting, James J. Delaney, a New York Democrat, was given the political plum of chairing the select committee. Until his death in 1952, Keefe corresponded regularly with Delaney from his bedside, advising him on the problems he felt should be addressed by the committee.[106]

As they developed, the Delaney Committee hearings centered less on Keefe's concerns and charges than on exploring the growing "chemicals-in-foods" problem overall and considering the proposal put forth during the first day of the hearings to shift the burden for toxicological testing of food chemicals from the government to private industry. The first witness, the eminent University of Chicago physiology professor Anton J. Carlson, had been advocating this for decades, ever since he had testified for the government in the cases against Monsanto beginning in 1917.[107] The government had been unsuccessful, after several legal attempts early in the century, in having saccharin removed from the market as "poisonous and deleterious to health," but Carlson had been an effective witness on behalf of the government's position. Relying heavily on the new drug provision in the 1938 law as a statutory precedent, the committee recommended that a mandatory system of premarket testing of new chemicals be enacted by Congress. In due course, the committee's counsel that "the barn door be closed" before new chemicals were allowed to escape into the marketplace was embodied in three new laws.[108] The Miller Pesticides Amendment of 1954, the Food Additives Amendment of 1958, and the Color Additives Amendment of 1960 all required manufacturers to shoulder the burden of testing new chemicals and submitting their test results to the FDA for approval before putting them on the market.[109] Although several witnesses testified to the dangers posed by carcinogens in foods during these hearings, such dangers were by no means a major focus of the hearings. The overuse of pesticides; the safety of POEMS; synthetic hormones in chickens, minks, mice, and humans; growing chemical sensitivities in humans; and the fluoride issue were the major problems addressed in nearly two years of hearings. Indeed, the hearings addressed many, if not most, of the major food chemical controversies to come in the "new chemical" age.[110]

Central to the hearings as well was Exhibit A, submitted by the FDA on the second hearing day: a master list of 842 chemicals that the agency claimed had been either used or proposed for use in food products. The entire list was privately circulated among committee members, but it was not made public, presumably because it contained sensitive trade-secret information which was protected under law. Instead a short list of sixty-two chemicals selected from the original list was included in the published hearing record. This abbreviated list reflected both broad and particular aspects of the growing chemicals in food products as addressed in the hearings.[111] The anonymous "list of 842" was the first of several "new chemical" lists compiled during this period. A public-relations tool in the coming decade, its numbers were reassessed periodically by industry to reassure the public of its sense of social responsibility and by the FDA to show progress in reducing dietary dangers. When presented to the Delaney Committee in 1950, the list was unquestionably effective both as a scare tactic and as an honest indicator of the growing need to exercise caution in the use of chemicals in food products.

In contrast to Harvey Washington Wiley, "crusading chemist" and "father of the 1906 Pure Food and Drugs Act," who had staunchly opposed the use of chemical preservatives in foods and advocated consumption of whole wheat flour over white flour a half century earlier, James Delaney, who emerged from the hearings as a new spokesperson for food safety, was no champion of natural foods. Far from advocating the elimination of chemicals from foods, he was instrumental in enacting a series of laws that implicitly acknowledged the utility, or perhaps the inevitability, of food chemicals in feeding an urbanizing population. Delaney and others did insist that the burden of proof demonstrating the safety of these chemicals as well as of other new ingredients should rest on the manufacturer rather than on the government. More than twenty-five bills were introduced between the Delaney hearings and the passage of the 1958 Food Additives Amendment. During this struggle to enact legislation, the "chemical fundamentalism" first advocated by the agency's founding father, Harvey Wiley, was largely replaced by "cancer fundamentalism" in the form of the Delaney anticancer clause. At the insistence of Delaney himself, this clause was included in the 1958 Food Additives Amendment and the 1960 Color Additives Amendment, which mandated FDA approval of new color and food additives prior to their use in food products.

The immediate events surrounding passage of the Delaney clause have been enumerated before, but in brief, an international scientific group known as the UICC (l'Union Internationale Contre le Cancer), which had been interested in the carcinogenic activities of food colors since 1939 and had been meeting in cities around the world since 1935, held its 1954 meeting in Rome. The German Research Council was formulating an additives policy at this time and recommended exploring the issue of environmental carcinogens. As Phyllis Meyer has noted, the Rome conference would have remained as obscure as earlier meetings if the scientists at the meeting had not named names.[112] Of the more than twenty colors not deemed safe, nine were used in the United States. The popular press picked up the debate over carcinogens and drew attention to a pending decision by the FDA on the pesticide Aramite, a known carcinogen, developed by U.S. Rubber. Delaney's insistence upon the clause as a price for legislation came after the FDA struggled mightily before finally setting a zero tolerance for the pesticide.[113]

The Delaney clause, which has been accurately labeled in subsequent decades "America's most famous public health statute," prohibits the approval of any substance for use in foods that is found "after tests which are appropriate for the evaluation of the safety of food additives, to induce cancer in man or animal."[114] Frequently condemned in subsequent decades as antiscientific and technologically stifling, it has been widely misunderstood. Scientists, in particular, accurately point out its poor grasp of theories of carcinogenicity and toxicity, especially in light of recent advances in toxicological testing procedures as they have developed over the past several decades.[115] Where once doses were calculated and measured in parts per million, they are now routinely studied in parts per billion, and it is possible to detect one part per trillion. Therefore, it is important to remember that the Delaney clause is a statement of public policy, not a scientific statement. A more fruitful line of inquiry would explore the extent to which the Delaney clause has forced scientists to clarify and communicate scientific information on carcinogens and carcinogenesis to the public in the later decades of the twentieth century. The Delaney clause has been the chalk line against which scientists have struggled.[116]

Also frequently overlooked is the fact that the Delaney clause accurately reflected the grass-roots experience and concerns of many scientists, especially toxicologists, during the period in which it was enacted. Experimenting on animals and adhering to a "hundredfold margin of safety" in evaluating new food additives, they had begun to discover carcinogenic effects as well as other adverse effects, not only from new food additives but also from small doses of older food additives long thought to be harmless.[117] In 1974, Philip Handler, President of the National Academy of Sciences, admitted his growing confusion over the safety of food additives at that time in a vivid anecdote recounting his crusade during the 1940s to convince southern legislators to introduce state legislation making it mandatory to fortify corn flour with nicotinic acid as a pellagra preventive. He admitted that although he had fed the fortified cornmeal to dogs and nothing had happened at the time, two years later he had fed a chemical variation to young rats at various levels in the diet. He reported that at 0.3 percent in their diet their growth was markedly impaired and they developed large, fatty livers that eventually became cirrhotic. Handler concluded, "Had I done that experiment first, I would never have had the courage to sell a program for the fortification of cornmeal with nicotinic acid. Nor am I certain what the Food and Drug Administration would have done had the program been challenged."[118]

The Delaney clause was an uncompromising effort to force farm and food capitalists to recognize and adhere to known scientific limitations regarding carcinogens. At that time, because no one knew how much of a carcinogen caused cancer nor by what mechanisms, the logical conclusion according to Delaney and the scientific community alike was that carcinogens simply could not be safely used. The absolute prohibition against the use of carcinogens embodied in the Delaney clause was clearly a populist response to the growing perception that an increasingly powerful chemical industry was willing to push past known scientific limits in pursuit of profit.

The debate itself highlighted both the advances that had taken place in the science of toxicology since the days of Harvey Wiley and the current limitations of toxicology, especially with regard to carcinogens. Scientifically, America was no more able at midcentury to say how much of a carcinogen might cause cancer than Harvey Wiley had been able to specify how much of a given preservative might interfere with digestion and disturb kidney function. The crucial difference, however, was a shift in power made evident in the chemogastric revolution. At the turn of the century, faced with only a few food chemicals and Harvey Wiley's determination to stretch the limits of science to declare very small amounts of all added chemical substances "poisonous and deleterious" under the law, the courts acted to restrain him from taking the law to extremes and interpreting its provisions in a manner inconsistent with the era's scientific knowledge and toxicological models.[119] During the Delaney Committee hearings a half century later, it had become clear that the burden for testing could no longer be assumed by the government, and soon Congress acted to force the chemical industry to act scientifically as well, performing adequate studies of the toxicological effects of new food chemicals before placing them on the market and adhering to known scientific limitations regarding carcinogens.

The Delaney clause remains the most visible and enduring populist legacy of the post-World War II world. Other postwar "pure food" advocates, including the organic farming movement and the counterculture with its critique of processed foods, rejected the chemicalizing world in favor of an alternative vision including a return to more "natural" foods.[120] During subsequent decades, however, the word "natural" was usurped by Madison Avenue, becoming one of the most misused terms in the food advertiser's lexicon. The Delaney hearings, the Delaney laws, and the Delaney clause all challenged chemical manufacturers, food producers, and food regulators more directly, both scientifically and politically. In particular, the controversies associated with Congressman Delaney's name focused attention on the largely unknown effects of the chemogastric revolution on public health and welfare and dramatized growing popular concerns about food safety in general, and carcinogens in particular.

The participants in the Beckman Center's 1990 conference on "Chemical Sciences in the Modern World" were very helpful in criticizing an early version of this paper. I am grateful to James Harvey Young, Aaron Ihde, Rima Apple, John Parascandola, John Swann, Naomi Rogers, and Dale Smith, as well as Robert Schuplein and William Horwitz of FDA's Center for Food Safety and Applied Nutrition, for their help in revising it. I am particularly indebted to Seymour Mauskopf for his gentle but persistent pushes for thematic focus.

Notes

1. J. H. Young, F. L. Lofsvold, W. F. Janssen, R. G. Porter, with B. J. Vos, O. G. Fitzhugh, E. P. Laug, G. Woodard, Oral History of the U.S. Food and Drug Administration: Pharmacology, 1980, transcript, History of Medicine Manuscripts Division, National Library of Medicine, Bethesda, Md., p. 123 (hereafter cited as Oral History, Pharmacology). The chemogastric revolution I refer to in this paper began before World War II and extended into the 1960s. Aspects of the changes engendered during this period are captured in almost all writings about food. For statistical confirmation, see George W. N. Riddle, "A Decade of Growth in the Food Industry, *Food Business* 13 (1965): 18–27. Riddle chose 1952–54 as his baseline and ends his study with averages obtained in 1963.

2. Christopher Hamlin, "Between Knowledge and Action: Themes in the History of Environmental Chemistry," this volume, pp. 295–321.

3. Alfred D. Chandler, Jr., *Scale and Scope: The Dynamics of Industrial Capitalism* (Cambridge, Mass.: Harvard University Press, 1990), p. 163.

4. Production and processing of food accounted for sales of $52.5 billion, as compared with $23.2 billion for the chemical industry, $19.3 billion for the automotive industry, and $14.8 billion for steel. *Chemical Industry Facts Book,* 4th ed., 1960–61 (Washington, D.C.: Manufacturing Chemists' Association, Inc.), p. 61.

5. George Brown Tindall, *America: A Narrative History* (New York: W. W. Norton, 1984), p. 1180.

6. Eric F. Goldman, *The Crucial Decade and After: America 1945–1960* (New York: Vintage Books, 1960), pp. 26–27; National Commission on Food Marketing, *The Structure of Food Manufacturing,* Technical Study no. 8, Washington, D.C., June 1966, p. 294.

7. Arnold Thackray, Jeffrey L. Sturchio, P. Thomas Carroll, and Robert Bud, *Chemistry in America, 1876–1976* (Dordrecht: D. Reidel, 1985); David A. Hounshell and John Kenly Smith, Jr., *Science and Corporate Strategy: DuPont R&D, 1902–1980* (Cambridge: Cambridge University Press, 1988); Aaron J. Ihde, *The Development of Modern Chemistry* (New York: Harper & Row, 1964); Hugh D. Crone, *Chemicals and Society: A Guide to the New Chemical Age* (Cambridge: Cambridge University Press, 1986); Alfred D. Chandler, Jr., *The Visible Hand: The Managerial Revolution in American Business* (Cambridge, Mass.: Belknap Press of Harvard University Press, 1977).

8. *Chemical Industry Facts Book,* (Note 4), 7. As late as 1958, the combined sales of the three largest chemical companies amounted to less than 16 percent of industry sales.

9. Charles M. Haar, ed., *The End of Innocence: A Suburban Reader* (Glenview, Ill.: Scott, Foresman & Co., 1972), p. 173.

10. William L. O'Neil, *American High: The Years of Confidence,* 1945–1960 (New York: Free Press, 1987).

11. Regarding difficulties in defining and discussing the American twentieth-century "consumer culture," see Richard Wightman Fox and T. Jackson Lears, eds., *The Culture of Consumption: Critical Essays in American History, 1880–1980* (New York: Pantheon, 1983), p. x. Elaine Tyler May, *Homeward Bound: American Families in the Cold War Era* (New York: Basic Books, 1988). May's is the most recent study of suburban family life, and she characterizes the domestic suburban household as "child centered and sexually charged" (p. 162). Exploring its larger symbolic value in the political sphere of the era, however, she concludes that the consumer-oriented suburban home was viewed as the locus of the "good life" during the period, even though the vision obliterated class distinctions and accentuated gender distinctions. The suburban kitchen, in particular, provided the "evidence of democratic abundance" and "American superiority" employed so skillfully by Vice President Richard M. Nixon in his "kitchen debate" with Soviet Premier Nikita Khrushchev in 1959.

12. David Riesman, "The Suburban Dislocation," in *The End of Innocence,* p. 33.

13. Alan M. Kraut, "The Butcher, The Baker, The Pushcart Peddler: Jewish Foodways and Entrepreneurial Opportunity in the East European Immigrant Community, 1880–1940," *Journal of American Culture* (Winter 1983): 80; United States Department of Agriculture, "Technology in Food Marketing: A Survey of Developments and Trends in the Processing and Distribution of Farm-Produced Foods, 1930–1950," USDA monograph no. 14 (Washington, D.C.: U.S. Department of Agriculture, 1952), pp. 65–66.

14. E. W. Williams, *Frozen Foods: Biography of an Industry* (Boston: Cahners Publishing Co., 1970), p. 197. Williams was the editor of the trade journal *Frozen Foods* from its inception following the war and was a keen observer and tireless promoter of frozen foods. *Design for Decision* (Wilmington, Del.: E.I. du Pont de Nemours and Company, 1949–50), p. 10.

15. Between 1947 and 1967, the value added by farmers *decreased* by nearly 27 percent while the value added during food processing *increased* by 124 percent. In 1947, value added in processing was $11.6 billion. By 1967, however, this had more than doubled to $26.0 billion. W. Smith Greig, *The Changing Structure of the Food Processing Industry: Description, Causes, Impacts, and Policy Alternatives,* Bulletin no. 827, College of Agriculture Research Center, Washington State University (Pullman, September 1976), p. 1.

16. American Marketing Association, "Report of the Definitions Committee," *Journal of Marketing* 13 (October 1948): 202–217.

17. Advertising played an important role in generating consumer acceptance of these "convenience foods." In one study of advertisements for these new foods, Mary Anne Anselmino found that convenience itself was not specifically touted as a virtue. Rather, the ads consistently emphasized the quality, wholesomeness, and flavor of the product as if to assuage guilt over the ease of preparation. Anselmino concluded that "During these years, women were being asked to accept new products and abandon some of their traditional attitudes toward food preparation." Mary Anne Anselmino, "Factors Influencing the Emergence and Acceptance of Food Innovations in Twentieth Century America" (Ph.D. diss., Teachers College, Columbia University, 1986), p. 259.

18. Several studies of food expenditures for "convenience foods" during this period seem to indicate that they actually cost slightly less than equivalent amounts of home-prepared foods. See Robert D. Buzzell and Robert E. M. Nourse, *Product Innovation in Food Processing, 1954–1964* (Boston: Harvard University Graduate School of Business Administration, Research Division, 1976), pp. 140–141. Cost considerations aside, there is still ample justification for James Turner's observation in 1970 that "making food appear what it is not is an integral part of the $125 billion food industry." James S. Turner, *The Chemical Feast* (New York: Grossman Publishers, 1970), p. v. George W. Larrick, "The Consumer Looks at Chemicals in Our Food," Third Annual Conference, Council on Consumer Information, St. Louis, Mo., 5 April 1957. Files of FDA History Office.

19. Bradshaw Mintener, "A Turning Point in Food and Drug Protection," *Food, Drug, and Cosmetic Law Journal* 11, 2 (February 1956): 54. Mintener attributes to Paul S. Willis, then President of the Grocery Manufacturers of America, the statistic that one-third of the grocery sales in 1956 consisted of items which "did not exist 10 years ago or were sold only in token quantities."

20. Emil M. Mrak, "Food Science and Technology: Past, Present, Future," *Nutrition Reviews* 34, 7 (1976): 195; Michael R. Taylor, "Food and Color Additives: Recurring Issues in Safety Assessment and Regulation," *75th Anniversary Commemorative Volume of Food and Drug Law* (Washington, D.C.: Food and Drug Law Institute, 1984), p. 198; "Boilable Bag Breakthrough," *Quick Frozen Foods,* July 1963, 53.

21. Williams, *Frozen Foods,* p. 65.

22. Minute Rice was the first precooked rice on the market. A palatable

version was available as early as 1942 as a result of cooperative work carried on between General Foods and a cousin of the King of Afghanistan, Ataullah K. Ozai Durrani. The Army used the precooked rice in its field rations, and in 1946 it was test-marketed in civilian markets, where it proved very popular. Buzzell and Nourse, *Product Innovation,* p. 63; James R. and Barbara G. Shortridge, "Patterns of American Rice Consumption: 1955 and 1980," *Geographical Review* 73, no. 4 (1983): 417–429.

23. Clarence Birdseye, "Progress of Quick-Freezing in the United States," *Frozen Foods* 89 (1935): 129–130; Harry Carlton, *The Frozen Food Industry* (Knoxville: University of Tennessee Press, 1941); Buzzell and Nourse, *Product Innovation,* pp. 47–54.

24. Wallace F. Janssen, "Changes in Food Production and Their Relationship to the Consumer" (Presentation to the Institute for College Health Educators on Advances in the Health Sciences, New York University, 30 October 1962). Files of the FDA History Office.

25. Buzzell and Nourse, *Product Innovation,* p. 7.

26. Oscar E. Anderson, Jr., *Refrigeration in America: A History of a New Technology and Its Impact* (Princeton, N.J.: Princeton University Press, 1953), pp. 273–274. Inaccurate rumors concerning the vitamin value of frozen foods had been largely dissipated before World War II. Carlton, *The Frozen Food Industry,* p. 114; Agnes Fay Morgan, "Interactions of Food Technology with Nutrition During the Last Twenty-Five Years," *Food Technology* 18 (September 1964): 70.

27. Williams, *Frozen Foods,* p. 21. Edgar P. Hawk, *Fruit and Vegetable Canning Industries,* 1934–45 (Washington, D.C.: Government Printing Office, 1945), p. 1.

28. Williams, *Frozen Foods,* p. 23. Although most observers in the early 1940s thought that frozen foods would compete most effectively with canned foods, in reality they nearly displaced fresh products by the mid-1960s. H. H. Mottern and A. H. Johnson, "Fifty Years of New-Product Development: 1939–1989, *Food Technology* 18 (September 1964): 87–90.

29. Buzzell and Nourse, *Product Innovation,* pp. 51–53. Williams, *Frozen Foods,* p. 43.

30. In 1930 farmers received only 39 percent of each dollar paid for farm food products, while processors and agents received 61 percent. In 1950, there was a more equitable distribution in which farmers received 48.5 percent and processors and distributors received 51.5 percent of the consumer dollar. "Technology in Food Marketing: A Survey of Developments and Trends in the Processing and Distribution of Farm-Produced Foods, 1930–1950," (Washington, D.C.: U.S. Department of Agriculture, 1952), p. 3.

31. Williams, *Frozen Foods,* pp. 56–57, 62.

32. Ibid., p. 104.

33. Ibid., p. 67. The frozen citrus concentrate industry succeeded only after years of research in perfecting low-temperature processing and equipment. The blending of concentrates themselves also required technical knowledge of colors, essential oils, acid ratios, and vitamin contents and retention characteristics. C. G. King, "The Chemist and Engineer in Fifty Years of Food Processing," *Industrial and Engineering Chemistry* 50 (January 1958): 91A.

34. Williams, *Frozen Foods,* p. 85. By the mid-1950s, frozen concentrated orange juice accounted for 24 percent of the total sales of household oranges. "Technology in Food Marketing" p. 11.

35. Anderson, *Refrigeration in America,* p. 279; Williams, *Frozen Foods,* p. 42.
36. Williams, *Frozen Foods,* p. 77.
37. Ibid., p. 165. Charles C. Slater, *Baking in America: Market Organization and Competition* (Evanston, Ill.: Northwestern University Press, 1956), 2:360.
38. Slater, *Baking in America,* pp. 75, 360; Williams, *Frozen Foods,* p. 104.
39. Williams, *Frozen Foods,* p. 86.
40. Principal accusers include Warren G. Belasco, *Appetite for Change: How the Counterculture Took on the Food Industry,* 1966–1988 (New York: Pantheon, 1989); Turner, *The Chemical Feast* (note 18); Beatrice Trum Hunter, *The Mirage of Safety: Food Additives and Federal Policy* (New York: Charles Scribner's Sons, 1975).
41. Anderson, *Refrigeration in America,* pp. 114–115.
42. An association of MSG with "Chinese Restaurant Syndrome" was first published in 1969. H. H. Schaumburg et al., "Monosodium L-Glutamate: Its Pharmacology and Role in the Chinese Restaurant Syndrome," *Science* 163 (1969): 826. This association, however, was not linked to any overall concerns about the safety of MSG as a food additive. Richard L. Hall, "GRAS—Concept and Application," *Food Technology* 29 (January 1975): 48–53.
43. *Chemical Industry Facts Book,* (note 4), p. 64.
44. 21 U.S.C., 52 Stat. 1040, 403(k).
45. 21 U.S.C., 52 Stat. 1040, Sec. 401, Definitions and Standards for Foods.
46. Walter H. Eddy, "The New Meaning of 'Food Standards,'" *Glass Packer* 17, 8 (1938): 477–505; Ole Salthe, "Food Standard Making—What Did Congress Intend?" *Food, Drug, and Cosmetic Law Journal* 1:1 (1946): 174–190; Charles W. Crawford, "Ten Years of Food Standardization," *Food, Drug, and Cosmetic Law Journal* 3, 2 (1948): 243–254; Henry A. Lepper, "The Evolution of Food Standards and the Role of the AOAC," *Food, Drug, and Cosmetic Law Journal* 8, 3 (1953): 133–189; Robert R. Williams, "Standards of Identity for Foods," *Food, Drug, and Cosmetic Law Journal* 7, 9 (1952): 565–572; William W. Goodrich, "Food Standards Past, Present, and Future," *Food, Drug, and Cosmetic Law Journal* 24, 10 (1969): 464–473; Richard A. Merrill and Earl M. Collier, Jr., "Like Mother Used to Make: FDA Food Standards of Identity," *Columbia Law Review* 74 (1974): 561–621.
47. Franklin C. Bing, "Chemicals Introduced in the Processing of Foods," *American Journal of Public Health* 40 (1950): 156–163; W. B. White, "Protection Afforded the Consumer Against Added Chemicals in Foods," *Food, Drug, and Cosmetic Law Journal* 4, 4 (1949): 478–496. W. B. White, "Addition of Chemicals to Foods," *Food, Drug, and Cosmetic Law Quarterly* 2, 4 (1947): 475–489.
48. Memorandum for the File on Monochloracetic Acid and Quaternary Ammonium Compounds, Box 1, RG 233, U.S. House of Representatives, House Select Committee on Chemicals in Food Products, 1950–1952, National Archives, Washington; White, "Addition of Chemicals to Foods," p. 477. The term "preservative" carried a stigma with the public as well as with the FDA. This stigma was rooted in Harvey Wiley's highly publicized and largely successful fight against preservatives such as benzoic acid and boric acid at the turn of the century. The beverage maker, undoubtedly, did not have such a long historical memory. The company was most likely not trying to dodge the term "preservative" so much as it was trying to promote its chemical product as a stabilizer. Hall substantiates that while the demand during the Progressive era was for preserving chemicals, most chemicals following World War II were used primarily as antioxidants or stabilizers, "also referred to as emulsifiers,

plasticizers, and thickening agents." Lloyd A. Hall, "Chemicals: Twenty-Five Years of Progress," *Food Technology* 18 (September 1964): 131–132.

49. *Federal Food, Drug, and Cosmetic Law Administrative Reports, 1907–1949* (New York: Commerce Clearinghouse, 1951), p. 1195 (hereafter cited as *Administrative Reports, 1907–1949*).

50. Bing, "Chemicals Introduced in Processing," p. 160.

51. Ibid.

52. E. B. Astwood, "Treatment of Hyperthyroidism with Thiourea and Thiouracil," *Journal of the American Medical Association* 122, 2 (1943): 78–80; Sally H. Dieke and Curt P. Richter, "Acute Toxicity of Thiourea to Rats in Relation to Age, Diet, Strain and Species Variation," *Journal of Pharmacology and Experimental Therapeutics* 83 (1945): 195–202; Bing, "Chemicals Introduced in Processing," p. 160. Astwood's studies were human drug studies, whereas Dieke and Richter summarize the animal studies. There was general agreement following these studies that because of its druglike hyperthyroid effects thiourea did not belong in foods. Nonetheless, there was considerable disagreement as to the real dangers presented to humans during the four years it remained on the market. The editor of *American Magazine* published a letter noting that "the army also used this chemical on tons of citrus fruit to prevent spoilage during the war period. No adverse effect was observed in the thousands of subjects under close supervision." George L. McNew to Mr. Sumner Blossom, 3 July 1951, Box 9, RG 233, U.S. House of Representatives, House Select Committee to Investigate the Use of Chemicals in Food Products, 1950–1952, National Archives, Washington.

53. O. Garth Fitzhugh and Arthur Nelson, "Liver Tumors in Rats Fed Thiourea or Thioacetamide," pp. 626–628; House Select Committee to Investigate the Use of Chemicals in Food Products, *Chemicals in Food Products,* 81st Cong., 2d sess., 73 (hereafter cited as House Select Committee, *Chemicals in Food Products*). *Administrative Reports, 1907–1949,* p. 575.

54. *Administrative Reports, 1907–1949,* pp. 1244–1245, 1264.

55. Microfilm of correspondence and files of the Division of Pharmacology, Food and Drug Administration, FDA History Office (hereafter cited as Microfilm, Division of Pharmacology, FDA History Office). Also Oral History, Pharmacology (note 1).

56. Oral History, Pharmacology, p. 15. *Administrative Reports, 1907–1949,* 848. The research on the use and misuse of glandular preparations was being conducted by Dr. W. T. McCloskey, a leading specialist on the pituitary gland. Wallace F. Janssen, "Washington Prepares for a New Food, Drug, and Cosmetics Act," *The Glass Packer* 15, 1 (1936): 26. George E. Farrar, Jr., "The Short Happy Life of the First Pharmacology Division of the FDA: 1935–1936," *Clinical Therapeutics* 11 (1989): 183–184.

57. Walter W. Piegorsch, "Quantification of Toxic Response and the Development of the Median Effective Dose (ED50)—A Historical Perspective," *Toxicology and Industrial Health* 5, 1 (1989): 55–62.

58. Stechl cites Joshua Burns in assessing Trevan's contribution. "If we first determine the characteristic curve for the action of any toxic substance on any species of animals, Trevan showed that the curve may then be used for determining the average lethal dose of any unknown sample." Peter Stechl, "Biological Standardization of Drugs Before 1928 " (Ph.D. diss., University of Wisconsin, 1969), p. 204. For related toxicological assessment models also in use, see Piegorsch, "Quantification of Toxic Response."

59. James Harvey Young, "Sulfanilamide and Diethylene Glycol," in *Chemistry and Modern Society*, ed. John Parascandola (Washington, D.C.: American Chemical Society, 1983), pp. 105–125. Charles Wesley Dunn, *Federal Food, Drug, and Cosmetic Act: A Statement of Its Legislative Record* (New York: G. E. Stechert & Co., 1938), pp. 1316–1327.

60. "Sulfanilamide—The Drug," *Reports by the Council on Pharmacy and Chemistry* (Chicago: American Medical Association, 1937), pp. 1–53.

61. White, "Addition of Chemicals to Foods" (note 47), 478.

62. Such long-term problems included that of chemical carcinogens. It was in connection with the glycol research that the first evidence linking industrial chemicals with cancer surfaced. Scientific interest within the FDA in food chemicals had long preceded the establishment of the Division of Pharmacology, however, and tumor-producing agents such as coal tar colors, saccharin, ergot, and selenium had attracted the special interest of researchers from the beginning. Oral History, Pharmacology.

63. W. B. White, "Protection Afforded the Consumer Against Added Chemicals in Foods," *Food, Drug, and Cosmetic Law Journal* 4, 4 (1949): 480; White, "Addition of Chemicals to Foods " (note 47), 478.

64. Edward P. Laug, Herbert O. Calvery, Herman J. Morris and Geoffrey Woodard, "The Toxicology of Some Glycols and Derivatives," *Journal of Industrial Hygiene and Toxicology* 21 (1939): 173–201.

65. I am indebted to Christopher Sellers for sharing with me his research on thresholds and tolerances obtained in the course of his study of early federal research into the chronic effects of chemicals.

66. J. H. Draize, G. Woodard, and H. O. Calvery, "Method for the Study of Irritation and Toxicity of Substances Applied Topically to the Skin and Mucous Membranes," *Journal of Pharmacology and Experimental Therapeutics* 82: 377–390; John Parascandola, "The Development of the Draize Test for Eye Toxicity," *Pharmacy in History* 33, 3 (1991): 111–117.

67. Scientists and leaders from what would become Union Carbide, Dow Toxicology Group, Hazleton Labs, and Food Research Labs all spent time in the FDA's Division of Toxicology. Oral History, Pharmacology, pp. 32–34.

68. Microfilm, Division of Pharmacology, FDA History Office. Also Oral History, Pharmacology.

69. Oral History, Pharmacology, pp. 123–124.

70. Arnold J. Lehman et al., "Procedures for the Appraisal of the Toxicity of Chemicals in Foods, Drugs, and Cosmetics," *Food, Drug, and Cosmetic Law Journal* 10, 10 (1955): 679–748.

71. Although objections were raised to the costs of testing, especially by smaller companies, the biggest complaint appears to have been the extended periods required to complete animal testing and the concern that competitors might not adhere to the same testing timetables and thereby gain advantage in the marketplace. Microfilm, Division of Pharmacology, FDA History Office. Also Oral History, Pharmacology.

72. Andrew Rowan notes that the LD-50 was used around the world by regulatory officials as a rough classification scheme designed to distinguish between very toxic, toxic, and nontoxic substances. Andrew N. Rowan, *Mice, Models, and Men: A Critical Evaluation of Animal Research* (Albany: State University of New York Press, 1984).

73. The U.S. Food and Drug Administration certainly used the LD-50 as a

screening device at a time when some information about a new chemical was better than no information. Important questions still remain, however: How important was the LD-50 to regulators? How important did commercial interests perceive it to be in securing approval of a new product? Is there a discrepancy between the two? The LD-50 became important because it was a pivotal point in the emerging struggle between regulators and companies to require food, drug, and cosmetic products to adhere to a higher standard of safety. Scientists, especially pharmacologists, certainly knew of the quantitative limitations of the test and, given any choice, preferred detailed qualitative data as well as pharmacological assessments and analyses of a substance's mode(s) of action to a mere assessment of the chemical's LD-50. From the division's correspondence, they appear to have accepted the LD-50 data, however, as a first step in pushing toward more extensive testing at a time in which it was not mandated for any product other than "new drugs." Many companies, on the other hand, felt that if the LD-50 proved their chemical to be low in toxicity, they had met their testing burden. Most were reluctant to invest money while risking market usurpation for voluntary extended toxicity testing. Threats of FDA action and/or product liability lawsuits sometimes persuaded them that the additional tests recommended in conference with FDA scientists were prudent, but many if not most companies sought consultations with FDA scientists, bringing the LD-50 data with them in the hope that substances with a low order of toxicity could merely be deemed "safe" without any further testing. Many companies had to be convinced of the need for their conducting more extensive and revealing toxicity tests, and thus the need for their obtaining the means to do so.

74. The new generation of organic or "persistent" pesticides, insecticides, and rodenticides remained at the top of the agency's list of concerns: problems with DDT, parathion, aldrin, and dieldrin were the most notorious examples. A. J. Lehman, "Pharmacological Considerations of Insecticides," *Bulletin of the Association of Food and Drug Officials of the United States* 13, no. 2 (April 1949): 65–70; Paul B. Dunbar, "The FDA Looks at Insecticides," *Food, Drug, and Cosmetic Law Quarterly* 4, 2 (1949): 233–239. Although FDA scientists criticized what they viewed as the pointless proliferation and overuse of agricultural chemicals, the FDA itself had no jurisdiction over backyard gardens and nearly negligible influence over farmers, especially since the U.S. Department of Agriculture was simultaneously promoting the use of pesticides. See James Whorton, *Before Silent Spring: Pesticides and Public Health in Pre-DDT America* (Princeton, N.J.: Princeton University Press, 1974); Christopher J. Bosso, *Pesticides and Politics: The Life Cycle of A Public Issue* (Pittsburgh: University of Pittsburgh Press, 1987). Passage of the Federal Pesticide, Fungicide, and Rodenticide Act of 1947 left the establishment of tolerances up to the FDA, but gave it no veto power over the registration of new pesticides. The FDA was only able to act to limit pesticide residues in foods and to control new uses in food packaging.

75. House Select Committee, *Chemicals in Food Products,* 12 December 1950, testimony of Morton Biskind, M.D., p. 709.

76. House Select Committee, *Chemicals in Food Products,* 28 Nov. 1950 testimony of Arnold Lehman, pp. 382–390; also 29 Nov. 1950 pp. 407–412.

77. W. B. White, "Addition of Chemicals to Foods" (note 47), 489.

78. Victoria A. Harden, *Inventing the NIH: Federal Biomedical Research Policy, 1887–1937* (Baltimore: Johns Hopkins University Press, 1986), p. 186.

79. Quotation is from Fritz Eichholtz, "The Chemical Preservation of Food and Its Importance for Human Health—Pharmacological Comments," *Food, Drug, and Cosmetic Law Journal* 10, 4 (1955): 207. Informal standards for bread were promulgated in 1923 under Food Inspection Decision no. 188. These standards were edited in 1936 with small changes, but they remained advisory until passage of the 1938 Food, Drug, and Cosmetic Act. 1941 and 1948 Bread Standard Hearing Transcripts, FDA Hearing Clerk's Office, Docket no. FDC-31 and 31 (a). Hereafter cited as Transcripts.

80. Exchange between William R. Goodrich, attorney for the government, and C. J. "Ten Ingredient" Patterson, *Baker's Letter,* no. 110, pt. 12, 15 April 1949, 7.

81. Transcripts, pp. 67, 79, 80, 81.

82. Ibid., p. 463.

83. Ibid.

84. Ibid., pp. 200–280.

85. Ibid., p. 282.

86. In a phone conversation on 7 March 1989, William Goodrich confirmed that he had been "hard nosed" at the hearing because he and Joe Callaway felt bakers were being sold a lot of unnecessary ingredients, especially "hocus pocus" for preventing rope and mold.

87. Transcripts, pp. 277–278.

88. Ibid., p. 330.

89. "Enrichment of Bread and Flour: A History of the Movement," *Bulletin of the National Research Council* 110 (November 1944); Russell M. Wilder, "A Brief History of the Enrichment of Flour and Bread," *Journal of the American Medical Association* 162, no. 17 (1956): 1539–1541. For an alternative view of enrichment, see William Longood, "White Bread—Enriched but Still Impoverished," *Poisons In Your Food* (New York: Grove Press, 1960), pp. 179–196.

90. The hearings transcripts for the 1948 bread hearings, FDC-31 (a), are so voluminous that the courts accepted summary accounts published by *Baker's Weekly* as part of the record for appeals. I also use and cite from these summaries.

91. "Bread Standards—Their Past and Their Future," *Baker's Helper,* 10 July 1948, 57.

92. *Baker's Letter,* no. 104, pt. 6, 4 February 1949, 9.

93. Ibid., 10. *Baker's Letter,* no. 105, pt. 7, 11 February 1949, 2.

94. Phone conversation on 7 March 1989 with William Goodrich. Goodrich noted that following the bread hearings, both Atlas and Glyco continued to seek to have their emulsifiers included in food standards, but they became more selective and ultimately were more successful in getting them included in other products, e.g., pickles.

95. *Proceedings of the Twenty-Ninth Annual Meeting of the American Society of Bakery Engineers,* Chicago, Ill., 2–5 March 1953, Tuesday afternoon session, pp. 143–144. Transcript courtesy of the American Institute of Baking, Manhattan, Kansas.

96. *Bakers' Letter,* no. 103, 22 December 1948, 12; *Proceedings of the American Society of Bakery Engineers* (1949), pp. 77. Files of the American Institute of Baking, Manhattan, Kansas; *Baker's Letter,* pt. 18, 17 June 1948, 3. On 6 June 1949, one bakery employee testified that the salesman had never told him that the product was a "shortening extender," but he had told him that if he used it, he could use less shortening in the bread recipe.

97. *Baker's Letter,* no. 111, pt. 12, 29 April 1949, 6–10.

98. C. M. Keyworth, "Recent Developments in the Use of POEMS in Baking," *Food,* August 1955, 291.

99. *Baker's Letter,* no. 100, 8 December 1948), 3. Ibid., no. 113, pt. 14, 13 May 1949, 7; Ibid., no. 115, pt. 17, 14 June 1949, 1–2. Arthur T. Joyce, "This Week in Washington: Noted Scientist Opposes Softeners," *Baker's Weekly,* 6 June 1949, 45.

100. *Bakers' Letter,* no. 121, pt. 23, 23 September 1949, 2.

101. Arthur T. Joyce, "This Week in Washington: Softener Experts Admit Need for More Research on Toxicity," *Baker's Weekly,* 14 February 1949, 38–42; *Baker's Letter,* no. 121, pt. 23, 23 September 1949, 3.

102. Typescript, 1948 Bread Standards Hearings, Records of the Hearings Clerk, U.S. Food and Drug Administration, Rockville, Md.; *Baker's Letter,* no. 104, 4 February 1949), 1–9; Ibid., no. 105, 11 February 1949, 4; Suzanne White, "POEMS, Glycol Globules and Good Bread," paper delivered at the Sixty-Second Annual Meeting of the American Association for the History of Medicine, Birmingham, Ala., 29 April 1989. Regulatory scientists at the FDA speculate, but cannot prove on the basis of later testing, that some of the early POEMs may have been chemically contaminated, which may have skewed some of the early rat studies and account for some of the inconsistent toxicological findings which were reported. "Preservatives Warning: Surface Active Compounds," *Newsweek,* 18 July 1949, 41–42; William Horwitz and Leo Friedman, "Heated Fats and the Chick Edema Factor," *Bureau By-Lines* 3, 2 (October 1961): 45–55; D. Firestone et al., "The Chick-Edema Factor," *Journal of The American Oil Chemists Society* 38 (1961): 253–418.

103. Robert S. Aries and William Copulsky, "The Marketing of Chemical Products" (New York: R. S. Aries and Associates, 1948), p. 6. Copyrighted manuscript in the possession of Woodruff Library, Emory University, Atlanta, Ga.

104. House Select Committee, *Chemicals in Food Products,* 81st Cong., 2d sess., 16 November 1950, 257–268.

105. Arthur T. Joyce, "This Week in Washington: Keefe Blasts Use of Chemicals in Bread and Other Foods, Demands Congressional Hearing," *Baker's Weekly,* 15 May 1949, 38–40.

106. Papers of the House Select Committee on the Use of Chemicals in Food Products, National Archives, Washington, D.C.

107. James Harvey Young, "Saccharin: A Bitter Regulatory Controversy," in *Research in the Administration of Public Policy,* ed. Frank B. Evans and Harold T. Pinkett (Washington, D.C.: Howard University Press, 1975), pp. 39–49.

108. This phrase was popular throughout the debate over the Delaney Committee's recommendation, appearing in both popular and technical periodical accounts of the hearings.

109. 1954 Pesticides Amendment, 68 Stat. 511; 1958 Food Additives Amendment, 72 Stat. 1785; 1960 Color Additives Amendment, 74 Stat. 399.

110. Hugh D. Crone, *Chemicals and Society: A Guide to the New Chemical Age* (Cambridge: Cambridge University Press, 1986).

111. In 1989, when the House changed its rules on access to House Committee records, the original "list of 850" became available in the committee documents.

112. Phyllis Anderson Meyer, "The Last Per-Se: The Delaney Cancer Clause in U.S. Food Regulation " (Ph.D. diss., University of Wisconsin, 1983), p. 144.

113. Ibid., 147–148, 170–176.

114. Food, Drug, and Cosmetic Act. 21 CFR 409 (c) (3) (A).

115. E.g., Vincent A. Kleinfeld, "The Delaney Proviso—Its History and Prospects," *Food, Drug, and Cosmetic Law Journal* 28, 9 (1973): 556–565; Philip H. Abelson, "Testing for Carcinogens with Rodents," *Science* 21, 249 (1990): 1357; *A Symposium on the Delaney Clause,* 26 March 1988 (Washington, D.C.: Food and Drug Law Institute, 1988); Roger D. Middlekauff, "200 Years of U.S. Food Laws: A Gordian Knot," *Food Technology,* June 1976, 52. *Food Chemical News,* Special Edition on Delaney Committee Report, 16 May 1960, 1–14.

116. For an informative debate on the major issues, see "Conspiracy of Silence," *American Council on Science and Health News and Views* 1, 2 (February 1980): 4,5,13. A report by the Institute for Science in Society reports that the Delaney clause has prevented some newer and safer pesticides from entering the market to replace older, more dangerous pesticides that were grandfathered under the law. Institute for Science in Society, "Unraveling Delaney's Paradox: Challenges for the 102nd Congress," 19 September 1991 (Washington D.C.: The Institute).

117. Maurice H. Seevers, "Perspective versus Caprice in Evaluating Toxicity of Chemicals in Man," *Journal of the American Medical Association* 153, no. 15 (1953): 1329–1333; National Academy of Sciences, *How Safe Is Safe? The Design of Policy on Drugs and Food Additives* (Washington, D.C.: National Academy of Sciences, 1974), p. 7. By the sixties, this increasing ability to test for even minute amounts of carcinogens, coupled with the knowledge that certain chemical substances could interact with DNA to cause mutations that increased the incidence of cancer, helped create widespread confusion among regulators and scientists. In the W. O. Atwater Memorial Lecture sponsored by the Agricultural Research Service of the Department of Agriculture, and delivered at the 5 April 1976 Centennial Meeting of the American Chemical Society, Emil M. Mrak observed from his experience: "It appears to me that, more often than not, industry scientists have been unaware of what is really expected of them with respect to safety testing, of what protocols are to be used and of what is meant by teratogenesis, mutagenesis, interaction and even carcinogenesis. This is more the result of governmental uncertainties than neglect on the part of food scientists. The question has been, and is, what to do about these matters and where to obtain information. Frequently there has been a lack of agreement in government and even a lack of knowledge regarding what protocols to follow in testing for these effects." Emil M. Mrak, "Food Science and Technology: Past, Present, Future," *Nutrition Reviews* 34, 7 (July 1976): 196.

118. National Academy of Sciences, *How Safe Is Safe?* p. 7.

119. Mastin G. White and Otis H. Gates, *Decisions of Courts in Cases Under the Federal Food and Drugs Act* (Washington, D.C.: U.S. Government Printing Office, 1934), pp. 443–577; *U.S. v. Lexington Mill and Elevator Co.,* 232 U.S. 399.

120. Belasco, *Appetite for Change* (note 40); Caroline E. Mayer, "Pressing On at Rodale Foods," Washington *Post,* El [n.d.]. FDA History Office Files, Rodale.

The Chemical Industries and Their Publics: How Can History Help?

E. N. Brandt

Having devoted most of my career as a public relations man to the relationship between the chemical industry and its publics, I have done an uncommon amount of thinking about that relationship. I am therefore beginning this essay with some of my reflections.

On most days, as we read our newspapers and watch television, it seems as though chemistry is being tried at the bar of public opinion. Almost any publication mentions the latest attack on the industry or its practices, or the latest chemical product being attacked, the "chemical of the week."

The case of alar, a product used on apples to keep them on the tree longer, allowing them to ripen fully, lengthening their storage life and improving their appearance, is fairly typical. For most of a week in February 1989, the nation, and especially its press, was in a panic about the dangers of our children eating apples. The U.S. government wheezed and huffed and said that alar should no longer be used. Schools pulled apples off their cafeteria menus, and, overnight, apples were treated as poison. The apple growers discussed substitutes for alar. In a scenario we have seen many times before, the scientific facts in the case were rather cavalierly brushed aside and virtually ignored, even though they solidly supported alar and demonstrated clearly the benefits of its use.[1] Thus alar became another perfectly good product tossed on the trash heap of chemical history.

We all recognize that chemicals can benefit us, just as chemicals can harm us. But this is true of just about everything else around us. Humans cannot survive without water, but we can also drown in it. The governing factor, as we know, is those upper and lower limits—how much is enough, how much is too much—and how the material is used.

So those of us in the chemical business are put constantly on the defensive. And we discovered long ago that the Nuremberg defense is not enough. For many years the chemical industry people took the attitude that they were only the servants of the public, that they devised and made and sold materials they believed to be beneficial. How these materials were used was the function of government, of the policy makers: "We don't make policy, we only make the chemicals." At Nuremberg those who had followed the policies set by Hitler were judged, and the defense that "we only made the gas for the gas chambers because Hitler ordered us to do it" was deemed insufficient. If you know better, ran the reasoning, it is your responsibility to oppose the orders of a tyrant. And over the years that rule has been extended beyond tyrants to government generally in the minds of a large portion of the populace.[2]

So in our generation we have had to take more responsibility for our products than had ever been the case before. This means not only that we require better scientific data to back up our products but also that we need to explain our products better.

It is paradoxical that, now that we have the communications devices to present and explain our case, the availability of these new devices only intensifies our problem. We have television and other remarkable means of communication available to us that were not available to previous generations. But these means of communication have been used to attack chemicals and their uses far more often than to explain or defend them. The horrors wrought with chemistry or by chemical means are easy to display on the television screen, but the good that chemistry does is oft interred with the bones of a product like alar. A revolution is indeed taking place in communications techniques, but we in chemistry are far more often the victims of it than the beneficiaries.

It should be observed that all the sciences have the same problem. The amount of new scientific knowledge available has grown at such a prodigious rate in our time that we have moved far beyond the possibility of the general public understanding more than a fraction of it. The press, which is the main means of reaching this public, doesn't understand very much more of it than the unwashed public, unfortunately. And the press therefore is increasingly vulnerable to the attacks of those who claim to know that chemicals are bad. As a consequence, attacks based on emotional appeals—"your child may become seriously ill if you continue to feed him apples treated with alar"—are often successful. The scientific data are ignored, either because they also are not understood, or believed or, I'm afraid, because the actual data are attacked by those who claim to know that chemicals are bad. Very often, these claimants produce conflicting data based on their own research.

This is the whole problem of the chemical industry and has been for the past thirty years, since about the time of Rachel Carson's *Silent Spring*.[3] The chemical industry has devised many ways to meet this problem. Dozens of ways can be cited in which chemists and chemical manufacturers are seeking to bring about dialogue with their critics, to educate the general populace about chemicals, and to present sound science more palatably and dramatically.

But somehow we have not used history very much in meeting this problem. It is still a much underused tool in the arsenal of science. It seems to me that the use of insecticides, for example, is marvelously clarified when viewed in terms of the history of humankind's efforts to feed itself, in competition with the insect world. In general, every problem is easier to understand if you understand the history of that problem. Therefore, chemistry, with all its problems, is one of the fields most critically needful of history.

I would like to get at the question posed in the title of my essay—How can history help?—in a rather elliptical fashion. I would like to discuss a couple of cases from my own experience that would have benefited greatly, I feel, from the spotlight of history. I would like also to tell you a little about the Dow Chemical Company and its history, because that is, I'm afraid, a little-known subject, and to give you some notion of what has been done with that company's history I would like to sketch for you what the company is doing now, and why it's doing it.

First, let us take up a case where history could have helped: the discovery of mercury in the Great Lakes and other bodies of water around the world, about twenty years ago.

In the late 1960s a Swedish graduate student named Norvald Fimreite had to write a paper. Being an unusually bright fellow, he decided to combine business and pleasure in the research for his paper. He proposed to test fish taken in a Swedish lake where mercury was present and to measure by the advanced techniques then coming into use whether the fish were taking up and storing mercury in their bodies. This gave him an excuse to go fishing while actually working on his degree. Now, this was a paper that should probably never have been written, and one I think most of us in the 1960s, if we were faculty advisors, would have discouraged. After all, the literature for the past two hundred years had shown that mercury was inert in water, that metallic mercury vaporizes in air and is injurious to humans, and the textbooks said the proper way to dispose of waste mercury, for this reason, was to wash it into the nearest moving stream. Why repeat research merely to confirm what we had already known for many generations? "Go find a better subject, Mr. Fimreite," we would have said.

What happened, of course, is history; it turned out that fish in certain circumstances do take up mercury from their environment and do store it in their flesh, and Norvald Fimreite was the first to discover this. Unfortunately, his unheralded student paper was then published in a rather obscure Swedish technical journal and lay there quietly for many months; papers in Swedish were not routinely translated into English or the other popular scientific languages—not, at least, from the journal in which the Fimreite paper appeared.

Many readers will remember the ensuing hubbub, triggered in large part by the Minamata disaster in Japan, for it quickly developed that the mercury problem was worldwide, and that a heavy diet of such fish could be quite harmful to human health. The sale and eating of fish from the Great Lakes was thereupon halted for several years. The U.S. Senate investigated. The Canadian government sued Dow Chemical of Canada. And the general public was once again astounded by the apparent stupidity of the chemical industry. How could they have been dumping mercury into the rivers and lakes; why didn't they know it could hurt people? Why hadn't they read Norvald Fimreite's paper and acted on it?

The point of recalling this bit of fairly recent history is that there was no way of repairing the damage once it was done. Once the emotional impact of the Minamata victims had been unleashed, the question was no longer quite rational, it was emotional, and there were many who wanted the chemical companies severely punished for their sins. And I'm sure there are people even today who believe the chemical industry flushed mercury into the streams in full knowledge of the hazards involved.[4]

Think of the problems that could have been avoided if there had been better general knowledge of the historical background of the element mercury, and why it was handled as it was by highly educated and highly responsible chemical personnel. It seems to me that would have made a great difference in the perception of chemical company behavior. by the public, which seemed to see only that the chemical industry had been caught red-handed in a practice that was contemptuous of the environment and of human health as well. Yes, history could have helped, I think.

Let me switch now and give you some historical background concerning the company I represent.

Herbert H. Dow, the founder of Dow Chemical, was the son of an American tinkerer best remembered as inventor of the turbine used by the U.S. Navy for many years to propel torpedoes. Herbert was rather a precocious boy and from the age of about ten worked with his father on his inventions. He went to a hometown university, the Case School of Applied Science in Cleveland, and graduated in 1888. He became interested in the brines of the Midwest while a student and traveled about Ohio, Michigan, and Pennsylvania obtaining samples of the local brines. There was a growing demand for bromine at the time, for medical and photographic uses, and he developed a new method of extracting bromine from the brine, the "blowing out" method, that didn't depend on heating it and in consequence was cheaper, requiring no heavy energy input. His research told him the brines in central Michigan were richest in bromine content, so he decided to go there to try to prove out his process. His first effort there failed, mainly because it was underfunded. In fact, the first three companies that he set up did not do well. When the third one became modestly profitable, Herbert Dow proposed to develop products from the other components of the brine, which at that time were just being dumped. He proposed to make bleach, using mainly chlorine derived from the discarded wastes of the brine stream. His financial backers said no, those earnings should be returned as dividends to his investors, not squandered in hazardous ventures. Eventually the young man left this company to set up a new one to make bleach. This new company he called the Dow Chemical Company, and it was a success almost from the beginning, established from the outset as a strongly research-oriented, new product-oriented firm. Within a half dozen years it bought out the profitable but stand-pat firm that Herbert Dow had left.

From those humble beginnings the Dow Chemical Company has become one of the great American success stories. It is now the number two U.S. chemical firm (after du Pont) and number six in the world, approaching $20 billion in annual sales.

Herbert Dow became a chemical company executive but never lost touch with the lab bench, and he is credited with 107 patents. In many respects he was far ahead of his time. We have an automatic control device that he and his brother-in-law worked on in 1899 to enable them to produce chemicals without constant human attendance. He and Tom Griswold finally had to give up on the project, and it was a great many years later that automatic controls finally came into common use in the chemical business.

A full description of Herbert Dow's many accomplishments is beyond the scope of this essay, but it is noteworthy that at his death in 1930 he was attempting to expand his research from mining chemicals from underground brines, which had occupied much of his career, to the mining of chemicals from sea water. The sea is full of chemical components, of course, and Herbert Dow reasoned that the brines he was working with had once been part of some prehistoric sea. The chemicals in the sea were just a little more dilute.[5]

As you may know, Herbert Dow's son Willard, who succeeded him at his death, completed the project, by mining first bromine and then magnesium from seawater. I believe these are still the only major chemical products produced with seawater as the raw material. Most of the U.S. magnesium supply today comes from a Dow plant based on the seawater process at Freeport, Texas. In the case of bromine it has proven more economical to produce the element from brines, and most of the bromine is produced nowadays in Arkansas, not in Michigan.

Herbert Dow is relatively little known, I think, in the galaxy of American chemists and business pioneers. I believe the main reason for this is that he did his life's work in what was then backwoods Michigan, far from the madding crowd, far from the attention of the press and the scholars. The company itself remained obscure, too, in large part because it became a chemical company's chemical company—a supplier of bulk chemicals to other chemical manufacturers. Until fairly recent years such companies did not come very much to the public notice and were certainly not household words.

For these reasons (and others), not a great deal of history has been written either about Herbert Dow (or, for that matter, any other of the brilliant men with whom he surrounded himself) or about the Dow company. The main historical sources are a 1949 biography of Herbert Dow, *Herbert H. Dow: Pioneer in Creative Chemistry*, and a 1969 history of the company, *The Dow Story,* by Don Whitehead.[6] There are a number of other volumes by and about Dow achievements, notably a history of Dow research and development, *Dow Research Pioneers, 1888–1949*, but by and large a great deal remains to be done.[7]

This brings me, logically enough, to the question, "What are you doing about it?" And I'm glad you asked that.

Like other large companies, the Dow company generates prodigious amounts of paper documents, photographs, videotapes, slides, and other potentially historical material. By 1973 the company was so swamped in paper that its board decreed that any piece of paper not needed for current business would be thrown out, and this resolution included a proviso that no historical records would be kept. Literally tons of paper were pitched out and burned at that time, records of all kinds. From a historical viewpoint this was a total disaster; there was no effort at all to sort out or salvage any historical materials. There was at the time no history function in the company, no one to defend history and become the shepherd of historical records.

Oddly, out of this major calamity came the first organized historical effort in the company. For every action there is an equal and opposite reaction. Well, I was among those who objected most vociferously to the destruction of all this history, and when I kept right on objecting I was given an additional duty—that of company historian. The original idea was that I should spend 5 or 10 percent of my time working on the problem of how history might be kept. I had a license, in effect, to squirrel away historical records that might be on the verge of being destroyed. And I was in the challenging position of being the historian of a company that doesn't keep historical records. That, in fact, is still the case today.

The eventual solution to this problem was to enroll the help of the Dow family foundation. We began by establishing a rule that our historical records became the property of the foundation, not of the company. This made them legal, and I am much indebted, as you can imagine, to the support of the foundation, which has continued unflagging ever since.

In 1987 we opened an archives building in an old brick schoolhouse belonging to the foundation, an 1876 one-room schoolhouse where Grace Ball taught school before she married a young chemical entrepreneur named Herbert Dow. We have remodeled it into a pleasant and quite serviceable archive, supported financially both by the company and by the foundation. And for the first time we are able to serve the needs of scholars and writers and others for information about the history of the Dow company.

In 1988 we embarked on another venture, the Herbert H. Dow Historical Museum, also in Midland. This is a project of our county historical society. The museum is a replica of the Evens Mill, originally a Main Street gristmill that had a brine well adjacent, the first site of Herbert Dow's operations in Midland and the ancestor, three companies later, of the Dow Chemical Company. The museum opened on August 14, 1990, at 2:30 P.M., the exact centennial of the date and time the young Herbert Dow got off the train in Midland and walked up the street to see about leasing the Evens Mill.

We have some gaping holes in the Dow history, as you might gather, and we have been attempting to plug them by engaging in a rather heavy program of oral history. We are working hard to compile a first-rate series of oral history interviews of dozens of Dow Chemical pioneers. I think it may be one of the most extensive oral history programs currently being conducted in an American company.

Let me return now to my main theme of how history can help. I hardly need convince the reader of the value of history, either in decision making or as a pattern for conduct. But I do want to emphasize my belief that chemical history can be a very present help in times of trouble.

In support of this premise I'd like to cite the napalm case at Dow Chemical during the war in Vietnam, 1966 to 1971. Many of you will remember—perhaps you were on campus at the time—that Dow became the company you loved to hate during that war. Dow was in an expansion mode at the time and was recruiting rather heavily on most large U.S. campuses. Vocal protests against the war erupted on most college campuses during that period, and the favorite pattern came to be a protest of the presence on campus of the Dow recruiter. He, poor fellow, was seen as some devil creature, bent on recruiting innocent, dewy-eyed young Americans to manufacture napalm, a horrible form of liquid fire used in Vietnam to bomb women and children as part of Lyndon Johnson's awful war there. "Napalm—Johnson's baby powder," the protest signs said. "Dow shalt not kill."

During a three-year period there were 226 major protests on American campuses focusing on Dow recruiters. In many of these there were arrests, students and police were injured, the Dow recruiter was held prisoner, campus buildings were blockaded. Harvard, by the way, still holds the school record for keeping a Dow recruiter prisoner in the interview room the longest—twelve hours. That is something of a poetic justice because napalm B, the form being used, was invented by Professor Louis H. Fieser of Harvard.

There was very little that could be done to halt all this, of course; it seemed then and still seems today to have had a life of its own, to have been an outward manifestation of the American psyche, an emotional outpouring that could not have been prevented by any conceivable amount of facts or data, no matter how cogently and dramatically presented.

But I think history could have helped. Most of the students were convinced that napalm had been invented by Dow in its own research laboratories; it of course had not. There is a whole history of the use of fire and flame throwers in warfare, dating back to Greek fire in the seventh century. The first napalm as such was developed in World War I in response to the development of the tank; the modern form was developed by Professor Fieser in World War II. In the latter war, which was a "good" war, it was used to dislodge the Japanese from their bunkers in the Pacific islands, and Americans then thought napalm was a splendid product. Thirty years later, in Vietnam, it was criminal. Ironically, medical teams determined later that napalm burns among the civilian population were extremely rare, and that most of the civilian burns came from black market gasoline. But by that time the evils of napalm had been too solidly established to be corrected.

The history of the use of fire in warfare, the history of napalm, if it had been disseminated—if it could have been delivered through the astounding depths of emotion that surrounded the debate on the Vietnam war—would possibly have brought some perspective to the problem. To be effective at all, of course, the information would have to have been known and communicated very early, before the napalm protests began. In the hands of key academics and public policy officials such history would have had a considerably damping effect, I believe, on the focus that napalm came to be.

That may be a key to the effectiveness of disseminating the history of chemistry and chemicals—doing it early enough and widely enough that the situation will be understood, at least by scattered specialists, before it is too late, before the product becomes notorious and the object of some emotional pileup.

Invariably, we in the chemical business are thrust on the defensive by some event. All too often we lack the data with which to defend the product. Perhaps someday we will reach the point where we know all there is to know about any chemical anyone can mention. But we certainly aren't there yet, and it will be very long before we are, if it ever happens at all. I suppose that's one of the main reasons for continuing the development of chemicals and of their contributions to humankind, and for recording the history of that activity.

It is key, I think, to the entire history of science.

Notes

1. Regarding the alar episode, see Kenneth Smith, "Alar: One Year Later," Special Report, American Council on Science and Health, New York, March 1990.

2. Dow Chemical was accused of using the "Nuremberg defense" during the napalm protests of the 1960s. See H. D. Doan (President, The Dow Chemical Company),"Dow Will Direct Its Efforts Toward a Better Tomorrow," in *Dow Diamond,* a publication of The Dow Chemical Company 30, 4 (1967).

3. Rachel Carson, *Silent Spring* (Boston: Houghton-Mifflin, 1962).

4. See LeRoy D. Smithers (Chairman, Dow Chemical of Canada, Ltd.), "Some Aspects of the Mercury Problem," in the *Dow Maple Leaf,* a publication of Dow Chemical of Canada (April 1971).

5. Murray Campbell and Harrison Hatton, *Herbert H. Dow, Pioneer in Creative Chemistry* (New York: Appleton-Century-Crofts, 1951).

6. Don Whitehead, *The Dow Story* (New York: McGraw-Hill, 1968).

7. Robert S. Karpiuk, *Dow Research Pioneers: Recollections 1888–1949* (Midland, Mich.: Pendell Publishing Co., 1981).

IV
Prospects

The Prospect from Here

Erwin N. Hiebert

What impressed me most during the conference is that there are so many different and yet interconnected and complementary ways in which to pursue the history of the chemical sciences and technology. In our concerted efforts to examine and elucidate the cognitive and contextual dimensions of the discipline of chemistry in an integrated way, methodological pluralism has been much in evidence. It has been recognized as such, and emphasized.

Historians, I believe, are uniquely aware of the astounding diversity intrinsic to the task of achieving excellence within the different branches and domains of the sciences and technology. In chemistry, more than in physics, for example, the essential and deep natural complexity of the subject matter is conspicuous. There are, it seems, no ultimate answers. One settles perforce for partial understanding. Rock bottom solutions are nowhere in sight. Wherever one digs, one can dig deeper. In this perennial grubbing for new information and insight into the nature of things, one can find creative artistry, esthetics, and myth in abundance—if one looks for it. Chemical laws are intuitively discovered, chemical compounds are imaginatively constructed, chemical technologies are ingeniously invented, chemical agendas are passionately and idealistically drawn up.

Reductionism fails miserably across the board as one gets into the fine structure of the chemical universe. Atoms presumably are all alike, but their combinations are near infinite. No two forms of any kind of life are alike. In biochemistry and the life sciences the complexity and the richness—and therefore the creative component requisite for engaging in such domains of science gainfully—unfolds so rapidly and radically that it is imperative to learn to live with diversity. This diversity, this unsounded depth of nature, this unmeasured and open-ended world in which the chemist works as architect (in areas that nature never knew) provides, I believe, an essential perspective for the well-proportioned methodological craft of the historian of the chemical sciences and technology.

So where does one go from here? One cannot take possession of the whole possible world of discourse in science history and ride off, Hegelian fashion, in all conceivable directions at the same time. On the other hand, from the more or less blurred general perspective on the larger landscape of science from which we normally launch our studies in a given direction, the historian, eventually, after some muddling, will be compelled to formulate meaningfully focused, that is, restricted, amenable, specific questions and issues worthy of exploration. The problem formulation, one assumes, will lie within the range of the historian's knowledge and skills; but the historian also must have access to tools and documents and other archival resources that are commensurate to the task.

Accordingly, the historian narrows down the focus by force of circumstance. But to what? There are many alternative routes to take and questions to pursue. Some are deemed not to be sufficiently significant or challenging. Others are not manageable or are too narrow or too boundless. But choices are mandatory. And will the choices end up being whiggish? Yes, I think so. Some whiggism is essential to making the choices. One cannot get rid of the "here and now," and in any case, even if one could do so, not all agendas and formulated issues merit the same intensity of study. I would even suggest that it will not suffice to take on a project that, as one might want to say, "has never been done before and needs to be done in order to set the historical record right once and for all." There are many records to lay open, and the inquiry into some of them is more demonstrative of significant science history than others. One can ask countless questions of an historical nature, but they cannot all be asked at the same time. Choices become inevitable, and this suggests priorities.

Let us leave it at that and not enter into what determines the historian's priorities. They rest mostly on intensely personal decisions, albeit decisions that are not entirely free from the historical fashions and the environmental constraints of one's time. As for myself, the most challenging problems worth exploring are such as emerge—take their point of departure if not all of their ingredients or significance—from within the substantive/cognitive content of science itself. Here the fundamental issues are given. One need not invent them. But the outcome of analysis, I suggest, is not totally determined by the historical process itself. Rather, the final product will be defined in part by the questions one seeks to answer and the issues one chooses to illuminate.

One could argue that for the discriminating historian the intrinsic elements of science, ab initio, will be seen to be coextensive with environmental factors such as epistemology, culture, language, geography, politics, and ideology. As for the way in which the growth of the disciplines takes place, science, like literature or art or religion, is contextually conditioned—even while some of us believe that the sciences and technology advance in a more unilinear fashion toward some kind of realistic target than do literature or art or religion.

This overlapping and interpenetration of the sciences and technology with each other and with the contextual circumstances in which they advance or retreat means, of course, that the historian will not be able to avoid becoming a near expert in those other scientific and nonscientific domains that are essential to the explication of the more narrowly chosen original problem. As an example, I think it obvious that a genuinely useful historical analysis of a problem in chemical physics or physical chemistry would demand far more than a perfunctory account of some elements of physics grafted onto chemistry or some elements of chemistry grafted onto physics. The same holds for physiological chemistry, or chemistry conditioned by Marxist thought, or chemistry in the service of the military, or the philosophical implications of quantum chemistry.

Historians of the chemical sciences and technology perhaps could use some new molds of thinking in order to shed light on the theme of these plenary sessions, namely, Where do we go from here? What I have wanted to suggest is that I cannot imagine that our new molds of thinking—if that is what we need—would want to include reductionism, or a uniform language of discourse, or identical educational backgrounds, or a monism of method for pursuing history. The discipline, I think, should remain open-ended, and this implies radical methodological pluralism. So let us celebrate it.

Bewitched, Bothered, and Bewildered

John W. Servos

Upon hearing of plans for the conference, I was full of doubts and reservations. Chemistry surely has a history, and that history has long been neglected, even by historians of science and technology. But would a gathering of its students help to remedy such neglect? Conferences can win resources and visibility for fields of inquiry, but they can also encourage specialists to talk to one another instead of to outsiders, thereby erecting barriers instead of building bridges. They can stimulate scholars to broaden their thinking, but can also promote particularistic sentiments that are inimical to growth. They can generate a kind of esprit de corps by revealing common denominators in scattered work, but can also accentuate the gaps that separate specialists in a given field from each other as well as from those with other interests and concerns. Is the history of chemistry better served by bringing historians of chemistry together to talk about their common interests or by encouraging them to reach out toward colleagues who currently find the phrase "history of chemistry" either archaic or extravagantly narrow?

These doubts have been allayed, but only partially, by the meeting that has resulted from the thoughtful work of Sy Mauskopf and his associates. Fine papers, excellent commentaries, lively discussions, and good company have made this conference an intense and extraordinarily useful one for me and, I suspect, for many of the others in attendance. Less certain, however, is the likelihood that nonspecialists shall find comparable rewards in these published proceedings, not for any lack of scholarly quality, but rather because of the inherent difficulty of speaking at once to multiple audiences. Nor is it clear how it could be otherwise. Hence the title of this brief commentary, "Bewitched, Bothered, and Bewildered."

Bewitched. I am happy to say that my remarks here could fill much more than the allotted space. Impressive, first of all, are the imaginative ways in which Mary Jo Nye, Alan Rocke, Bob Friedel, and Larry Holmes are reworking the history of chemistry in the nineteenth century, a period much in need of ordering principles. Rocke's "quiet revolution," Nye's analysis of chemists' shifting philosophical commitments, Friedel's polarities (analysis and synthesis, craft and science) and Holmes's "complex multiscale narratives" all promise to reveal to us new and unanticipated novelties in familiar but underanalyzed materials.

Many of the papers on the twentieth century are equally impressive. They do not converge on common sources as did the contributions treating the nineteenth century. Perhaps they could not. Chemistry has diversified in so many ways during the past century as to overwhelm the most complicated of multiscale narratives. Nevertheless, it is easy to recognize, to borrow Sy Mauskopf's term, "developmental profiles" in these contributions: in the history of instruments, a crucial part of the history of chemistry that we can all agree has been narrowly construed and unjustly overlooked; in the history of the intricate dance of chemists and physicians in pharmacological research; and in the history of environmental chemistry, a subject that provoked some of the broadest and most intense discussion of the meeting.

As valuable as the papers were the informal discussions and table talk. One example may suffice. At one of those now-storied breakfasts at which Reese Jenkins sat, Reese tossed off the suggestion, between gulps of coffee, that the continuous flow processes of twentieth-century chemical engineering might be better understood if viewed against the backdrop of the continuous production processes of nineteenth-century mechanical engineering. He went on to suggest that the chemical engineering program at MIT might be a critical site in the transfer of this mode of thought and practice from one field to the other, both because of the important role of mechanical engineering courses in MIT's early chemical engineering program and because MIT was the training ground for the chemical engineers at Jersey Standard who were most aggressive in developing and refining continuous-process technologies. As someone who has done work on the history of both MIT and chemical engineering, I found this more than a little interesting. I tried to sit near Reese at other meals as well but found that his table filled too rapidly. When Reese speaks, people listen.

Bothered. Here let me begin with a confession that may mark me as a hopeless reprobate. I have a high regard for the history of ideas. I enjoy works that reconstruct in graceful and sensitive ways the thoughts of others; that analyze the logical and temporal relations among those thoughts; that give us a connected, albeit always incomplete, account of how ideas about the natural world have evolved over time; and that exhibit affinities, sympathies, and antipathies among the concepts of our predecessors.

Frank Sulloway might explain this curiously conservative taste by reference to birth order. But I think there is another reason. I spend most of my time teaching undergraduates—marvelous, bright young men and women whose imaginations and capacities often exceed my own. Bright as they are, however, they often have minds that, in that nineteenth-century phrase, are poorly furnished. A few know something of the work of Aristotle, Galileo, Newton, Lavoisier, and Darwin. But the names of Boyle, Berthollet, Dumas, Liebig, van't Hoff, or Lewis are as ciphers.

Students who barely know the name of Lavoisier and who have never heard of Liebig are not likely to appreciate debates over the social construction of their theories. Students who can associate chemistry only with Spandex or hazardous wastes are not prepared to appreciate the differences between medical and chemical approaches to pharmacological research, much as specialists may savor the discrimination.

I am not telling you anything new. But a comment that was made in the discussion of William Jensen's paper is worth repeating. We are not serving our audiences well. And by failing them, we endanger our future. To this I would only add that our most important audience is composed of liberal arts students (a category that does include science majors). Such students, of course, constitute our pool of students and recruits, but, more important, they purchase and read books and often (if not so often as we would wish) become adults with enduring tastes for serious writing. It is they who ultimately sustain the professional communities to which we belong.

Yet I see an enormous gulf between the literature that is accessible to such students and the esoteric writings that we produce. This gulf is so forbidding that it discourages not only students from studying the history of chemistry but also historians from teaching it. In conversation with others who write about the history of chemistry, I have again and again been struck by how few teach the subject on a regular basis. I, for one, have not offered a course on the history of chemistry in over ten years, preferring to concentrate on topics in which the literature is better suited to undergraduate needs, such as the scientific revolution and the development of evolutionary biology.

Let me not be misunderstood. I am not advocating a moratorium on professional research. Far from it. We want all sorts of good books. Yet our most desperate need, it seems to me, is for an infusion of books and articles that are inviting to readers intermediate in sophistication between Robert Bud's "bright twelve-year-olds" and Ph.D.s in the history of science. Nor do I have in mind the slickly packaged and lightweight surveys, anthologies, and textbooks that sometimes pass for undergraduate readings. Our students, at least the ones we most want to reach, are too discerning to be satisfied by such fare. Though their minds may be poorly stocked, they are open to and readily engaged by ideas. They are interested in politics; they are intrigued by the relationship between knowledge and power; they are receptive to inquiries into epistemological issues so long as those issues are framed in reasonably accessible terms; they can learn to enjoy the study of the past for the reasons most of us enjoy it—the satisfaction of discovering hidden relationships and the joy of learning that the strange can seem familiar and the familiar strange. The trick, of course, is in finding ways to make the history of chemistry into a vehicle for satisfying these interests and curiosities.

Here, as you may already suspect, is where this writer becomes bewildered. How are we to meet the needs of this, our most important and demanding audience? Perhaps guidance is to be found in the examples offered by other fields that have been more successful in engaging undergraduates. Why is it so much easier to interest intelligent liberal arts students in courses on the scientific revolution or the Darwinian revolution, or—to pick topics that have become foci of undergraduate interest in more recent years—the history of science in America, or the history of American technology? Surely not because these subjects are closer to the personal concerns of students than the history of chemistry. No science has been entwined with politics, profit, and power for so long as chemistry; none has had a greater impact on the material world that surrounds us. Nor are these other subjects inherently more accessible. The undergraduate is as likely to be befuddled by the history of rational mechanics as by the history of the atomic theory and as likely to become lost in the names and missions of our welter of federal scientific agencies as in the properties of phlogiston.

I think the answer has much to do with the nature of the secondary literature in these fields. The student (or teacher) entering upon the study of the scientific revolution or the Darwinian revolution can readily find inviting works that both inform and tease the imagination—books that are written in vigorous and supple prose, that express the personalities of the authors as well the character of their subjects, and that manage to use narratives to impart a sense of the past but also show us how the past, while having an integrity of its own, has shaped our present. I have in mind books and articles that exhibit the grace, precision, and passion that Alexandre Koyré brought to the history of the scientific revolution; the meticulous archival knowledge, technical proficiency, and clarity of Richard Westfall; the lively prose and broad vision of Herbert Butterfield; the imagination and vigor of Thomas Kuhn's historical writings; and the craftsmanship of Daniel Kevles, whose work subordinates the history of ideas to the history of social interests, without disparagement. It is books such as these that attracted many of us to the study of the history of science and technology; they continue to do duty in undergraduate classrooms.

The history of chemistry offers few examples of works that could stand comparison with the literature just cited. It is as if the field has skipped a stage in its development, leaping from retrospectives by professional chemists and vapid popularizations to the monographs of professional historians, without producing the broad-scale narratives and ambitious biographies that would attract liberal-arts students. We need such books—books that reach out rather than look inward. I do not know how to write in such a manner about the history of chemistry. Perhaps some of you will show us the way. I hope so. It is books like these that create and sustain vigorous scholarly communities.

Women in Research Schools: Approaching an Analytical Lacuna in the History of Chemistry and Allied Sciences

Pnina G. Abir-Am

From Great Chemists and Obscure Chemical Practitioners to Research Schools as Relational Units of Socio-Historical Analysis

The history of chemistry, much as the history of other sciences, has long been polarized between a focus on great chemists as innovators or originators of great chemical ideas, theories, and concepts (Nye 1972; Bensaude-Vincent 1979, 1983, 1993; Holmes 1985; Rocke 1993) and a later emphasis on the social composition of chemical societies and related professional forums, whose changing membership provided vital clues for understanding chemistry's shifting place in society, culture, and history (Morrell 1972; Bud, Carroll, Sturchio, and Thackray, 1981; Klosterman 1985; Nye 1990; Fruton 1990; Servos 1990; Jensen, this volume).

However, the rise of a cognitive turn in the sociology of science since the mid-1970s, periodically and critically examined by historiography and social theory buffs (Shapin 1982; Whitley 1983; Fuller 1984; Abir-Am 1987a; Shinn 1988; Golinsky 1990; Clarke 1991), has profoundly shifted the terms of the above debate (usually referred to as the "internalist-externalist" debate) toward an ongoing rapprochement, a process being also evident in this volume.

By now it is a truism that neither intellectual nor social approaches are methodologically satisfactory when deployed in isolation from each other, or even in a crudely additive fashion. The current major methodological problem for historians of science, including historians of chemistry, is how to select socio-historical units of analysis that best enable the science analyst to explore the interaction between the origins of ideas, their reception (or rejection), and their subsequent impact on both neighboring disciplines and social institutions at large.

Indeed, the terms of the great internalist-externalist debate have shifted so dramatically as to include the very constitution of scientific facts (and artifacts) as a topic of investigation rather than as an epistemological given. In seeking to comprehend why and how certain scientific innovations succeed or fail to become accepted or institutionalized as facts, discoveries, or laws, most historians of science have ceased appealing to the inherent truth of accepted innovations (or to the "inborn error" associated with discarded ones).

As this comprehensive collection demonstrates, increasing emphasis is being given to recovering the detailed sequence of social processes involved in stabilizing scientific information as objective knowledge (an outcome usually referred to as a "success" or "truth" or no longer in need of social explanation); or in destabilizing, dissolving, or deconstructing such potentially objectifiable knowledge into contingent opinion (an outcome usually referred to as a "failure" or "error," and hence as the object par excellence of social analysis qua excuse for deviation from the path to progress) (Callon, Law, and Rip, eds. 1986; Latour 1987; Star and Gerson 1986; Clarke and Gerson 1989).

The acceptance or rejection of scientific and technological innovations are of major concern not only for historians but also for other science analysts, even economists and business strategists. In the aftermath of the Manhattan project and related World War II collaborative ventures, innovation has come to depend on opportunities for collaboration, opportunities that are often associated with participation in transitory social groupings at the rapidly shifting and increasingly interdisciplinary frontier of science. This dependence stems from a unique property of transitory social formations, namely that of mediating between the subjectivity, fluidity, and idiosyncrasy of individuals and the objectivity or transcontextuality of institutions or reproducible social formations (Giddens 1984; Abir-Am 1987b).

As a result, many innovations are both enabled and constrained by the properties of the social context in which they originate. Indeed, the formal authorship of, and hence intellectual property rights to, many innovations is an outcome of specific processes of "primitive accumulation" and subtle appropriation of scientific credit, reputation, and authority on the part of scientists who are well positioned on the fast lanes of the intersecting social and cognitive hierarchies of science (Shinn 1988; Abir-Am 1991b).

For example, the papers in this volume by Larry Holmes on Justus Liebig in mid-nineteenth-century Germany and Robert Kohler on George Beadle in mid-twentieth-century United States attempt to situate successful scientist entrepreneurs in the social context of their time while combining meticulous scientific detail with intriguing evidence of the role of disciplinary politics in their careers. Yet, despite this promising combination of both internalist and externalist ingredients in the makeup of these leading scientific figures, both papers refrain from exploring how the scientific standing of their great scientist subjects derived from elaborate social strategies of scientific collaboration with large numbers of mostly subordinate "disciples," in almost industrially organized research groups. Conceivably, the organization of huge, productive, and innovative research groups was the ultimate achievement of a chemist such as Liebig or a biochemical geneticist such as Beadle, since the results on which their respective fame is based, including publicity-generating controversies, could not possibly have been reached by them alone, or without access to a wider infrastructure of interdependent cognitive and institutional hierarchies.

This stance of avoiding the study of scientist leaders in the constitutive social context of their collaborators is even more surprising since historians of chemistry and allied fields, especially biochemistry and molecular biology, were pioneers in the study of intellectual circles, disciplinary clubs, and other transient social formations, most notably research schools, that link great figures with a social infrastructure of collaborators in an interdependent outcome of collective creativity (Morrell 1972; Nye 1979, 1990; Geison 1981; Klosterman 1985; Fruton 1990; Abir-Am 1991b).

Work on the history of research schools is perhaps the single most promising approach to reconciling the traditional internalist and externalist approaches to the history of science; it offers a "natural" unit of analysis for the study of making or missing discoveries, making or breaking careers, and the pervasive rise of transdisciplinary collaborative innovation, the quintessential feature of twentieth-century molecular science. Such a focus on research schools needs to be amplified, especially in view of the recent availability of computerized relational databases and network analysis software that greatly facilitate the measurement of various aspects of collective creativity in science and its social locus in informal, intergenerational, and spatially mobile groupings (Abir-Am 1991c).

A new theoretical and empirical focus on units of collaborative innovation, especially research schools that combine coherent research programs, technical choices, social/institutional networks, and policy initiatives in a comparative, cross-national context, holds a far greater promise for the history of science, including the history of chemistry, than studies of great individuals such as a Liebig or a Beadle, even when such studies combine scientific detail with political maneuvering.

Indeed, the question persists why veteran historians of chemistry seeking to deploy social analysis in examining historical phenomena such as shifts in organisms as research tools (as in Kohler's analysis of Beadle) or changes from philosophically relevant to politically charged research programs (as in Holmes's analysis of Liebig), limit themselves to a Latourian fashionable gloss of abstract social discourse on allies and counter-allies (Latour 1987; Callon, Law, and Rip, eds. 1986) rather than deploying the more relevant perspective of the pragmatist and symbolic interactionist tradition of sociology of work (Strauss 1978, 1988; Gerson 1983; Star 1985; Star and Gerson 1986; Fujimura 1988; Clarke and Gerson 1989).

Finally, one cannot but wonder whether names such as Knorr-Cetina, Amsterdamska, Fujimura, or Abir-Am are either too exotic or too genderized or both, and hence invisible for established historians of chemistry in search of social-theoretical guidance. Their avoidance of any engagement with the work of theoretically oriented, often less established female scholars in social and historical studies of science, coupled with their insistence on tolerating only the most fashionable social-theoretical work by male scholars, even when that work is less relevant to their tasks, holds the history of chemistry captive to possibly unconscious forms of non-scholarly or sexist, ethnic, and ageist bias.

These multiple biases lead at "best" to treating as invisible the contributions of women and younger scholars of both sexes (but especially younger women scholars); at worst they directly marginalize and obstruct those scholars' careers. This observation brings us directly to the other single conspicuous omission in the present otherwise comprehensive and balanced collection: the omission of the role of gender and of women scientists in the growth of modern chemical sciences and its associated professional structure.

From Historical Case Studies to a Science Policy Agenda for Women in Chemistry and Allied Sciences

Though widely acknowledged as a major variable in history and society (the index of successive annual meetings of the American Historical

Association suggests that "gender" is perhaps the fastest growing category of historical analysis and one of the largest on any program), the role of gender in the rise of the chemical establishment does not feature among this collection's diverse topics. Ironically, a recent (April 1990) double session of the American Chemical Society on Great Women Chemists, which featured ten speakers and included papers by women historians of science (Johnson 1990; Abir-Am 1990a) on Nobelist women chemists Maria Goeppert Mayer and Dorothy Crowfoot Hodgkin respectively, suggests that there is no lack of qualified researchers or subjects for research on the role of women in modern chemical sciences and allied fields.

Indeed, another speaker at the above meeting was Margaret Rossiter, the leading historian of women in American science (Rossiter 1982), currently MacArthur Fellow and Professor at Cornell University, who started her career as a historian of agricultural chemistry. Furthermore, Rossiter, Helena Pycior, Ruth Sime, and Pnina Abir-Am were among the participants in sessions on the history of women in science at the 1989 and 1990 Annual Meetings of the History of Science Society; this focused on contributions to chemistry, including those of Marie Curie, Lise Meitner, Ida Noddack, Irene Joliot-Curie, Maria Goeppert Mayer, Gerty Cori, Kathleen Lonsdale, Dorothy M. C. Hodgkin, Rosalind Franklin, along with other, less well-known women chemists (Abir-Am and Outram, eds. 1987, 1989).

Chemistry is a field in which relatively large numbers of women have been active, with a considerable number, including the above mentioned mostly Nobelist contingent, achieving excellence. Yet there are no book length biographical studies of great women chemists (with the exception of the legendary Marie Curie, the subject of several biographies in both French and English, the most recent one being Pflaum 1989). This invisibility is particularly disturbing for both historical and social reasons. The popularity of the autobiographical genre among scientists has been in ascendance through the last decade (Abir-Am 1991a). Socially, the lack of biographies on women scientists deprives young women of suitable role models in developing career aspirations, while further reinforcing the image of chemistry as a field inhospitable to women throughout both science and society.

Indeed, clarifying the role of chemists and other scientists as mentors or obstructors of women's careers is not only a question of potential historical and sociological interest, it is also an urgent matter of science policy as well as social policy. In addition to the obvious and primary issue of gender equality, the projected shortage of scientists in the 1990s and beyond has serious implications for a country's competitiveness in the international economic order. It can be best solved by improving the recruitment and retention of women scientists at all stages of the pipeline (Dix 1987, 1991; Abir-Am 1990c, 1992b).

Historical work on the obstacles faced by women in science, pioneered by Margaret Rossiter in *Women Scientists in America* (1982) and further developed in a cross-national context by the twelve authors of *Uneasy Careers and Intimate Lives: Women in Science, 1789–1979* (Abir-Am and Outram, eds., 1987, 1989) suggests that certain recurrent problems in women's careers can serve as guidelines for devising an urgently needed gender-responsive science policy.

For example, the cultural and social constraints on cross-gender mentorship and patronage, crucial ingredients in the making of any career, and the greater complexity for women in combining career with family responsibilities stemming from the deeply entrenched gender asymmetry with regard to primary responsibility for child care, household management, and social life, are topics that require further attention from science analysts. In this connection, the performance of chemical versus other professional organizations (as discussed, for example, in a special session at the American Chemical Society's 200th meeting in Washington, D.C. in August 1990) should be of direct interest to historians and sociologists of chemistry (Georg 1990).

One can enumerate a few other topics that inevitably slipped from the net of this collection, for example, the role of chemists such as the American James Conant (Hershberg 1990); the German Fritz Haber (Gay 1989); the French Frederic Joliot-Curie (Pflaum 1989); or the Israeli Chaim Weitzman (Reinhartz, 1985) in both science and public policy; or the role of chemistry in recent mega projects such as the Human Genome Project as both a language and a tool for decoding biomedical information (Nova, 1990).

But perhaps the most useful way to conclude this commentary on past neglected topics and future strategic research opportunities is to exemplify a synthesis of three key topics identified as lacunae in the history of modern chemistry—the role of informal social formations in scientific innovation; the role of women and gender in chemical practice and knowledge; and the interaction between chemists and science policy—by presenting an outline of work-in-progress on the rise of the research school of chemical crystallography at Oxford, under the leadership of Nobelist Dorothy M. C. Hodgkin, in the period 1934–70. This study focuses on the role of gender, post-World War II policy, and the discipline of chemical crystallography in sustaining a renowned research school that further included a substantial number of women scientists.

Toward a Synthesis of Neglected Topics in the History of Chemistry: Dorothy M. C. Hodgkin and the Oxford School of Chemical Crystallography, 1934–70

Dorothy Mary Crowfoot Hodgkin, hereafter D. C. Hodgkin, is primarily known as a pioneer in the application of X-ray crystallography to bioorganic chemistry and as the architect of the solution of several, increasingly complex, molecular structures of biomedical import, most notably the antibiotic drug penicillin (in the 1940s); vitamin B-12 (in the 1950s); and the hormone insulin (in the 1960s). In 1964, D. C. Hodgkin was the sole recipient of the Nobel Prize for chemistry.

Of equal interest for historians of science should be the less well-known role of D. C. Hodgkin as leader of a large research school of international renown, with about 100 members in over a dozen countries (Dodson, Glusker, and Sayre, eds. 1981). Based at Oxford University, D. C. Hodgkin's school was unique in its relatively large proportion, about one third, of female scientific progeny. Scientific leadership on this scale is rather rare among all scientists, but especially among women scientists, who tend to be "lone stars" even when successful (Sayre, 1975; Abir-Am and Outram, eds. 1987, 1989; Levi-Montalcini 1988).

The question thus persists how D. C. Hodgkin could create a research school of her own at Oxford, given the difficulties most women faced in pursuing a scientific career, either before or after World War Two. A wide variety of factors—historical, social, familial, educational, institutional, political, ideological, personal, and scientific—appear to have converged in sustaining her dual career as both a great scientist and a leader of a research school, in a professional world that has remained, until recently, thoroughly patriarchal.

From an historical viewpoint D. C. Hodgkin grew up at the time of great social reform in the position of women brought about by World War I and the suffrage movement. She was eight years old when the vote was granted to women in 1918, so she missed the most militant phase of the suffrage movement (Vicinus 1985, chapter 7; Marcus, ed. 1987). In 1921, seven years before her arrival as a student, Oxford University voted to grant degrees to women. The "gay" 1920s, when she was an adolescent, were a period of relaxation from both the Victorian standards of tight social control of gender relations and the Edwardian obsession with class distinctions.

Thus, by the time she came to Oxford in 1928, D. C. Hodgkin had experienced a period of increased personal freedom and educational opportunities for women. While somewhat older women colleagues, for example, Dorothy Wrinch (1894–1976) (Abir-Am 1987b), tended to remember the militant postures of the suffragettes in the 1910s and to cultivate a confrontationist outlook toward the mostly male scientific and academic establishment, D. C. Hodgkin seemed particularly capable of working smoothly within it. This capacity is even more remarkable because she held socially progressive views, such as a concern for the welfare of the British working class and the colonized would-be Third World, that were at odds with the conservative values prevailing in the scientific and academic establishments.

From the social viewpoint D. C. Hodgkin was born into an upper middle class professional family whose social position was further enhanced by colonial service for the British Empire (in Egypt and Sudan). This class origin meant not only access to useful social connections but also, and chiefly, freedom from the obsession with social climbing that consumed so many men and women of talent born into the lower strata of the middle classes. Though D. C. Hodgkin came to hold socially progressive views, still, her social origins, forever inscribed (as for all Britons) in her mannerisms and speech patterns, spared her the need to waste time, energy, and emotions on social respectability and acceptance, surely not a negligible advantage for any human being, professional or otherwise, in highly ritualized, class obsessed, regionally and ethnically stratified, conformist British society.

From the familial viewpoint D. C. Hodgkin's background seems to have been conducive (or at least ideologically and pragmatically unhindering) to her career as an academic chemical crystallographer. On the one hand, by her own account, she had a role model in her mother, an expert in ancient textiles who pursued her professional interests, which required extensive traveling, with the apparent blessing of her archaeologist husband, four daughters, and related family. On the other hand, D. C. Hodgkin's sisters not only were available as a "backup" system but on occasion kept a joint household with her, an arrangement that greatly facilitated the care of her three children during her own extensive professional trips.

Furthermore, her parents-in-law provided a free residence since they occupied the Master's Lodge at Queen's College, Oxford, thus freeing the young couple from financial worries about lodging in an accessible location. Other relatives, some of whom feature in Lord Noel Annan's "intellectual aristocracy" network (Annan 1959) provided an extended familial context of both academic excellence and professionally informed contacts. Last but not least, D. C. Hodgkin had a non-conventional husband, the African scholar Thomas Hodgkin (1909–1988), of Quaker and communist persuasions, who at the time of their marriage in 1937 had just transferred from a position with the colonial service in Palestine (which he quit in protest over the repression of the 1936 Arab uprising) to a position motivated by his social consciousness, namely, evening adult education for the working classes.

Such ideological commitments enabled the Hodgkins to pioneer an egalitarian marriage in the 1930s, an arrangement essential for her to maintain her career despite a family of three children. Child care and household maintenance duties, which often neutralized married women scientists or greatly reduced their capacity to retain a high-powered career, appear to have been shared among the Hodgkins; indeed some observers have indicated that Tom Hodgkin shouldered the lion's share of them. His work with adult education in the evenings meant that he was often available during the day, or when she worked long hours in the laboratory, especially at a time when their children were young (before his emergence in the 1950s as an African scholar with increasing traveling duties). Many household responsibilities were also delegated to a live-in Welsh nanny (whose affordability was ensured, needless to say, by her origin in an economically depressed and ethnically oppressed social group).

Of special importance was the fact that D. C. Hodgkin did not need to relocate after marriage (relocation to the place of the husband's job has often adversely affected the career prospects of women scientists, including those who married scientists), since her husband was able to travel to his work from Oxford, where, as a graduate of Balliol College and nephew of a former Balliol master, he had a solid basis for an academically oriented social life. In this way, D. C. Hodgkin both benefited from the inbreeding habits at Oxford, habits that secured for her a donship in the mid-1930s, and learned the ropes of university politics conveniently at the dinner table in the homes of various academic relatives.

From the educational-institutional viewpoint D. C. Hodgkin's lifelong association with Somerville College as student, tutor, and Fellow was significant in terms of her absorbing a value system and image that combined expectations for female intellectual excellence with social reformism and non-conformism (for example, Somerville was the first women's college at Oxford to accept dissenters, Catholics, and "Jewesses").

At the same time, the still precarious and marginal position of women at Oxford (for example, the Chemical Club at Oxford, a student-sponsored organization, which invited D. C. Hodgkin, a tutor in chemistry at Somerville, to give a talk, did not have women members) meant that the women students' new sense of independence and personal and social power, stemming from their experience with success in competitive collegiate examinations, sports, and dramatic productions, was constrained by the old male tradition at Oxford that still treated women as objects of toleration. Women students were subject to greater discipline and had lower standards of living than men students.

As Vicinus put it, academic women's freedom was bought "at the price of political timidity, a frequent fear of change and a dislike of innovation" (Vicinus 1985, p. 135). Many women dons led controlled, passionless and pliable inner as well as outer lives (Leonardi 1989, p. 49).

However, the Principal of Somerville in D. C. Hodgkin's time as student was Margery Fry, sister of the Bloomsbury painter and biographer of Virginia Woolf Roger Fry, who provided a new role model as a noted social reformer (it was at a college event organized by Principal Fry in 1929 that D. C. Hodgkin met her future husband, whose family was related to the Frys by marriage) (Huwes Jones 1977, p. 146).

Another unconventional role model was the flamboyant medical scientist Janet Vaughn, who became Principal during D. C. Hodgkin's time as a young don (Caldecott 1984). While former Principals had been chiefly preoccupied with maintaining strict disciplinary and scholarly standards so as to secure tolerance of their "girlies" at Oxford, those in D. C. Hodgkin's formative time as a don were independent women who provided an inspiring role model and a strong sense of social and professional mission.

Yet another institutional advantage that accrued to D. C. Hodgkin over other women scientists was the early offer of a position as tutor and later Fellow in her own college, the only way for a woman to have professional and economic security at Oxbridge at that time. Furthermore, by the late 1930s, the social control of married women was more relaxed at Oxford than at Cambridge; thus D. C. Hodgkin was allowed to retain her Fellowship upon marriage.

From the political viewpoint D. C. Hodgkin's formative years coincided with the rise to power of the Labour Party, and with the increasing appeal of socialist politics among Oxbridge students and intellectuals. Socialist ideology was particularly appealing to women scientists because of its vague but nevertheless progressive and egalitarian position on women's status. Though D. C. Hodgkin chose to devote the bulk of her energy and time to science, assuming a modest, indirect role in political activism, nevertheless she supported the communist activism of her Ph.D. adviser, mentor, and collaborator J. D. Bernal; her husband Thomas Hodgkin; and many other colleagues and friends. Her main benefit from her adherence to this leftist orientation and company appears to have been access to a supportive social environment, which also provided legitimation and tolerance for her career aspirations despite the quiet but steady deviation they represented from women's traditional roles.

Leftist ideology, being the most progressive in the 1930s (the alternatives were the declining liberalism of appeasement, conservativism, or fascism) proved inspiring for many scientists not only because of its goal of a just political order but also because it advocated a more central role for science in the planning of the utopian socialist society. D. C. Hodgkin's mentor and collaborator J. D. Bernal was a major spokesman for this sort of scientistic Marxism. Leftist scientists in the 1930s thus enjoyed the dual gratification of believing that their scientific career was not only a quest for truth but also a key step in assisting society to undergo an imminent social revolution.

From the scientific viewpoint D. C. Hodgkin's early interest in chemical crystallography, channeled by J. D. Bernal into the X-ray crystallography of key biological compounds, especially sex hormones and proteins, enabled her to pioneer, together with the visionary Bernal and his few collaborators, the new field of biomolecular crystallography. As collaborators they complemented each other well. He preferred to pursue too many new ideas rather than completing any particular project, a goal which would have involved routine rather than exploratory work; in contrast, she possessed the temperament and tenacity required to bring projects to completion. Indeed, when she realized the utopian dimensions of Bernal's vision for solving the problem of protein structure, D. C. Hodgkin boldly chose to focus on less grandiose goals, ones that could possibly be successfully approached by the then available X-ray theory and technology.

This tendency for pragmatism, also informed by her passion to demonstrate the then yet unrecognized usefulness of X-ray crystallography to organic chemists (who still resisted such "indirect" proofs of molecular structure), and by her responsibility as a tutor for students who needed manageable research problems, led D. C. Hodgkin to start her own "production line" of solving the structure of "smaller" biological compounds such as cholesteryl iodide and later penicillin and vitamin B12. In this way she anticipated the solution of the first protein structures, made possible only by the advent of the early computers in the 1950s. However, in the 1960s she returned to "her own protein," the bi-peptide insulin, with a large team and powerful equipment, a long journey from her encounter with insulin crystals in 1935 (Dodson, Glusker, and Sayre, eds.' 1981; Abir-Am 1992c).

D. C. Hodgkin's research school thus revolved around a specific scientific vision that she shared with Bernal, namely that molecular structure is the clue to biological function. But she also adapted the utopian problem of protein structure of the 1930s (at that time proteins were considered to be the most important biomolecules, hence decoding their structure was often described as equivalent to discovering the "secret of life") to the structurally feasible goals of chemical crystallographic technology in the 1940s and later, and to the institutional requirements of tutorials and degrees at Oxford.

D. C. Hodgkin's research school formed around an innovative, nonconformist subject matter that combined a physical technique, a chemical goal, and a biological meaning long before molecular biology became the heroic field of the late twentieth century. A charismatic personality and an attractive institutional basis, the other key ingredients of a research school, appear to have been amply present in D. C. Hodgkin's vision turned international laboratory; indeed, tens of former students and research associates provided testimony to her gentle but firm guiding presence in their careers (Dodson, Glusker, and Sayre, eds. 1981).

Other key elements sustaining D. C. Hodgkin's remarkable career pertained to *mentorship* and patronage by her Ph.D. adviser and collaborator, J. D. Bernal, who not only supported her career for reasons of both merit and egalitarian conviction but further enlisted the support of W. H. and W. L. Bragg, the founders of X-ray crystallography, who allowed her access to rare equipment at the Royal Institution laboratory in London and wrote crucial letters of recommendation on her behalf. Bernal's and D. C. Hodgkin's pioneering collaboration on the X-ray crystallography of sex hormones and proteins in the 1930s, during which they obtained and interpreted the first X-ray photo of a globular, or biologically active protein, an event recalled fifty years later as the beginning of molecular biology (Abir-Am 1992c), was facilitated by Bernal's habit of delegating responsibility to his associates, thus minimizing the adviser-student distance.

Furthermore, at the outset of World War II, when Bernal was recruited as scientific adviser to various government ministries, he transferred to D. C. Hodgkin his equipment, staff, and support from the Rockefeller Foundation. These resources boosted the capacity of her modest laboratory, which remained active at a time when Bernal's had suffered from the London blitz. At the end of the war, Bernal's and Sir Lawrence Bragg's letters were decisive in securing D. C. Hodgkin a university lectureship at Oxford, successfully competing against a male leading candidate.

It was during World War II that D. C. Hodgkin, by then a mother of two and who took in refugees from blitzed London (Edsall, Bearman, and Abir-Am 1979), got the opportunity to expand her research and coordinate the work of a team that was part of the Oxford penicillin effort of Howard Florey and Ernst Chain. Eventually, her team solved the structure while demonstrating that X-ray crystallography could anticipate classical organic synthesis in proving a complex structure. The war not only conferred social relevance on her X-ray structural efforts on penicillin but provided crucial experience in team management, an experience essential for her later ventures into the larger projects of Vitamin B12 and insulin.

The work on penicillin also led to D. C. Hodgkin's election to the Royal Society in 1947, a year after the birth of her third child, and at age 37, making her the youngest woman to have been elected (the Royal Society elected its first women members in 1945; those elected were veteran scientists, often ten years or more senior to D. C. Hodgkin).

Another valuable experience gained during the war was D. C. Hodgkin's grantsmanship with the Rockefeller Foundation, which continued to support equipment and research assistance for her work into the late 1940s, when she declined further grants on the grounds that Oxford University should by then have assumed this responsibility. By the late 1940s, she toured the United States extensively at the invitation of the Rockefeller Foundation to see the early computers and was well prepared to launch the next phases, well described by her many students and collaborators in a festschrift produced on the occasion of her seventieth birthday (Dodson, Glusker, and Sayre, eds. 1981).

While many questions remain to be sorted out before a full biography of D. C. Hodgkin, let alone a collective biography of her school, can be completed, there is no doubt that a study of D. C. Hodgkin's school combines all the most exciting topics in the history of modern science, world history, and gender history. Such a study, currently under way with support from the Visiting Professorships for Women Program of the National Science Foundation, may not fill all current lacunae in this or other collections but may add totally new dimensions to the history of chemistry.

References

Abir-Am, Pnina G. 1987a. "The Biotheoretical Gathering, Transdisciplinary Authority, and the Incipient Legitimation of Molecular Biology in the 1930s: New Perspective on the Historical Sociology of Science." *History of Science* 25: 1–70.

———. 1987b. "Synergy or Clash: Disciplinary and Marital Strategies in the Career of Mathematical Biologist Dorothy Wrinch." In Pnina G. Abir-Am and Dorinda Outram, eds., 1987, 1989, *Uneasy Careers and Intimate Lives: Women in Science, 1789–1979,* 388–394. New Brunswick, N. J. and London: Rutgers University Press.

———. 1990a. "Science Policy or Social Policy for Women in Science: Lessons from Historical Case Studies." *EASST Newsletter* 9: 14–17.

———. 1990b. "Leadership and Gender in Science: D.C. Hodgkin and the Oxford School of Chemical Crystallography." Paper read at the American Chemical Society Meeting, Boston, 24 April. Session on "Great Women Chemists" organized by Jane Miller.

———. 1991a. "Nobelesse Oblige: Lives of Molecular Biologists." *Isis* 82: 326–343.

———. 1991b. "Research Schools of Molecular Biology in the UK, US, and France: A Study of Collaborative Innovation, Leadership, and Policy." Paper read at the Symposium for the Philosophy of Science, Boston University, 16 April.

———. 1991c. "A Relational Database for the Study of Research Schools with Special Emphasis on Molecular Biology in the UK, US, and France." Paper read at the History of Science Society Annual Meeting, Madison, 1 November. Also read at a Colloquium in History and Philosophy of Science, University of Maryland at College Park, 10 October.

———. 1992a. "The Politics of Macromolecules: Molecular Biologists, Biochemists and Rhetoric." *Osiris* 7: 210–237.

———. 1992b. "Science Policy or Social Policy for Women in Science: From Historical Case Studies to an Agenda for the 1990s." *Science and Technology Policy* (April) : 11–12.

———. 1992c. "A Historical Ethnography of a Scientific Anniversary in Molecular Biology: The First Protein X-ray Photo, 1984, 1934." *Social Epistemology* 6: 323–354; 363–364; 371–372; 380–387.

Abir-Am, Pnina G. and Dorinda Outram, eds. 1987, 1989. *Uneasy Careers and Intimate Lives: Women in Science, 1789–1979.* New Brunswick, N.J. and London; Rutgers University Press.

Amsterdamska, Olga. 1987. *Schools of Thought: The Development of Linguisitics from Bopp to Saussure,* Dordrecht and Boston: Reidel.

Annan, Noel G. "The Intellectual Aristocracy." In J. H. Plumb, ed., *Studies in Social History: A Tribute to G. M. Trevelyan,* 214–287. London: Longmans, Green.

Bensaude-Vincent, Bernadette. 1979. "Le Mandarinat des chimistes, français au XIX siècle." *Bulletin de 1'Union des Physiciens* (December): 383–392.

———. 1983. "A Founder Myth in the History of Science: The Lavoisier Case." In Loren Graham, Wolf Lepenis, and Peter Weingart, eds., *Functions and Uses of Disciplinary Histories. Sociology of Science Yearbook* 7: 53–78. Dordrecht and Boston: Reidel.

———. 1993. *Lavoisier et ses historiens.* In press.

Bud, Robert, Thomas Carroll, Jeffrey Sturchio, and Arnold Thackray. 1981. *Chemical Indicators: A Study of American Chemists, 1870–1970.* Philadelphia: University of Pennsylvania Press.

Caldecott, Leonie. 1984. *Women of Our Century.* London: British Broadcasting Corporation.

Callon, Michel, John Law, and Arie Rip, eds. 1986. *Mapping the Dynamics of Science and Technology.* London: Sage.

Clarke, Adele, E. "Social Worlds/Arenas Theory as Organizational Theory." In D. R. Maines, ed., *Social Organization and Social Process: Essays in Honor of Anselm Strauss,* 119–158. New York: Aldine de Gruyter.

Clarke, Adele E. and Elihu M. Gerson, 1989. "Symbolic Interactionism in

Social Studies of Science." In Howard S. Becker and Michal McCall, eds., *Symbolic Interaction and Cultural Studies,* 1–52. Chicago: University of Chicago Press.

Collins, Randall. 1987. "A Micro-Macro Theory of Intellectual Creativity: The Case of German Idealist Philosophy." *Sociological Theory* 5: 47–69.

Dix, Linda S., ed. 1987. *Women: Their Underrepresentation and Career Differentials in Science and Engineering.* Washington D.C.: National Academy Press.

———, ed. 1991. *Women in Science and Engineering: Increasing Their Numbers in the 1990s, a Statement on Policy and Strategy.* Washington, D.C.: National Academy Press.

Dodson, Guy, Jenny P. Glusker, and David Sayre, eds. 1981. *Structural Studies on Molecules of Biological Interest: A Volume in Honour of Dorothy Hodgkin.* Oxford: Oxford University Press.

Edsall, John T., David Bearman, and Pnina G. Abir-Am. 1979. "Report on the Archives of J. D. Bernal, W. T. Astbury and the Braggs." *Newsletter of the Survey of Sources for the History of Biochemistry & Molecular Biology* 10: RA 2–6, Philadelphia: American Philosophical Society Library.

Fujimura, Joan. 1988. "The Molecular Biological Bandwagon in Cancer Research: When Social Worlds Meet." *Social Problems* 35: 261–283.

———. 1991. "On Methods, Ontologies, and Representation in the Sociology of Science: Where do We Stand?" In D. R. Maines, ed., *Social Organization and Social Process: Essays in Honor of Anselm Strauss,* 207–248. New York: Aldine de Gruyter.

Fuller, Steve. 1984. "The Cognitive Turn in Sociology." *Erkenntnis* 21: 439–450.

Fruton, Joseph S. 1990. *Contrasts in Scientific Style: Research Groups in Chemical and Biochemical Sciences.* Philadelphia: American Philosophical Society Library.

Gay, Peter. 1990. Paper read at the conference on which this volume is based, Chemical Sciences in the Modern World, Beckman Center for the History of Chemistry, Philadelphia, 17–20 May 1990.

Geison, Gerald L. 1981."Scientific Change, Emerging Specialties and Research Schools." *History of Science* 19: 20–40.

Georg, Gunda, organizer. 1990. "The Impact of Science Policy on Women Chemists." Session at the 200th Meeting of the American Chemical Society, Washington, D.C., 28 August.

Gerson, Elihu M. 1983. "Scientific Work and Social Worlds." *Knowledge: Creation, Diffusion, Utilization* 4: 357–377.

Giddens, Anthony. 1984. *The Constitution of Society: Outline of the Theory of Structuration.* Cambridge: Cambridge University Press.

Golinsky, Jan. 1990. "The Theory of Practice and the Practice of Theory: Sociological Approaches in the History of Science." *Isis* 81: 492–505.

Hershberg, James G. 1990. "Aloof but not Oblivious: James B. Conant, Manhattan Project Scientists, and Controlling the Atomic Bomb, 1944–1945." Paper read at the Annual Meeting of the History of Science Society, Seattle, 31 October.

Hodgkin, D. M. C. 1980. "J. D. Bernal, 1901–1971." *Biographical Memoirs of Fellows of the Royal Society* 26: 1–80.

———. 1987. "Patterson and Pattersons." In Jenny P. Glusker et al., eds., *Patterson and Pattersons: Fifty Years of the Patterson Function,* 167–193. Oxford: Oxford University Press and International Union of Crystallography.

Holmes, Frederic L. 1985. *Lavoisier and the Chemistry of Life: An Explanation of Scientific Creativity.* Madison: University of Wisconsin Press.
Huwes Jones, Enid. 1977. *Margery Fry.* Oxford: Oxford University Press.
Johnson, Karen, F. 1990. "Maria Goeppert Mayer and the Shell Model." Paper read at the American Chemical Society Meeting, Boston, 24 April. Session on "Great Women Chemists" organized by Jane Miller.
Klosterman, L. 1985. "A Research School of Chemistry in the 19th Century: Jean-Baptiste Dumas and His Research Students." *Annals of Science* 43: 1–80.
Knorr-Cetina, Karin D. 1981. *The Manufacture of Knowledge, An Essay on the Constructivist and Contextual Nature of Science.* New York: Pergamon.
Knorr-Cetina, Karin D. and Aaron Cicourel, eds. 1981. *Advances in Social Theory and Methodology: Toward an Integration of Micro- and Macro-Sociologies.* Oxford: Oxford University Press.
Knorr-Cetina, Karin D. and Michael Mulkay, eds. 1983. *Science Observed.* London: Sage.
Latour, Bruno. 1987. *Science in Action.* Cambridge, Mass,: Harvard University Press.
Leonardi, Susan J. 1989. *Dangerous by Degrees: Women at Oxford and the Somerville College Novelists.* New Brunswick, N.J. and London: Rutgers University Press.
Levi-Montalcini, Rita. 1988. *In Praise of Imperfection: My Life and Work.* New York: Basic Books.
Marcus, Jane, ed. *Suffrage and the Pankhursts.* London: Routledge and Kegan Paul.
Morrell, Jack, B. 1972. "The Chemist Breeders: The Research Schools of Liebig and Thomas Thomson." *Ambix* 19: 1–46.
Nova (Public Television Series). 1990. *The Book of Life: The Story of the Human Genome Initiative.* Videocassette.
Nye, Mary Jo. 1972. *Molecular Reality, A Perspective on the Scientific Work of Jean Perrin.* New York: American Elsevier.
———. 1979. "The Boutroux Circle and Poincaré's Conventionalism." *Journal of the History of Ideas* 15: 107–120.
Pflaum, Rosalynd. 1989. *Grand Obsession: Marie Curie and Her World.* New York: Doubleday.
———. 1990. "Chemical Explanation and Physical Dynamics: Two Research Schools at the First Solvay Chemistry Conferences." *Annals of Science* 46: 461–480.
Reinhartz, Yehuda. 1985. *Chaim Weitzman.* Oxford: Oxford University Press.
Rocke, Alan J., 1993. *The Quiet Revolution: Hermann Kolbe and the Science of Organic Chemistry.*
Rose, Hilary, 1990. "Talking About Science in Three Colors: Bernal and Gender Politics in the Social Studies of Science." *Science Studies* 3: 5–20.
Rossiter, Margaret W. 1982. *Women Scientists in America: Struggles and Strategies to 1940.* Baltimore, Johns Hopkins University Press.
Sayre, Anne. 1975. *Rosalind Franklin and DNA.* New York: Norton.
Servos, John W. 1990. *Physical Chemistry from Ostwald to Pauling: The Making of a Science in America.* Princeton, N.J.: Princeton University Press.
Shapin, Steven, 1982. "History of Science and Its Sociological Reconstructions." *History of Science* 20: 157–211.
Shinn, Terry. 1988. "Hierarchies des chercheurs et formes des recherches." *Actes de la Recherches des Sciences Sociales* 74 (September): 2–22.

Star, Susan Leigh. 1985. "Scientific Work and Uncertainty." *Social Studies of Science* 15: 391–427.
Star, Susan Leigh and Elihu M. Gerson. 1986. "The Management of Anomalies in Scientific Work." *Sociological Quarterly* 28: 147–169.
Strauss, Anselm. 1978. *Negotiations, Varieties, and Contexts, Processes and Social Order.* New York: Jossey Bass.
———. 1988. "The Articulation of Project Work: An Organizational Process." *Sociological Quarterly* 29: 163–178.
Vicinus, Martha. 1985. *Independent Women.* London: Virago.
Whitley, Richard. 1983. "From the Sociology of Scientific Communities to the Study of Scientists' Negotiations and Beyond." *Social Science Information* 22: 681–720.

Bibliography

Abir-Am, Pnina G. "The Politics of Macromolecules: Molecular Biologists, Biochemists, and Rhetoric." *Osiris* 7 (1992): 210–237.

———. "A Historical Ethnography of a Scientific Anniversary in Molecular Biology: The First Protein X-ray Photo, 1984, 1934." *Social Epistemology* 6 (1992): 321–387.

———. "Science Policy or Social Policy for Women in Science: From Historical Case Studies." *Science and Technology Policy* 5 (1992): 11–12.

———. "Nobelesse Oblige: Lives of Molecular Biologists." *Isis* 82 (1991): 326–343.

———. "The Biotheoretical Gathering, Transdisciplinary Authority, and the Incipient Legitimation of Molecular Biology in the 1930s: New Perspective on the Historical Sociology of Science." *History of Science* 25 (1987): 1–70.

Abir-Am, Pnina G. and Dorinda Outram, eds. *Uneasy Careers and Intimate Lives: Women in Science, 1789–1979.* New Brunswick, N.J. and London: Rutgers University Press, 1987, 1989.

Alexander, Philip N. and Helen W. Samuels. "The Roots of 128: A Hypothetical Documentation Strategy." *American Archivist* 50 (Fall 1987): 518–531.

Allen, Garland E. *Thomas Hunt Morgan: The Man and His Science.* Princeton, N.J.: Princeton University Press, 1978.

Anschütz, Richard. *August Kekulé.* Berlin: Verlag Chemie, 1929.

Aulie, Richard P. "Boussingault and the Nitrogen Cycle." Ph.D. dissertation, Yale University, 1968.

Bachelard, Gaston. *La pluralisme cohérent de la chimie moderne.* Paris: Vrin, 1931. Reprint. Paris: Éditions du Seuil, 1969.

Backman, Jules. *The Economics of the Chemical Industry.* Washington, D.C.: Manufacturing Chemists Association, 1970.

Barnes, Barry. *Interests and the Growth of Knowledge.* London: Routledge and Kegan Paul, 1977.

Barnes, Barry and David Edge, eds. *Science in Context: Readings in the Sociology of Science.* Cambridge, Mass.: MIT Press, 1982.

Barnes, Barry and Steven Shapin, eds. *Natural Order: Historical Studies of Scientific Culture.* Beverly Hills and London: Sage, 1979.

Bearman, David. *Archival Methods.* Pittsburgh: Archives and Museum Informatics, 1989.

Beckman Center for the History of Chemistry. *Final Report: Polymers and People.*

NSF Project no. SES 84-21278. Philadelphia: Beckman Center for the History of Chemistry, University of Pennsylvania, 1988.

Beer, John J. *The Emergence of the German Dye Industry.* Urbana: University of Illinois Press, 1959.

Belasco, Warren J. *Appetite for Change: How the Counterculture Took on the Food Industry, 1966–1988.* New York: Pantheon, 1989.

Benjamin, Walter. "The Work of Art in the Age of Mechanical Reproduction." In Benjamin, *Illuminations,* edited by Hannah Arendt, translated by Harry Zohn, 219–254. London: Fontana, 1973.

Bensaude-Vincent, Bernadette. "A Founder Myth in the History of Science: The Lavoisier Case." In *Functions and Uses of Disciplinary Histories,* edited by Loren Graham, Wolf Lepenies, and Peter Weingart, 53–78. Dordrecht and Boston: Reidel, 1983.

Berenson, Conrad, ed. *The Chemical Industry: Viewpoints and Perspectives.* New York: Interscience, 1963.

Bijker, Wiebe E., Thomas P. Hughes, and Trevor Pinch, eds. *The Social Construction of Technological Systems: New Directions in the Sociology and History of Technology.* Cambridge, Mass.: MIT Press, 1987.

Bloor, David. *Knowledge and Social Imagery.* London: Routledge and Kegan Paul, 1976.

Borscheid, Peter. *Naturswissenschaft, Staat und Industrie in Baden (1848–1914).* Stuttgart: Klett, 1976.

Bosso, Christopher J. *Pesticides and Politics: The Life Cycle of a Public Issue.* Pittsburgh: University of Pittsburgh Press, 1987.

Bovet, Daniel. *Une chimie qui guérit: histoire de la découverte des sulfamides.* Paris: Éditions Payot, 1988.

Bradley, John. "On the Operational Interaction of Classical Chemistry." *British Journal for the Philosophy of Science* 6 (1955–1956): 32–42.

Brannigan, Augustine. *The Social Basis of Scientific Discoveries.* Cambridge: Cambridge University Press, 1981.

Brooke, John. "Organic Synthesis and the Unification of Chemistry—A Reappraisal." *British Journal for the History of Science* 5 (1971): 363–392.

———. "Avogadro's Hypothesis and Its Fate." *History of Science* 19 (1981): 235–273.

Brooke, John Hedley. "Laurent, Gerhardt, and the Philosophy of Chemistry." *Historical Studies in the Physical Sciences* 6 (1975): 405–429.

———. "Methods and Methodology in the Development of Organic Chemistry." *Ambix* 34 (1987): 147–155.

Bruemmer, Bruce H. and Sheldon Hochheiser. *The High-Technology Company: A Historical Research and Archival Guide.* Minneapolis: Charles Babbage Institute, University of Minnesota, 1989.

Bud, Robert. "Biotechnology in the Twentieth Century." *Social Studies in Science* 21 (1991): 415–457.

———. "The Myth and the Machine: Seeing Science Through Museum Eyes." In *Picturing Power: Visual Depictions and Social Relations,* edited by Gordon Fyfe and John Law, 34–59. Sociological Review Monographs 35. London: Routledge and Kegan Paul, 1988.

———. *The Uses of Life: A History of Biotechnology.* Cambridge: Cambridge University Press, 1993.

Bud, Robert and G. K. Roberts. *Science Versus Practice: Chemistry in Victorian Britain.* Manchester: Manchester University Press, 1984.

Bud, Robert, Thomas Carroll, Jeffrey Sturchio, and Arnold Thackray, eds. *Chemical Indicators: A Study of American Chemists, 1870–1970*. Philadelphia: University of Pennsylvania Press, 1981.
Bugos, Glenn. "Managing Cooperative Research and Borderland Science in the National Research Council, 1922–1941." *Historical Studies in the Physical and Biological Sciences* 20, 1 (1989): 1–32.
Burian, Richard M., Jean Gayon, and Doris Zallen. "The Singular Fate of Genetics in the History of French Biology, 1900–1940." *Journal of the History of Biology* 21 (1988): 357–402.
Butterfield, Herbert. *The Origins of Modern Science, 1300–1800*. 1957. Revised edition, New York: Free Press, 1965.
Caldin, E. F. *The Structure of Chemistry in Relation to the Philosophy of Science*. London and New York: Sheed and Ward, 1961.
Callon, Michel, Arie Rip, and John Law, eds. *Mapping the Dynamics of Science and Technology: Sociology of Science in the Real World*. London: Sage, 1986.
Campbell, Murray and Harrison Hatton. *Herbert H. Dow, Pioneer in Creative Chemistry*. New York: Appleton-Century-Crofts, 1949.
Cannadine, David. "The State of British History." *Times Literary Supplement*, 10 October 1986, 1139–1140.
Carson, Rachel. *Silent Spring*. Boston: Houghton-Mifflin, 1962.
Chandler, Alfred D., Jr. *The Visible Hand: The Managerial Revolution in American Business*. Cambridge, Mass.: Belknap Press of Harvard University Press, 1977.
Chandler, Alfred D., Jr., with Takashi Hikino. *Scale and Scope: The Dynamics of Industrial Enterprise*. Cambridge, Mass.: Harvard University Press, 1990.
Chandler, Alfred D., Jr., and Herman Daems, eds. *Managerial Hierarchies: Comparative Perspectives on the Modern Industrial Enterprise*. Cambridge, Mass.: Harvard University Press, 1980.
Channell, David. *The Vital Machine: A Study of Technology and Organic Life*. Oxford: Oxford University Press, 1991.
Colebrook, Leonard. "Gerhard Domagk 1895–1964." *Biographical Memoirs of Fellows of the Royal Society* 10 (1964): 39–50.
Coleman, William and Frederic L. Holmes, eds. *The Investigative Enterprise: Experimental Physiology in Nineteenth-Century Medicine*. Berkeley: University of California Press, 1988.
Coley, Noel G. "Physicians, Chemists, and the Analysis of Mineral Waters: The Most Difficult Part of Chemistry." In *The Medical History of Waters and Spas*, edited by Roy Porter, 56–66. Medical History, supplement 10. London: Wellcome Institute for the History of Medicine, 1990.
Collins, Harry M. "Stages in the Empirical Program of Relativism." *Social Studies of Science* 11 (1981): 3–10.
———. *Changing Order: Replication and Induction in Scientific Practice*. London: Sage Publications, 1985.
Collins, Randall. "A Micro-Macro Theory of Intellectual Creativity: The Case of German Idealist Philosophy." *Sociological Theory* 5 (1987): 47–69.
Corbin, Alain. *The Foul and the Fragrant: Odor and the French Social Imagination*. Cambridge, Mass.: Harvard University Press, 1986.
Cox, Richard J. "A Documentation Strategy Case Study: Western New York." *American Archivist* 52 (Spring 1989): 192–200.
———. "Professionalism and Archivists in the United States." *American Archivist* 49 (Summer 1986): 229–247.

Crone, Hugh D. *Chemicals and Society: A Guide to the New Chemical Age*. Cambridge: Cambridge University Press, 1986.

Dagognet, François. *Tableaux et langage de la chimie*. Paris: Éditions du Seuil, 1969.

Debus, Allen G. *The Chemical Philosophy: Paracelsian Science and Medicine in the Sixteenth and Seventeenth Centuries*. 2 vols. New York: Science History, 1977.

Dix, Linda S., ed. *Women: Their Underrepresentation and Career Differentials in Science and Engineering*. Washington, D.C.: National Academy Press, 1987.

Doan, H. D. "Dow Will Direct Its Efforts Toward a Better Tomorrow." In *Dow Diamond* (publication of The Dow Chemical Company) 30, 4 (1967).

Dodson, Guy, Jenny P. Glusker, and David Sayre, eds. *Structural Studies on Molecules of Biological Interest*. A volume in honour of Dorothy Hodgkin. Oxford: Oxford University Press, 1981.

Donovan, Arthur. *Philosophical Chemistry in the Scottish Enlightenment: The Discoveries of William Cullen and Joseph Black*. Edinburgh: Edinburgh University Press, 1975.

Douglas, Mary. *Risk Acceptability According to the Social Sciences*. London: Routledge and Kegan Paul, 1985.

———. *Natural Symbols: Explorations in Cosmology*. New York: Vintage, 1973.

———. *Purity and Danger: An Analysis of the Concepts of Pollution and Taboo*. London: Routledge and Kegan Paul, 1966.

Douglas, Mary and Baron Isherwood. *The World of Goods: Towards an Anthropology of Consumption*. New York: Basic Books, 1979. Reprint. New York: Norton, 1982.

Douglas, Mary and Aaron Wildavsky. *Risk and Culture: An Essay on the Selection of Technical and Environmental Dangers*. Berkeley: University of California Press, 1982.

Dowling, Harry F. *Fighting Infection: Conquests of the Twentieth Century*. Cambridge, Mass. and London: Harvard University Press, 1977.

Du Bois, J. Harry. *Plastics History U.S.A.* Boston: Cahners, 1972.

Durkheim, Émile and Marcel Mauss. *Primitive Classification*, translated by Rodney Needham. Chicago: University of Chicago Press, 1963.

Eliade, Mircea. *The Forge and the Crucible*, 2d ed., translated by Stephen Corrin. Chicago: University of Chicago Press, 1978.

Everett, Kenneth G. and Will S. DeLoach. "Who is Teaching the History of Chemistry?" *Journal of Chemical Education* 64 (1987): 991–993.

Forbes, Gilbert B. and Grace M. Forbes. "An Historical Note on Chemotherapy of Bacterial Infections." *American Journal of Diseases of Children* 119 (1970): 6–11.

Fox, Richard Wightman and T. J. Jackson Lears, eds. *The Culture of Consumption: Critical Essays in American History, 1880–1980*. New York: Pantheon, 1983.

Fox, Robert and George Weisz, eds. *The Organization of Science and Technology in France, 1808–1914*. Cambridge: Cambridge University Press, 1980.

Friedel, Robert. *Pioneer Plastic: The Making and Selling of Celluloid*. Madison: University of Wisconsin Press, 1983.

Fruton, Joseph S. *Contrasts in Scientific Style: Research Groups in the Chemical and Biochemical Sciences*. Philadelphia: American Philosophical Society, 1990.

Fujimura, Joan. "The Molecular Biological Bandwagon in Cancer Research: Where Social Worlds Meet." *Social Problems* 35 (1988): 261–283.

Fuller, Steve. "The Cognitive Turn in Sociology." *Erkenntnis* 21 (1984): 439–450.

Garnham, P. C. C. "History of Discoveries of Malaria Parasites and of their Life Cycles." *History and Philosophy of the Life Sciences* 10 (1988): 93–108.

Geison, Gerald L. "Scientific Changes, Emerging Specialties and Research Schools." *History of Science* 19 (1981): 20–40.

Gerson, Elihu M. "Scientific Work and Social Worlds." *Knowledge: Creation, Diffusion, Utilization* 4 (1983): 357–377.

Giddens, Anthony. *The Constitution of Society: Outline of the Theory of Structuration.* Berkeley: University of California Press, 1984.

Giere, Ronald. *Explaining Science: A Cognitive Approach.* Chicago: University of Chicago Press, 1988.

Gillespie, Brendan, Dave Eva, and Ron Johnston. "Carcinogenic Risk Assessment in the USA and the UK: The Case of Aldrin/Dieldrin." In *Science in Context: Readings in the Sociology of Science,* edited by B. Barnes and D. Edge, 303–335. Cambridge, Mass.: MIT Press, 1982.

Golinsky, Jan. "The Theory of Practice and the Practice of Theory: Sociological Approaches in the History of Science." *Isis* 81 (1990): 492–505.

Gooding, David, Trevor Pinch, and Simon Schaffer, eds. *The Uses of Experiment: Studies in the Natural Sciences.* Cambridge: Cambridge University Press, 1989.

Goodman, D. C. "Chemistry and the Two Organic Kingdoms of Nature in the Nineteenth Century." *Medical History* 16 (1972): 113–130.

Graham, Loren. "Why Can't History Dance Contemporary Ballet or Whig History and the Evils of Contemporary Dance." *Science, Technology and Human Values* 6 (1981): 3–6.

———. "Epilogue." In *Functions and Uses of Disciplinary Histories.* Sociology of the Sciences Yearbook no. 7, edited by Loren Graham, Wolf Lepenies, and Peter Weingart, 291–295. Dordrecht: Reidel, 1983.

Grimaux, Edouard and Charles Gerhardt, Jr. *Charles Gerhardt: sa vie, son oeuvre, sa correspondance 1816–1856.* Paris: Masson, 1900.

Haas, Joan K., Helen Willa Samuels, and Barbara Trippel Simmons. *Appraising the Records of Modern Science and Technology: A Guide.* Cambridge, Mass.: Massachusetts Institute of Technology, 1985; distributed by the Society of American Archivists.

Haber, Ludwig F. *The Chemical Industry, 1900–1930: International Growth and Technological Change.* Oxford: Clarendon Press, 1971.

Hacking, Ian. *Representing and Intervening.* Cambridge: Cambridge University Press, 1983.

Hackman, Larry J. and Joan Warnow-Blewett. "The Documentation Strategy Process: A Model and Case Study." *American Archivist* 50 (Winter 1987): 12–47.

Hall, A. Rupert. "On Whiggism." *History of Science* 21 (1983): 45–59.

Hamerow, Theodore S. *Reflections on History and Historians.* Madison: University of Wisconsin Press, 1987.

Hamlin, Christopher Stone. *What Becomes of Pollution: Adversary Science and the Controversy on the Self-Purification of Rivers in Britain, 1850–1900.* New York: Garland, 1987.

———. *A Science of Impurity: Water Analysis in Nineteenth Century Britain.* Berkeley: University of California Press, 1990.

———. "Robert Warington and the Moral Economy of the Aquarium." *Journal of the History of Biology,* 16 (1986): 131–145.

———."Chemistry, Medicine, and the Legitimization of English Spas, 1740–1840." *The Medical History of Waters and Spas,* edited by Roy Porter, 67–81. Medical History, Supplement 10. London: Wellcome Institute for the History of Medicine, 1990.

Hannaway, Owen. *The Chemist and the Word: The Didactive Origins of Chemistry.* Baltimore: Johns Hopkins University Press, 1975.

Harden, Victoria A. *Inventing the NIH: Federal Biomedical Research Policy, 1887–1937.* Baltimore: Johns Hopkins University Press, 1986.

Harris, J. and W. H. Brock. "From Giessen to Gower Street: Toward a Biography of Alexander W. Williamson." *Annals of Science* 31 (1974): 95–130.

Hayes, Peter. *Industry and Ideology: I. G. Farben in the Nazi Era.* Cambridge: Cambridge University Press, 1987.

Haynes, Williams. *American Chemical Industry: A History.* 6 vols. New York: Van Nostrand, 1945–54.

———. *Cellulose: The Chemical That Grows.* Garden City, N.Y.: Doubleday, 1953.

Heitmann, John A. and David Rhees. *Scaling Up: Science, Engineering, and the American Chemical Industry.* Philadelphia: Beckman Center for History of Chemistry, University of Pennsylvania, 1984.

Herbert, Vernon and Attilio Bisio. *Synthetic Rubber: A Project That Had to Succeed.* Westport, Conn.: Greenwood Press, 1985.

Hickel, Erika. "Die neue Alchemisten-(Un)kultur." *Forum Wissenschaft* 6, 3 (1989): 28–31.

Hochheiser, Sheldon. *Rohm and Haas: History of a Chemical Company.* Philadelphia: University of Pennsylvania Press, 1986.

Hodgkin, D. C. "Patterson and Pattersons." In *Patterson and Pattersons, Fifty Years of the Patterson Function,* edited by Jenny P. Glusker, Betty K. Patterson, and Miriam Rossi, 167–193. Oxford: Oxford University Press, 1987.

Hofmann, August Wilhelm. *Zur Erinnerung an vorangegangene Freunde,* 3 vols. Brunswick: Vieweg, 1888.

Holmes, Frederic L. "Liebig, Justus von." In *Dictionary of Scientific Biography,* vol. 8, 333–338. New York: Charles Scribner's Sons, 1970–80.

———. "The Complementarity of Teaching and Research in Liebig's Laboratory." *Osiris* 5 (1989): 121–164.

Holton, Gerald. *The Scientific Imagination: Case Studies.* Cambridge: Cambridge University Press, 1978.

Hounshell, David A. and John Kenly Smith, Jr. *Science and Corporate Strategy: Du Pont R&D, 1902–1980.* New York: Cambridge University Press, 1988.

Hughes, Thomas P. "Technological Momentum in History: Hydrogenation in Germany 1898–1933." *Past and Present* 44 (1969): 106–132.

———. *Networks of Power: Electrification in Western Society 1880–1930.* Baltimore: Johns Hopkins University Press, 1983.

Hull, David L. "In Defense of Presentism." *History and Theory* 18 (1979): 1–15.

———. *Science as a Process: An Evolutionary Account of the Social and Conceptual Development of Science.* Chicago: University of Chicago Press, 1988.

Ihde, Aaron J. *The Development of Modern Chemistry.* New York: Harper and Row, 1964.

Johnson, Jeffrey. "Academic Chemistry in Imperial Germany." *Isis* 76 (1985): 500–524.

Jordanova, Ludmilla. "Objects of Knowledge: A Historical Perspective on Museums." In *The New Museology,* edited by Peter Vergo, 22–40. London: Reaktion Books, 1989.

Karpiuk, Robert S. *Dow Research Pioneers, 1888–1949*. Midland, Mich.: Pendall Publishing, 1981.

Kauffman, George B., ed. *Teaching the History of Chemistry: A Symposium*. Budapest: Akadémiai Kiadó, 1979.

Kay, Lilly E. "Selling Pure Science in Wartime: The Biochemical Genetics of G. W. Beadle." *Journal of the History of Biology* 22 (1989): 73–101.

Kennedy, Paul M. *The Rise and Fall of the Great Powers: Economic Change and Military Conflict from 1500 to 2000*. New York: Random House, 1987.

Kevles, Daniel J. *The Physicists: The History of a Scientific Community in Modern America*. New York: Knopf, 1978.

Klosterman, Leon. "A Research School of Chemistry in the 19th Century: Jean-Baptiste Dumas and His Research Students." *Annals of Science* 43 (1985): 1–80.

Knight, David M. *The Transcendental Part of Chemistry*. Folkestone, Kent: Dawson, 1978.

Knorr-Cetina, Karin D. *The Manufacture of Knowledge: An Essay on the Constructivist and Contextual Nature of Science*. New York: Pergamon, 1981.

Knorr-Cetina, Karin D. and Aaron V. Cicourel, eds. *Advances in Social Theory and Methodology: Toward an Integration of Micro- and Macro-Sociologies*. Oxford: Oxford University Press, 1981.

Knorr-Cetina, Karin D. and Michael Mulkay, eds. *Science Observed: Perspectives on the Social Study of Science*. London: Sage, 1983.

Kohler, Robert E. *Lords of the Fly: Drosophila Genetics and the Nature of Experiment*. University of Chicago Press, 1994.

Koyré, Alexandre. *Metaphysics and Measurement: Essays in Scientific Revolution*. London: Chapman and Hall, 1968.

Kopp, Hermann. *Entwicklung der Chemie in der neueren Zeit*. Munich: Oldenbourg, 1873.

Krizack, Joan. "Hospital Documentation Planning: The Concept and the Context." *American Archivist* 54 (Summer 1991).

Kuhn, Thomas S. *The Copernican Revolution: Planetary Astronomy in the Development of Western Thought*. Cambridge, Mass.: Harvard University Press, 1957.

Latour, Bruno. *Science in Action: How to Follow Scientists and Engineers Through Society*. Cambridge, Mass.: Harvard University Press, 1987.

———. *The Pasteurization of France*, translated by Alan Sheridan and John Law. Cambridge, Mass.: Harvard University Press, 1988.

Latour, Bruno and Steve Woolgar. *Laboratory Life: The Social Construction of Scientific Facts*. London: Sage, 1979. 2d ed. Princeton, N.J.: Princeton University Press, 1986.

Laudan, Larry. "The Pseudo-Science of Science?" *Philosophy of the Social Sciences* 11 (1981): 173–198.

———. *Progress and Its Problems: Toward a Theory of Scientific Growth*. Berkeley: University of California Press, 1977.

Leach, Edmund, ed. *The Structural Study of Myth and Totemism*. London: Tavistock Publications, 1967.

Lenoir, Timothy. "A Magic Bullet: Research for Profit and the Growth of Knowledge in Germany Around 1900." *Minerva* 26 (1988): 66–88.

Leonardi, Susan J. *Dangerous by Degrees: Women at Oxford and the Somerville College Novelists*. New Brunswick, N.J. and London: Rutgers University Press, 1989.

Levenstein, Harvey A. *Revolution at the Table: The Transformation of the American Diet.* New York: Oxford University Press, 1988.
Levere, Trevor H. and William R. Shea, eds. *Nature, Experiment and the Sciences,* 207–223. Amsterdam: Kluwer, 1990.
Liebenau, Jonathan, Gregory H. Higby, and Elaine C. Stroud, eds. *Pill Peddlers: Essays on the History of the Pharmaceutical Industry.* Madison, Wis.: American Institute of the History of Pharmacy, 1990.
Lowenthal, David. *The Past is a Foreign Country.* Cambridge: Cambridge University Press, 1985.
———. "The Timeless Past: Some Anglo-American Historical Preconceptions." *Journal of American History* 75 (1989): 1263–1280.
MacDonald, D. K. C. "Physics and Chemistry: Comments on Caldin's View of Chemistry." *British Journal for the Philosophy of Science* 11 (1960): 222–223.
May, Elaine Tyler. *Homeward Bound: American Families in the Cold War Era.* New York: Basic Books, 1988.
McCall, Nancy and Lisa Mix. *A Model for Archives in the Health Field.* Baltimore: Johns Hopkins University Press, 1993.
McNeill, William H. *Mythistory and Other Essays.* Chicago: University of Chicago Press, 1986.
———. *The Pursuit of Power: Technology, Armed Force and Society Since A.D. 1000.* Chicago: University of Chicago Press, 1982.
Meyer-Thurow, Georg. "The Industrialization of Invention: A Case Study from the German Chemical Industry." *Isis* 73 (1982): 363–381.
Morrell, Jack B. "The Chemist Breeders: The Research Schools of Liebig and Thomas Thomson." *Ambix* 19 (1972): 1–46.
Munz, Peter. *The Shapes of Time: A New Look at the Philosophy of History.* Middletown, Conn.: Wesleyan University Press, 1977.
Nye, Mary Jo. "The Nineteenth-Century Atomic Debates and the Dilemma of an 'Indifferent Hypothesis'." *Studies in the History and Philosophy of Science* 7 (1976): 245–268.
———. "Berthelot's Anti-Atomism: A 'Matter of Taste'?" *Annals of Science* 38 (1981): 585–590.
———. "Chemical Explanation and Physical Dynamics: Two Research Schools at the First Solvay Chemistry Conferences. *Annals of Science* 46 (1990): 461–480.
Oesper, Ralph E. "Gerhard Domagk and Chemotherapy." *Journal of Chemical Education* 31 (1954): 188–191.
Olby, Robert C. *The Path to the Double Helix.* Seattle: University of Washington Press, 1974.
Olesko, Kathryn M. *Physics as a Calling: Discipline and Practice in the Königsberg Seminar for Physics.* Ithaca, N.Y.: Cornell University Press, 1991.
Parascandola, John, ed. *Chemistry and Modern Society: Historical Essays in Honor of Aaron J. Ihde.* Washington, D.C.: American Chemical Society, 1983.
Parascandola, John and Ronald Jasensky. "Origins of the Receptor Theory of Drug Action." *Bulletin of the History of Medicine* 48 (1974): 199–220.
Paul, Harry W. *The Sorcerer's Apprentice: The French Scientist's Image of German Science, 1840–1919.* Gainesville: University of Florida Press, 1972.
Paul, Karen D., Project Director. *The Documentation of Congress: Report of the Congressional Archivists Roundtable Task Force on Congressional Documentation.* Senate Publication 102-20. Washington, D.C.: United States Government Printing Office, 1992.

Plumpe, Gottfried. *Die I. G. Farbenindustrie AG: Wirtschaft, Technik und Politik 1904–1945*. Berlin: Duncker and Humblot, 1990.
Price, Derek De Solla. "Is Technology Historically Independent of Science . . .?" *Technology and Culture* 6 (1965): 550–568.
Proceedings of the Evangelical Archives Conference. Wheaton, Ill.: Billy Graham Center, 1988.
Rabkin, Yakov. "Technological Innovation in Science." *Isis* 78 (1987): 31–54.
———. *Science Between the Superpowers*. New York: Priority Press, 1988.
Reader, William J. *Imperial Chemical Industries: A History*. 2 vols. London: Oxford University Press, 1970–75.
Reid, Donald. *Paris Sewers and Sewermen: Realities and Representations*. Cambridge, Mass.: Harvard University Press, 1991.
Reingold, N. "History of Science Today. 1. Uniformity as Hidden Diversity: History of Science in the United States, 1920–1940." *British Journal for the History of Science* 19 (1986): 243–262.
Reynolds, Terry S. *75 Years of Progress: A History of the American Institute of Chemical Engineers, 1908–1983*. New York: American Institute of Chemical Engineers, 1983.
Rocke, Alan J. "Subatomic Speculations and the Origin of Structure Theory." *Ambix* 30 (1983): 1–18.
———. *Chemical Atomism in the Nineteenth Century: From Dalton to Cannizzaro*. Columbus: Ohio State University Press, 1984.
———. "Kolbe Versus the 'Transcendental Chemists': The Emergence of Classical Organic Chemistry." *Ambix* 34 (1987): 156–168.
———. "Kekulé's Benzene Theory and the Appraisal of Scientific Theories." In *Scrutinizing Science*, 45–161, edited by Arthur Donovan, Larry Laudan, and Rachel Laudan. Dordrecht: Kluwer, 1988.
———. *The Quiet Revolution: Hermann Kolbe and the Science of Organic Chemistry*. Berkeley: University of California Press, 1993.
Roth, Paul A. *Meaning and Method in the Social Sciences*. Ithaca, N.Y.: Cornell University Press, 1987.
Russell, Colin A. *The History of Valency*. Leicester: Leicester University Press, 1971.
———. "Rude and Disgraceful Beginnings: A View of History of Chemistry from the 19th Century." *British Journal for the History of Science* 21 (1988): 273–294.
Samuels, Helen Willa. *Varsity Letters: Documenting Modern Colleges and Universities*. Chicago: Society of American Archivists; Metuchen, N.J., and London: Scarecrow Press, 1992.
———. "Who Controls the Past." *American Archivist* 49 (Spring 1986): 109–124.
Sapp, Jan. *Beyond the Gene: Cytoplasmic Inheritance and the Struggle for Authority in Genetics*. Oxford: Oxford University Press, 1987.
Schmauderer, E., ed. *Der Chemiker im Wandel der Zeiten*. Weinheim: Verlag Chemie, 1973.
Sepper, Dennis. *Goethe Contra Newton: Polemics and the Project for a New Science of Color*. Cambridge: Cambridge University Press, 1988.
Shapin, Steven. "History of Science and Its Sociological Reconstructions." *History of Science* 20 (1982): 157–211.
Shinn, Terry. "Orthodoxy and Innotation in Science: The Atomist Controversy in French Chemistry." *Minerva* 18 (1980): 539–555.

Silverstone, Roger. *The Message of Television: Myth and Narrative in Contemporary Culture*. London: Heinemann, 1981.

Smith, Edgar F. "Observations on Teaching the History of Chemistry." *Journal of Chemical Education* 2 (1925): 533–555.

Smith, John Graham. *The Origins and Development of the Heavy Chemical Industry in France*. Oxford: Clarendon Press, 1979.

Smith, Kenneth. "Alar: One Year Later." Published as Special Report, American Council on Science and Health. New York, March 1990.

Spilker, Bert. *Multinational Drug Companies: Issues in Drug Discovery and Development*. New York: Raven Press, 1989.

Spitz, Peter H. *Petrochemicals: The Rise of an Industry*. New York: John Wiley and Sons, 1988.

Star, Susan Leigh and James R. Griesemer. "Institutional Ecology, 'Translations,' and Boundary Objects: Amateurs and Professionals in Berkeley's Museum of Vertebrate Zoology, 1907–1939." *Social Studies of Science* 19 (1989): 387–420.

Stocking, George W. and Myron W. Watkins. *Cartels in Action: Case Studies in International Business Diplomacy*. New York: Twentieth Century Fund, 1946.

Sturchio, Jeffrey L. "Artifact and Experiment." *Isis* 79 (1988): 369–372.

Suckling, Colin J., Keith E. Suckling, and Charles W. Suckling. *Chemistry Through Models: Concepts and Applications of Modelling in Chemical Science, Technology, and Industry*. Cambridge: Cambridge University Press, 1978.

Swann, John P. *Academic Scientists and the Pharmaceutical Industry: Cooperative Research in Twentieth-Century America*. Baltimore: Johns Hopkins University Press, 1988.

Thackray, Arnold. *Atoms and Powers: An Essay on Newtonian Matter-Theory and the Development of Chemistry*. Cambridge, Mass.: Harvard University Press, 1970.

Thackray, Arnold, Jeffrey L. Sturchio, P. Thomas Carroll, and Robert Bud. *Chemistry in America, 1876–1976: Historical Indicators*. Dordrecht: Reidel, 1985.

Travis, Anthony S. "Perkin's Mauve: Ancestor of the Organic Chemical Industry." *Technology and Culture* 31 (1990): 51–82.

Trescott, Martha. *The Rise of the American Electrochemicals Industry, 1880–1910: Studies in the American Technological Environment*. Westport, Conn.: Greenwood Press, 1981.

Turner, James S. *The Chemical Feast: The Ralph Nader Study Group Report on Food Protection and the Food and Drug Administration*. New York: Grossman Publishers, 1970.

Turner, Steven, Edward Kerwin, and David Woolwine. "Careers and Creativity in Nineteenth-Century Physiology: Zloczower Redux." *Isis* 75 (1984): 523–529.

Volhard, Jacob. *Justus von Liebig*. Leipzig: Barth, 1909.

Warnow, Joan et al. *AIP Study of Department of Energy National Laboratories*. Three reports: *A Study of Preservation of Documents at Department of Energy Laboratories; Guidelines for Records Appraisal at Major Research Facilities;* and *Files Maintenance and Records Disposition: A Handbook for Secretaries at Department of Energy Laboratories* by Jane Wolff. New York: American Institute of Physics, 1982; revised 1985.

Warnow-Blewett, Joan et al. *AIP Study of Multi-Institutional Collaborations, Phase I: High-Energy Physics. Report No. 1: Summary of Project Activities and Findings/Project Recommendations; Report No. 2: Documenting Collaborations in High-*

Energy Physics; Report No. 3: Catalog of Selected Historical Materials; Report No. 4: Historical Findings on Collaborations in High-Energy Physics. New York: American Institute of Physics, 1992. *Report No. 5: Sociological Analysis of Collaborations in High-Energy Physics* will be published in 1993.

Weil, B. H. and Victor J. Anhorn. *Plastics Horizons.* Lancaster, Pa.: Cattell Press, 1944.

Weinberg, Alvin. "Science and Transcience." *Minerva* 10 (1972): 209–222.

Westfall, Robert S. *Never at Rest: A Biography of Isaac Newton.* Cambridge: Cambridge University Press, 1980.

White, Hayden. *Metahistory: The Historical Imagination in Nineteenth-Century Europe.* Baltimore: Johns Hopkins University Press, 1973.

Whitehead, Don. *The Dow Story: The History of the Dow Chemical Company.* New York: McGraw-Hill, 1969.

Whitley, Richard. "From the Sociology of Scientific Communities to the Study of Scientists' Negotiations and Beyond." *Social Science Information* 22 (1983): 681–720.

Whorton, James. *Before Silent Spring: Pesticides and Public Health in Pre-DDT America.* Princeton, N.J.: Princeton University Press, 1974.

Willett, John. *Brecht in Context: Comparative Approaches.* London: Methuen, 1984.

Williamson, Harold F. et al. *The American Petroleum Industry: The Age of Energy, 1899–1959.* Evanston, Ill.: Northwestern University Press, 1963.

Woolgar, Steve, ed. *Knowledge and Reflexivity: New Frontiers in the Sociology of Knowledge.* London: Sage, 1988.

———. *Science: The Very Idea.* London: Tavistock Publications, 1988.

Pnina G. Abir-Am is NSF Visiting Associate Professor (1991–1993) in the Department of the History of Science, Medicine, and Technology at Johns Hopkins University. She has published on the history of molecular life sciences in *History of Science, Social Studies of Science, Isis, Minerva,* and more recently in *Osiris* and *Social Epistemology.* Her contribution to *Uneasy Careers and Intimate Lives: Women in Science, 1789–1979* (1987, 1989) received a 1988 award for outstanding research from the History of Science Society. She is the author of *Research Schools of Molecular Biology in the US, UK, and France, 1930–1970: A Comparative Study of Collaborative Innovation in a Transnational Context* (forthcoming).

Ellis N. Brandt is Historian of The Dow Chemical Company and Director of the Post Street Archives, the company's historical repository, in Midland, Michigan. He retired from Dow in 1986 following a 33-year career as a public relations executive with the company. The E. N. Brandt Professorship in Public Relations at Michigan State University is named in his honor.

He is vice president of the Midland County Historical Society, a trustee of the Historical Society of Michigan, vice president and secretary of the R. M. Gerstacker Foundation, and a former vice president of the Midland Foundation.

Robert Bud is Head of Life and Environmental Sciences at the Science Museum, London, where he has been responsible for galleries on petroleum, plastics, and the chemical industry. His main curatorial responsibility is now the biosciences, and he has recently published *The Uses of Life: A History of Biotechnology* (1993). With Susan Cozzens, he has also edited a volume on historial and policy issues associated with medium scale scientific instruments, *Invisible Connections: Instruments, Institutions, and Sciences* (1992), and has been the coauthor of volumes on the history of British and of American chemistry. He is project director and coeditor of the *Guide to the History of Technology in Europe* (Science Museum, 1992).

Robert Friedel is Associate Professor in the History Department of the University of Maryland, College Park, where he has taught the history of technology and science since 1984. Prior to that he was Director of the IEEE Center for the History of Electrical Engineering in New York City. His books include *Pioneer Plastic: The Making and Selling of Celluloid* (1983), *Edison's Electric Light* (1986), and *A Material World* (1988). He is currently completing a study of technological novelty as exemplified by the invention and marketing of zippers.

Christopher Stone Hamlin is Associate Professor in the Department of History and the Program in the History and Philosophy of Science at the University of Notre Dame, where he teaches the history of medicine, the history of technology, and environmental history. He has recently published *A Science of Impurity: Water Analysis in Nineteenth Century Britain* (1990).

Erwin N. Hiebert is Professor of the History of Science Emeritus, Harvard University. He has current research interests in the history of science, nuclear physics between the two world wars, low temperature physics, chemical physics, and quantum chemistry.

Frederic L. Holmes is Avalon Professor of the History of Medicine, Section of the History of Medicine, Yale University. He is a former President of the History of Science Society. His recent monographic publications include *Lavoisier and the Chemistry of Life* (1985), *Hans Krebs, Vol. I: The Formation of a Scientific Life, 1900–1933* (1991), *Vol. II: Architect of Intermediary Metabolism, 1933–1937* (in press). Current projects include the historical reconstruction of the Maselson-Stahl experiment on the replication of DNA; Liebig and the formation or organic chemistry; and the early physiological career of Helmholtz (with Kathryn Olesko).

William B. Jensen is Oesper Professor of the History of Chemistry and Chemistry Education in the Department of Chemistry of the University of Cincinnati, where he teaches a history of chemistry course at the senior-graduate level directed at chemistry majors. A former Chair of the Division of the History of Chemistry of the American Chemical Society and current editor of the *Bulletin for the History of Chemistry,* his historical interests center on the history of nineteenth- and early twentieth-century physical and inorganic chemistry and the history of chemical education.

Robert E. Kohler is Professor of the History and Sociology of Science at the University of Pennsylvania. His most recently published book is *Partners in Science: Foundations and Natural Scientists, 1900–1945* (1991), and he has recently finished *Lords of the Fly: Drosophila Genetics and the Nature of Experiment,* a book on experimental practics. His current interests include experimental practice in laboratory and field biology, and the social history of natural science in the modern era.

John E. Lesch is Associate Professor of History at the University of California, Berkeley. He is the author of *Science and Medicine in France: The Emergency of Experimental Physiology 1790–1855* (1984) and other publications on the history of the life sciences and medicine.

Seymour H. Mauskopf is Professor of History at Duke University. He was the first Edelstein International Fellow in the History of Chemical Sciences and Technology (1933–1989). His relevant research interests include the history of atomic and molecule theories (*Crystals and Compounds,* 1976) and his current project, the scientific study of explosives and munitions, 1775–1900.

Mary Jo Nye is George Lynn Cross Research Professor of the History of Science and Chair of the History of Science Department at the University of Oklahoma. Her new book *From Chemical Philosophy to Theoretical Chemistry: The Dynamics of Matter and the Dynamics of Disciplines, 1800–1950* is scheduled for publication in 1993.

Yakov M. Rabkin is Professor at the University of Montreal. His scholarly interests are divided mainly between two problem areas: (a) cultural, national, and transnational aspects of science, and (b) relations between science and technology. He is the author of *Science Between the Superpowers* (1988) and of several dozen book chapters and journal articles. He is currently co-editing a book on the interaction between Jewish and scientific cultures, and is preparing a French-language textbook in the history of science.

Alan J. Rocke is Professor of History at Case Western University in Cleveland, Ohio, where he teaches the history of sciences. In addition to a number of scholarly articles, he is the author of *Chemical Atomism in the Nineteenth Century: From Dalton to Cannizzaro* (1984) and *The Quiet Revolution: Hermann Kolbe and the Science of Organic Chemistry* (1993). The latter book develops several of the themes adumbrated in his essay in this collection.

Helen Willa Samuels has been Institute Archivist and Head of Special Collections in the MIT Libraries since 1977. She was trained as a librarian at Simmons College after receiving her bachelor's degree from Queens College in New York City and has worked with both as an archivist and a librarian. Her earlier writings dwell on the documentation of science and technology and the methods, such as documentation strategies, that can be used to improve the record of modern society. Her most recent publication is *Varsity Letters: Documenting Modern Colleges and Universities* (1992). She is a Fellow of the Society of American Archivists.

John W. Servos is Professor of History at Amherst College, where he teaches the history of science and medicine. He is author of *Physical Chemistry from Ostwald to Pauling: The Making of a Science in America* (1990), and is currently writing a history of the industrial sponsorship of scientific research in American universities.

John Kenly Smith, Jr. is associate professor of history at Lehigh University. He is coauthor of *Science and Corporate Strategy: Du Pont R&D, 1902–1980.*

Index

acetyl theory of organic radicals, 127
acetylene, 144, 147
acid anhydrides, 95, 96
acridine compounds, 171, 180, 194, 196–197, 198, 203
acrylic resins, 150, 152
aetherin theory of organic radicals, 126
affinity, 7–8, 10, 13, 15, 16, 17
agendas: chemical, 367; historical, 368
agricultural chemicals, 153, 154
alar, 356, 357
alchemy, 290
Allied Chemical and Dye Co., 147, 152
Ambix, 268, 271, 272, 273, 275
American Baking Association, 337
American Chemical Society (Division of the History of Chemistry), 262, 263, 268, 275
American Cyanamid, 144, 147, 152
American Institute of Physics, Center for History of Physics, 238, 254, 255–259
American Journal of Medical Sciences, 315
American Medical Association, Council on Foods and Nutrition, 339
amines, 96, 97
ammonia, 144–145
analogy, 6, 7, 8, 11, 13
analysis, 217, 218–219, 223–224, 226
animal experiment, 166–167, 168, 170, 171, 172, 173–174, 175, 177–178, 179–183, 188, 190–191, 194, 202, 205–206
Annalen der Chemie und Pharmacie (Liebig's Annalen), 97, 106
Appleton, John, 224
aramite, 343
archival strategies, xviii
Arrhenius, Svante, 16
arts, 221, 222
Atabrine, 171, 174, 194, 197, 199, 200
Atlas Powder Company, 337–338
atom, 3, 4, 8, 9, 10, 11, 12, 13–14, 15
atomic weights, 11, 13, 14, 91–94, 95, 99–101, 103–104
Attfield, John, 232
Avogadro, Amadeo, 4, 11
azo compounds, 180–181, 185, 192, 194, 195, 197–200, 203, 208

Bachelard, Gaston, 4
Baconian tradition, 220
Baekeland, Leo, 149
Baeyer, Adolf, 96, 98, 108
Bakelite, 149
balance, 28–30
Balard, Antoine, 104
Barclay, Alexander, 282
Barker, John, 301–304, 316
Barner, Jacques, 7
Barthes, Roland, 282, 284
BASF, 145, 154
basic alkylation, 194, 196–199, 202–203, 208
Baumé, Antoine, 7
Bayer, 160–161, 163, 164, 170, 171, 179, 194, 196, 204, 206–208
Bayer 205 (Germanin), 195
Bayer Co., 154
Beadle, George W., xiv, 44–78
Beckman Center for the History of Chemistry (Chemical Heritage Foundation), xii–xiii, 242, 244, 250, 255, 259–260, 262, 268, 275

Bensaude-Vincent, Bernadette, 4
benzene theory, 14
benzoyl radical, 126, 133
Bergman, Torbern, 299, 301, 302, 303–304, 316
Berichte der Deutschen Chemischen Gesellschaft, 91–92
Bernal, John Desmond, 384, 385, 386
Bernard, Claude, 132
Berthelot, Marcelin, 4, 12, 13, 15
Berzelius, Jöns Jakob, 10, 12, 13, 14, 31, 91, 92, 94, 105, 109, 123, 126, 127, 129–130, 132, 299; aetherin theory of, 126
Billy Graham Center, 242
biochemical genetics, 44–45, 49–50, 70–76
biochemistry and nutrition, 60, 63–65, 68–69, 70–71, 73–76
biotechnology, 282–283, 290
bond, chemical, 5, 15, 17
Bosch, Carl, 145
Bouasse, Henri, 4
Boullay, Polydore, 126
Boussingault, Jean-Baptiste, 305
Bovet, Daniel, 199–201
Boyle, Robert, 299
Bradley, John, 4
Brannigan, Augustine, 133
Brecht, Berthold, 290
Bridges, Calvin, 47–48
British Association for the Advancement of Science, 310
Brodie, Benjamin C., Jr., 95, 96
Brown, Alexander Crum, 108
Bulletin for the History of Chemistry, 263, 268, 271, 272
butadiene, 151
Butenandt, Adolf, 65, 67
Butlerov, Aleksandr Mikhailovich, 91, 98, 108
Bunsen, Robert Wilhelm, 31, 94, 97–98, 105, 108
Butterfield, Herbert, 374

Caldin, E. F., 3
caloric, 10
Calvery, Herbert, 331, 332
Cannadine, David, 279
Cannizzaro, Stanislao, 4, 93
Cantor, Geoffrey, 131
carbon, tetrahedron representation of, 12, 14, 18
carbonic acid theory, 106–107
carcinogens, 322, 341, 343–345
Carius, Ludwig, 96, 98
Carlson, Anton J., 341
Carothers, Wallace H., 149
Carson, Rachel, 358
catalysis, catalysts, 10, 139, 144, 148
causes, 7, 9, 10, 14, 15, 16
cellophane, 149
celluloid, 143, 146
cellulose, 143, 149
Chadwick, Edwin, 306
Chandler, Alfred D., Jr., 138
change, chemical, 223, 232
Channel, David, 282
Charles Babbage Institute for the History of Information Processing, 255, 256
"chemical century," xi, xvi
chemical crystallography, 380, 385
chemicals, 282, 289, 291
chemical engineering, 137, 139, 142, 148, 371
chemical industry, 137–138, 140, 141, 151–153, 154; role in diffusion of instruments, 26, 28, 32–33, 34, 36, 38; American, value of in 1989, xi
chemical instrumentation, xiv; cost, 29–30; development of, 37–38; images, 33, 38, 39; measuring, 29; impact on research orientation, 27–29, 30, 32, 38–39; production of, 31–32, 35–36; as a basis of chemical profession, 25, 28, 33–34, 39; as a source of inequality, 34–37, 39–40
Chemical News, 308
chemical processes, 137, 138, 139, 141–142, 144
Chemical Society of London, 97, 104
chemico-theology, 305–308
chemistry, 141; as "science" and "art," 218–220, 226, 232; changing public image, 216; definitions, 222–223, 230–233
"chemists' war" (World War I), xii
chemists: audience for history of chemistry, 264–269; reading habits, 269; writing habits, 270–272
chemogastric revolution, xx, 322–323, 326, 329, 333, 335, 344–345

chemotherapy, 160, 167, 169, 170–171, 174–177, 179, 185–193, 195–204, 204–208
chlorine, 144
classification, 9, 11, 13, 14, 186–187, 205–206
clinical trials, 167, 170, 171, 172, 190–191
coal tar dye industry, 224
Cohen, I. Bernard, 3
Collins, Harry M., 131
Color Additive Amendment (1960), 341, 342
complexity, 367
composition, 222–223, 230, 231, 232, 233
Comstock, John, 223–224, 230–231
Conant, James Bryant, 281
Condillac, [Abbé] Étienne Bonnot de, 9
Conservatoire des Arts et Métiers, 277, 283
consumer products, 153
contextual (historical) analysis, 369
convenience foods, 325–326
conventional(ism), 5–6, 9, 10, 11, 12, 14, 19
Cook, Josiah P., Jr., 225
Cooper, Thomas, 220
copula theory, 96, 105–106
Corfam, 153
Corfield, William, 307–308
Couper, Archibald Scott, 4, 91, 98, 103
critical problems, 175, 205, 207
Crookes, William, 144, 308–310
culture: categories, 282; popular, 279, 291; scientific, 280–281, 291

Dagognet, François, 4
Dalton, John, 4, 7, 10, 11, 14
Darby, William J., 339
Darwinism, 226
Daubeny, Charles G. B., 304
Davy, Humphry, 7, 8, 9, 30, 225
DDT, 334–335, 340
decomposition, organic, 307
Delaney, James J., 341, 342, 344, 345
Delaney anti-cancer clause, 343–345
DeLoach, Will S., 269
densities, vapor, 11
Department of Energy National Laboratories, AIP documentation research project on, 255–256
Descrozilles, François Antoine Henri, 32
design, divine, 225–226
Deutsche Chemische Gesellschaft, 91, 108
Deutsches Museum, 277, 283, 288
Deville, Henri Étienne Sainte-Claire, 4, 104
Dickinson, Henry Winram, 284
diethylene glycol, 332
diglycerides, 336, 338
discipline history centers, 254–255, 259–260
diversification, 146, 147–148, 153, 154
diversity, 367
Döbereiner, Johann Wolfgang, 10
documentation: research projects, 254, 255–260; strategies, 240–244, 251
Dodge, Bernard O., 69–70
Domagk, Gerhard, 158–159, 160–161, 163, 177–193, 194–195, 199, 202, 204, 205, 206, 208–209
Douglas, Mary, 284, 290, 296–297
Dow Chemical Company, xx, 144, 147, 152, 153; archives, 362; history, 358, 359, 360, 362
Dow, Herbert H., 359–361, 362
Dow, Willard H., 361
Draize test, 333
Drosophila, xiv, 44–47, 72, 76–77; transplantation method, 48–60; transfusion method, 61–62; feeding method, 61–66; and insect nutrition, 67–69
Duhem, Pierre, 3, 16, 267
Duisberg, Carl, 163
Dulong, Pierre Louis, 10
Dumas, Jean-Baptiste André, 7, 8–9, 10, 11–12, 14, 34, 90, 92, 94, 95, 96, 100, 101, 102, 103, 104, 105, 109, 123, 126–130, 305, 308
Du Pont Corporation, xii, 143, 145, 146, 147–151, 153, 154–155, 324
Durkheim, Émile, 284
dyes, 160, 163, 165, 166, 171, 173, 174, 182, 187, 194, 195, 199, 203, 204, 208; synthetic, 138, 142–143
dynamite, 143, 145

Eastman, George, 143
Eco, Umberto, 283, 289, 290

Edison, Thomas Alva, 174, 175
Ehrlich, Paul, 161, 167, 170, 195, 197, 206–207
Elberfeld, 161, 163, 165, 172, 178, 194–197, 202, 209
electrochemical dualism, 91, 92, 96, 99, 102, 105, 108
electrochemicals, electrochemistry, 30–31, 143–144, 146
electrons, 5, 17–18, 19
Elixir Sulfanilamide, 332
emulsifiers, 336, 337–338
energy, 16
engineering companies, 152
enriched bread/flour, 337
environmental chemistry, 371
environmental policy, xix–xx
Ephrussi, Boris, 44–66, 67, 72–74, 77, 78
epistemology, 3, 5
equivalents, 10, 13, 91, 94, 98, 100, 103
Erdmann, Otto, 99
Erlenmeyer, R. A. C. Emil, 96, 98, 108
ether(s), 94–96, 102–103
ethyl theory of organic radicals, 126, 128
Everett, Kenneth G., 269
experimental practice, 43–44, 59–60, 76–78

Faraday, Michael, 30
fermenters, 290–291
Fieser, Louis H., 363
Fimreite, Norvald, 358–359
fine chemical industry, 92
Fitzhugh, O. Garth, 333
Florida, 327
Florida Citrus Commission, 331
fluoride, 341
Food Additives Amendment (1958), 341, 342
Food and Drug Administration, U.S., 329–335, 341–343; Division of Pharmacology, 331–335
Food, Drug, and Cosmetic Act (1938), 329, 332, 333, 341
food standards, 329, 337, 339–340
forces, 7–8, 9–10, 11, 12, 13, 15, 16, 17
Fortune Magazine, xi
Foster, George Carey, 98
Fourcroy, Antoine, 299
Fourier, J. B. Joseph, 16
Fownes, George, 224
Fox, Cornelius, 315
Franco-Prussian War, 108
Frankland, Edward, 13, 15, 91, 96, 102, 105, 106, 107, 110, 309, 312–314
Frassen, Bas Van, 5
Friedel, Charles, 98, 104
frozen food industry, 325, 326–328, 330–331
function, chemical, 6, 9, 10, 12, 13

Gairdner, William, 304
gasometer, 30
Gay-Lussac, Joseph-Louis, 31, 33, 123
genetics, developmental, 48–51, 65–66
genome, 289, 292
geometry, spatial, 6, 12, 18
Gerhardt, Charles, xv, 13, 91–92, 93–105, 107, 109–111
Gerhardt-Laurent reform, 91–107, 110–111
Gesellschaft Deutscher Chemiker (Fachgruppe Geschichte der Chemie), 268
Gibbs, J. Willard, 16
Glyco Products Inc., 337–338
Gmelin, Leopold, 94, 96
Goethe, Johann Wolfgang von, 282
Goldschmidt, Richard, 45, 49, 57
Gooding, David, 121, 122, 131
Graham, Thomas, 10, 14–15

Haas, Otto, 146
Haber, Fritz, 145
Haber-Bosch process, 144
Hales, Stephen, 30
Hall, Charles Martin, 144
Handler, Philip, 343–344
Hannaway, Owen, 6, 227
Harcourt, Vernon, 16
Hazen, Edward, 231
Helmholtz, Hermann von, 16
Henry, William, 219–220, 222, 223, 225, 226
Hérault, Paul-Louis-Toussaint, 144
heritage business, 279
Hessler, John C., 223, 232–233
hexitol, 337
Hirschfelder, Joseph, 17
historical research, objectivity of, xx
history: academic, 277–279; annals, 288
history of chemistry, xii, xviii–xix, xx; lit-

erature of, 372–374; and undergraduates, 372–374; gender in, xxi, 379–380
history of ideas, 372
history of science, 280–281; internalist-externalist debate in, 119, 375–376; logical constructionism and, 119–121
Hodgkin, Dorothy M. Crowfoot, 379, 380, 381, 382, 383, 384, 385, 386, 387
Hoechst, 154
Hörlein, Heinrich, 159, 161, 163–176, 177, 178–179, 187, 194, 195, 204–209
Hofmann, August Wilhelm von, 90, 93, 95–96, 98, 100, 105–106, 299, 304
Hoffmann, Friedrich, 299, 302
Holton, Gerald, on science in the making, 131
hormones, 168, 169, 172, 173, 174
Horowitz, Norman H., 72, 75–76
House Select Committee to Investigate the Use of Chemicals in Food Products (Delaney Committee), xx, 340–342, 344
hundredfold margin of safety, 334, 343
Hyatt, John Wesley, 143
hypothesis, method of, 5, 7, 14

I. G. Farbenindustrie, xvii, 148, 149, 154, 158, 160–163, 164, 170, 174, 178–179, 201, 204, 207–208
Ihde, Aaron, 25, 127
Imperial Chemical Industries, Ltd. (ICI), 148, 149, 150, 154
in vitro trials, 177, 179–181, 185, 188, 189, 191, 194, 202
industrial research, xvi, 140–141, 142, 148, 152–153
Industrial Revolution, 141
industrialized invention, 159–160, 204
Ingen-housz, Jan, 305
Ingold, Christopher K., 18
Institute of Electrical and Electronic Engineers, Center for the History of Electrical Engineering, 259
Institute of Shortening Manufacturers and Edible Oils, 338
Institut Pasteur, 199
instrumental(ism), 5, 6
instrumentation, 139
instruments: animals as, 44–45, 76–78; history of, 371
intellectual aristocracy, 382
investigative pathways, as locus of individual scientific activity, 122, 131–132
ions, 5
Isis, 271, 272, 273, 275

Jacques, Jean, 4
Japanese Society for the History of Chemistry, 268
Jefferson, Thomas, 220–221
Jenkins, Reese V., 371
Joint Committee on Archives of Science and Technology, 256
Johnston, James Finlay Weir, 305, 307
Journal für praktische Chemie, 99, 108
Journal of Chemical Education, 264, 268, 271, 272, 275

Kagakushi, 268
kala-azar, 167, 171, 173, 175
Karlsruhe Congress, xv, 91, 93, 99, 100, 101, 104
Keefe, Frank B., 335, 340–341
Kekulé, F. August, 4, 14, 26, 90, 91, 93, 95, 96, 98, 100, 102, 103, 104, 106, 108–109
Kennedy, Paul, 286
Kenwood, Henry, 315
Kevles, Daniel J., 374
Khouvine, Yvonne, 60, 62, 63, 65
Kikuth, Walter, 171, 194, 202
kinetic theory, 15, 16
Kirchhoff, Gustav Robert, 31
Klarer, Joseph, 159, 161, 177, 179, 181, 185, 191, 192, 193, 194, 197–199, 205, 208, 209
Knorr, Ludwig, 163
Kolbe, Hermann, xv, 13, 14, 90, 91, 93, 95, 97–101, 103, 105–110, 112
Kopp, Hermann, 96, 98, 99, 106, 107, 267
Koyré, Alexandre, 374
Krantz, John C., Jr., 339
Krebs, Hans A., 123, 130, 132
Kühn, Alfred, 45, 65, 78
Kuhn, Thomas S., 374

Ladenburg, Albert, 267
Langmuir, Irving, 18
language, 6, 9, 12
Lapworth, Arthur, 18
Latour, Bruno, 120, 124, 129, 130

Laudan, Larry, 120
Laurent, Auguste, 13, 91, 92, 93, 101, 104, 105, 109, 130
Lavoisier, Antoine-Laurent, 3, 6, 8–10, 28–30, 31, 34, 108–109, 123, 132, 220, 299, 316
laws, chemical, 8, 11, 16
LD-50, 332–334
Le Blanc process, 138, 140, 141–142
LeBel, Achille, 12
Leeds, A. R., 315
Leeuwenhoek, Antoine van, 30
Lespieau, Robert, 4
Leverkusen, 161, 163, 194
Levi, Primo, 284
Lewis, Gilbert Newton, 18
Libavius, Andreas, 6–7
Liebig, Justus von, xv, 10, 12, 31, 34, 90, 92, 93, 94, 95, 97, 98–99, 100, 101, 104, 105, 108, 123–133, 305, 307, 316, 377
life, 282, 289, 291
Limpricht, Heinrich, 96, 98
Linde Air Products, 147
"list of 842", 342
Little, Arthur D., 137
Lodge, David, 285
Lowenthal, David, 279
Lucas, Charles, 300

MacDonald, D. K. C., 18
Macquer, Pierre-Joseph, 7
malaria, 167, 171, 175, 194, 196–197, 199, 202
mannitol, 337
Mansfield, Charles, 307
Margenau, Henry, 5
Marx, Karl, 280
Massachusetts Institute of Technology 227–228, 371; documentation of research, 238–239, 243
Mauss, Hans, 171, 194, 197
McNeill, William, 286
mechanisms, reaction, 16, 18
Melanchthon, Philipp, 6
Mendeleev, Dmitri, 18, 29, 36
mercury, 358–359
metaphors, 6, 11
methylcellulose, 328
Metzger, Hélène, 3
Meyer, Ernst von, 107
Meyer, Lothar, 15, 93, 98
Meyerson, Émile, 3
microscopy, 30, 32; field emission, 26
Mietzsch, Fritz, 159, 161, 171, 177, 179, 181, 185, 191, 192, 193–204, 205, 208, 209
Miller Pesticide Amendment (1954), 341
Minamata (Japan) disaster, 359
mineral theory of plant nutrition, 305
mineral water: analysis, 298–304; artificial, 302
Mitteilungen Fachgruppe Geschichte der Chemie, 268
models, 4, 6, 19
molecular structure, 381
molecules, 4, 6, 11, 14, 15, 16, 17, 18, 19
Monsanto, 341
Morgan, Thomas Hunt, Caltech Drosophila group of, 46–48, 49–51, 57, 59, 72
Morrison, A. Cressy, 282
Muir, M. M. Pattison, 232, 267
Mulliken, Robert, 18
Munz, Peter, 286–287
museums: artifacts in, 279, 285, 288, 290, 292; chemical displays in, 277, 281–282, 292; labels in, 288–289
Myrj, 337–338, 339
myth, 283–290

napalm, 363–364
Naquet, Alfed, 104, 111
narratology, 287
National Institutes of Health, 335
National Museum of American History, 279
National Research Council, 339; Committee on Cereals and Food of, 339; Food and Nutritution Board of, 335
Native Guano Co., 308–310
natural theology, 225; chemico-theology, 305–308
Nelkin, Dorothy, 298
Nelson, Erwin E., 331
neoprene, 150
Nernst, Walther, 16
Neurospora, xiv, 66–76, 77, 78
Newsletter for the Historical Group of the Royal Society of Chemistry, 268
Newton, Isaac, 7–8, 305, 316
Newtonian, 7, 10

Nickles, Thomas, 121
nitrocellulose, 143, 146
nitrogen fixation, 144–145, 306
Nitti, Frédéric, 199
Nobel, Alfred, 143
Nobel Prize, 158–160, 204, 208–209, 228
Noyes, William A., 222, [233]
Nuremberg defense, 357
nylon, 149–150, 151, 152

Odling, William, 93, 95–96
Oesper Collection in the History of Chemistry, 264, 268
Olmsted, Denison, 222, 223
orbitals, atomic and molecular, 18
Oren, Amos, 285
organic chemicals, 142
organic chemistry, movement of field in 1830s, 130
organic farming, 345
organic synthesis, 224 (13)
Ostwald, Wilhelm, 144, 267
Oxford University, 380, 381, 382, 383, 384, 386

Paneth, Erwin, 3
Paracelsus, P. A. T. Bombastus von Hohenheim, 3, 7
Parkes, Samuel, 225
Partington, John Riddock, 12
Pasteur-Pouchet debate, 291
patenting, 161, 163, 174, 185, 210
Pauling, Linus, 18
Pebal, Leopold von, 96, 98, 100
pellagra, 344
Pelouze, Théophile Jules, 128, 130
"pencil-and-paper" chemistry, 13, 14
Perkin, William Henry, Sr., 98, 142
petrochemicals, 151, 152, 154
pharmaceutical industry, 244–251
pharmaceuticals, 153, 154
pharmacology, 166–167, 170, 172, 181, 196, 371
philosophical chemistry, 7, 10, 16, 226
philosophy of chemistry (chemical philosophy), xiv, 3–6, 8–9, 11–12, 17–19, 89–90, 222
philosophy of science, 87–90, 110–112
photosynthesis, 305
physics and allied sciences, documentation projects, 254, 255–259
physiological chemistry, 168–169, 171–173, 174, 176, 206
Plasmoquine, 171, 173, 194, 196–197, 202
plastics, 140, 143, 149
Playfair, Lyon, 306–307
pluralism, methodological, 367, 369
polarimetry, 32
polyester, 152
polyethylene, 149, 150, 152
polymers, 140, 148–150, 151–152, 154–155
polyoxyethylene monostearate (POEMS), 337–338, 339–340, 341
polypropylene, 152
polystyrene, 149
polyurethane, 152
positiv(ism), 5–6, 13, 14, 16
preservatives, 329, 330, 333, 342, 344
Prest-o-Light, 147
Priestley, Joseph, 3, 299, 305, 308
Prigogine, Ilya, 4
principles, chemical, 7, 9
Procter and Gamble, 323, 336, 338
product responsibility, 357
promoters (of chemistry), 218
Prontosil, 158, 177, 181–185, 189, 191, 194, 197, 199, 201, 208
public health, 314
Pure Food and Drugs Act (1906), 322, 342

quantum mechanics, 6, 17, 19
Quarterly Journal of Science, 308
quinine, 171, 194, 196–197

R. T. Vanderbilt Co., 338, 340
radical theories in organic chemistry, 95–96, 99, 102, 105–106, 126–127, 128–129
Ramus, Petrus, 7
Randall, John Herman, Jr., 283–284
Ravetz, Jerome, 296, 316
rayon, 143, 148, 149
realism, 4, 5, 14
reductionism, 367, 369
refractometry, 32, 36
Regnault, Henri Victor, 104
Remsen, Ira, 226–227, 228

research schools, 377, 378, 381, 385, 386
reticuloendothelial system, 178, 185, 189
reverse salients, xvii, 175, 176, 204, 207
Richards, Theodore William, 237–238
Rifkin, Jeremy, 283
Robinson, Robert, 18
Robiquet, Jean-Pierre, 125, 131
Robison, John, 10
Rockefeller Foundation, 46, 58–59, 73, 386, 387
Roehl, Wilhelm, 161, 171, 195, 196, 202
Rohm and Haas Co., 145–146, 150
Rohm, Otto, 145–146
Roscoe, Henry, 96, 98, 100, 223, 231–232
Rothschild Institute, 51, 58, 60
Royal College of Chemistry, 15, 304
Royal Institution (London), 8
Royal Society of Chemistry (Historical Group), 268
Royal Society of London, 387
Russell, Colin, 267

Sabatier, Paul, 4
saccharin, 341
Saunders, William, 304
Saussure, Nicholas Théodore de, 305
scale, xvi, 138–139, 141, 142–143
Schönhöfer, Fritz, 194, 196, 202
Schrödinger equation, 19
Schulemann, Werner, 194, 196, 202
Schultz, Jack, 47–48, 50–51, 58, 59, 66, 77
Schweber, Silvan S., 17
science holiday movement, 282
Science Museum (London), 277, 281, 284, 288, 291
scope, xvi, xvii, 138–139, 141, 142–143, 144, 154
Semenov, N.N., 37
Serres, Michel, 4
Servos, John W., xxi
sewage treatment, 308, 309
Shapin, Steven, 120, 131
Shaw, Peter, 7
Silliman, Benjamin, 221–222, 230, 232
Silverstone, Roger, 287
Slater, Charles, 232
Slosson, Edwin, 282
Smith, Albert L., 223, 232–233
Smith, Edgar Fahs., 264
Smith, John Graham, 139
social construction of science, xv, 87–90, 110–112, 119–121, 122, 129, 130–131
Société Chimique, 104
Society for the History of Alchemy and Chemistry, 268
Society of Public Analysts, 314
sodium, 144
sodium thiosulfate, 337
Somerville College, 383, 384
sorbitol, 337
soy lecithin, 336, 338
Spans, 337–338
specialty chemicals, 153
specificity, biological, 186–188, 205–206
specificity, chemical, 188, 204, 205–206, 208
spectroscopy, 31, 32; infra-red, 32, 36, 37
Spilker, Bert, 246–247
Sprengel, Karl S., 305
stabilizers, 330
Standard Oil of New Jersey, 371
Stanford University, 59–60, 68, 71–72, 74
Staudinger, Hermann, 148
Stengers, Isabelle, 4
Stewart, Alfred W., 17
Stewart, Potter, 338
Stine, Charles M. A., 148
Stöckhardt, Julius A., 231
Strecker, Adolf, 96, 98, 99
streptococcal infections, 158, 179–180, 181–185, 191, 192, 194, 197, 199
strong program in the sociology of knowledge, 87–90, 110–112
structure, theory of, 12, 17, 18–19, 91–92, 96, 98, 102, 103, 107–110, 112
Sturtevant, Alfred H., 46–48, 50–51, 58, 59, 66, 77
substitution theory, 12, 91, 96, 105–106
subtle fluid, 302
sulfa drugs, xvii, 158–160, 177, 193, 204, 207, 208, 209, 210
sulfanilamide, 185, 199–201
sulfonamides, 158, 192, 195, 196, 197, 199, 201, 202, 203, 208
sulfuric acid, 138, 139, 141–142
Sulloway, Frank, 372
surfactants, 333, 337, 339
Swift and Company, 336, 338
symbols: algebraic, 6; chemical, 13, 17

synthesis, 217, 219, 223, 224, 226; organic, 94–95
synthetic: fibers, 149–150; rubber, 149, 151
"system of production," 7

table, periodic, 18
Tatum, Edward L., 45, 60–61, 62, 63–71, 73–75
Taylor, Alfred Swaine, 298
technological momentum, xvii, 204, 207–208
Teflon, 149–151
Thenard, Louis Jacques, 10, 94, 96, 104
thermodynamics, 5, 16
thiourea, 329–331
Thomsen, Julius, 15
Thomson, Joseph John, 18
Thresh, John Clough, 314–315
titrimetry, 32–33
toxicology, 331, 333, 341, 343–344
transgenic mouse, 291
Tréfouel, Jacques, 199
Tréfouel, Thérèse, 199
Tweens, 337–338, 339
type theory, 11–12, 13, 15, 91–92, 96, 99, 102–103, 106–108, 110

Union Carbide, 144, 146–147, 150–151
unit operations, 137

valence theory, 17–18, 91, 96, 105, 108, 109
Van't Hoff, Jacobus Henricus, 12, 13, 16
Vietnam, 363
Vieweg, Eduard, 97, 98, 105, 106
vinyl resins, 150
vitamins, 168, 172–173, 174, 175, 190, 192
Volhard, Jacob, 99
Volta, Alessandro, 30

Wanklyn, James Alfred, 312–314
water analysis, 311–315
Weinberg, Alvin, 296, 316
Weltzien, Karl, 96, 100, 104
Westfall, Richard, 374
whiggism, 272, 368
White, Hayden, 278, 288, 289
White, W. B., 335
Wigner, G. W., 309
Wiley, Harvey Washington, 322, 342, 344
Will, Heinrich, 93, 98–99, 105
Williams, R. R., 339
Williamson, Alexander William, xv, 4, 15, 92, 94–97, 98, 99, 102–103, 105, 107, 109, 110
Williamson synthesis, 94–96, 99, 102, 105
Wilson, E. Bright, 17
Wingler, August, 194, 196, 202
Wislicenus, Johannes, 12, 13
Wöhler, Friedrich, 12, 94, 97–98, 105, 107, 108, 123, 125, 126, 132–133
Woodward, Robert Burns, 237–238
Woolgar, Steve, 120, 124, 129, 130
World War I, 146
World War II, 140, 150, 151, 154
Wurtz, Charles Adolphe, 15, 35, 90, 93, 95–96, 97, 98, 101–104, 105–106, 107, 108–110, 112
Wurtz reaction, 95, 102, 105

X-ray crystallography, 381, 385, 386, 387

Zeiss Jena, 32, 36

This book was set in Baskerville and Eras typefaces. Baskerville was designed by John Baskerville at his private press in Birmingham, England, in the eighteenth century. The first typeface to depart from oldstyle typeface design, Baskerville has more variation between thick and thin strokes. In an effort to insure that the thick and thin strokes of his typeface reproduced well on paper, John Baskerville developed the first wove paper, the surface of which was much smoother than the laid paper of the time. The development of wove paper was partly responsible for the introduction of typefaces classified as modern, which have even more contrast between thick and thin strokes.

Eras was designed in 1969 by Studio Hollenstein in Paris for the Wagner Typefoundry. A contemporary script-like version of a sans-serif typeface, the letters of Eras have a monotone stroke and are slightly inclined.

Printed on acid-free paper.